Earth Days Reprised

William Dritschilo

Samizdat

Created using OpenOffice Writer 6.2.4.2 & 6.3.1.2

Cover design: Jamie Dritschilo
Front cover: sunlight through trees, Redwood National Park
Back cover: salt marsh next to Rachel Carson's home.

2019

Printed by Kindle Direct Publishing, an Amazon.com Company

© Copyright held by William Dritschilo, Proctor, Vermont. No republication, dissemination in any form, or other re-use, in part or in whole, is allowed without permission from the author.

Much of the material in this book was previously published in 2004 as *Earth Days: Ecology Comes of Age as a Science*. Some of it was later incorporated into an article published in *The Bulletin of the Ecological Society of America* and served as a jumping off point for *Magnificent Failure: Frank Egler and the Greening of American Ecology*. It was fictionalized in much less technical detail in my novel, *Ecologists*.

Table of Contents

Forward..4
Philadelphia, April 22, 1970, Earth Day...........................5
The Art of Earth Day..19
"Perhaps I Have Said Too Much Already"......................33
Great AEPPS..49
An Ecologist Reconsiders Silent Spring after 17 Years............61
Autecologists, Synecologists, and Genecologists....................74
A Tale of Two Brothers and an Uncle...............................90
Butterfly Collectors...107
The Curious Case of the Superorganism.....................126
Save the Cat!...140
Buried with Full Honors..156
A Mathematician Who Knew His Warblers................168
The Climax of the Ecosystem...188
Do Cats Eat Clover?...203
Diversity-Stability and Other Scams..............................214
Chaos Comes to Ecology..227
Nutches in Nitches, Atoms in Apples...........................250
A Doctor for a Sick Science..262
It Wasn't a Barroom Brawl. It Was an Athletic Contest..........285
Br'er Rabbit and the Tar Baby..313
Life is Simpler on an Island..321
SLOSS: Islands in the Wilderness..................................337
Do Owls Live on Islands?...349
Still Counting..361
Up Against Limits..367
Potatoes Made of Oil..377
What Are Nature's Services Worth?..............................403
Ecological Engineers and Environmental Doctors................419
A Brook Runs Through It...446
MacArthur's Ghost...464
Calculus and Commoner...472
Afterward...484
Published Sources of Quoted Material.........................496
Index..520

Forward

This work is the labor of a lifetime. I have been doing research for this book essentially from the day that the ideas of ecology took hold of me as an undergraduate. I was lucky that that happened during the excitement of Earth Day and at an age at which I could follow up on my enthusiasm for it.

What I learned in watching ecology transforming from a point of view to a science, I have tried to communicate to as broad an audience as possible. The specialist will read this work and benefit or not from it in the form it is written, but the reader that I have targeted has required me to skip some conventions that the specialist will take me to task over. First, I have tried to write informally. There is no set of notes and no index. My preference has been for a narrative that should be read sequentially, rather than perused. I have documented only the sources of material used directly out of a sense of ethics more than anything else.

A word needs to be said about the "samizdat" nature of this publication. The 21st century is not a time, I think, to have publishers and agents pussyfooting around works that need to be published. In this case, they did so out of concern that the subject matter would not appeal to the non-specialist, while the writing style was too "chatty," as one editor put it, for an academic audience. Having no academic position to nurture with a stodgy publication, however, I could take advantage of the ease and economy of electronic publication and put out exactly the type of book I would like to read on the subject. I hope you enjoy it.

A large number of people have helped or encouraged me in my efforts over the years. If I tried to mention them all, I am sure I would inadvertently leave someone one.

I thank all of them now. There are some who have been especially helpful and supportive. I am sure you already know who you are and have been shown my gratitude.

William Dritschilo
November 9, 2004

Philadelphia, April 22, 1970, Earth Day

I wasn't quite sure what I was doing in that crowd at the time. We were not about to march on a ROTC building, there were no hand-rolled smokes going around—at least not past me—and if there had been a rock band there, I had clearly missed it. What there was that was stirring the crowd was a rather dramatic looking man whose magnificent mustache and Scottish brogue stayed in my memory longer than his words. It was Ian McHarg, a professor of landscape architecture at the University of Pennsylvania and he was charging up the crowd as well as any political activist could. McHarg had just published *Design with Nature*, which called on planners to conform to nature, rather than do battle with it. He was the antithesis, I later learned, of New York City's Robert Moses, the creator of the Long Island and Cross Bronx Expressways. It was on Moses' principles that the Interstate Highway System had been built.

"You have no future," McHarg shouted, perhaps in contemplation of dreary stretches on those "I-something" highways. "You have no future. Why am I to tell you the bad news?"

I thought I had a future. I had plans for it.

Also a speaker in Philadelphia on that first Earth Day was counterculture personality Ira Einhorn. His courses in Penn's Free University—"Basic Ira Einhorn" was one title—promised exploration of the self. (It was mainly his own.) Scores of attractive women enrolled in the course. The Free University, in which courses with "relevance" were offered, was a concession to students that University of Pennsylvania administrators made in the turmoil of the 1960s. As in similar efforts sprouting throughout the US, interest as evidenced by enrollment seemed to be the only criterion for a course to be offered.

Other speakers in Philadelphia included ecologists Paul Ehrlich and Kenneth Watt, Senators Edmund Muskie and Hugh

Philadelphia, April 22, 1970, Earth Day

Scott, biologists René Dubos, Luna Leopold, and Nobelist George Wald, consumer activist Ralph Nader, poet Allen Ginsberg, and Dune author, Frank Herbert. The Philadelphia program had stretched for a full Earth week, and was covered by Walter Cronkite in a one-hour special news program.

Archdruid David Brower, another Philadelphia speaker, demanded less real estate development and more wilderness, a word heard as often on that day as "ecology."

"We're not borrowing from our children, we're stealing from them," he said. "And it's not even considered a crime."

Throughout the country, that first Earth Day celebration drew millions of people. "Teach ins," they were called then. Crowds stopped traffic on New York's Fifth Avenue. Congress was shut down for the day. By all reckoning, it was a turning point in national thinking on the environment. It's originator, Gaylord Nelson, then Senator from Wisconsin, jubilantly hopped from celebration to celebration. For years, Nelson had been trying to find a way to put the environment on the nation's political agenda. When the day of the teach-in finally came, the magnitude of the response to it overwhelmed him, bringing tears to his eyes.

Some 20 million people participated. Everywhere, as if in the air at the celebrations, there was concern for "the ecology." It was an old term and an old idea, but it was new on that day. Young people especially were discovering it, much as they were discovering marijuana, sex, or ways to get out of the draft. They made the concept their own.

There were speeches and clean-up activities in many cities, but in Washington, DC, college students and schoolchildren poured oil on the sidewalk in front of the Department of the Interior to protest oil spills. Other protests included dumping 5 tons of roadside trash (presumably picked up for the event) at a West Virginia courthouse and burying a car in San Jose, California. It was the era of protest. Ecology gave people of all ages something to protest.

In Los Angeles, a hitherto unremarkable engineering professor at UCLA found himself in the makeup room of a television studio. His extracurricular activities caused him to be invited to Pasadena for Cal Tech's Earth Day celebration. A devout fly fisherman and wilderness devotee, he had found a satisfying niche as an environmentalist. It seemed only fair to him to be one of three people being taped for television coverage to air that evening.

"What I remember is undergoing full makeup, as if I were to be a television star!" was his strongest memory of Earth Day.

That professor, who had started out his career as a petroleum chemist, enjoyed his stardom only briefly; other professors were propelled to something like superstardom. Along with them, an arcane, little known science was catapulted into the public limelight.

Not all speakers were equally concerned with "the ecology," if at all. At the University of Illinois, economist Julian Simon was asked to give a speech in order to balance points of view. Simon saw a rosier future than others.

"Of the 2000 people in attendance, fewer than a dozen concluded that anything I said made sense," he remembered years later. "My ten-minute talk so enraged people that it led to a physical brawl with another professor."

Earth Day was an emotional event, even for usually staid college professors. They vented before crowds at gatherings such as those I attended in Philadelphia.

That first Earth Day caught me in the final semester of a chemical engineering curriculum, and it was not hard to see where most engineers stood on the issues of that day. It was not in those crowds listening to speakers rail against polluting industries. They were instead on the side of products like napalm and companies like

Philadelphia, April 22, 1970, Earth Day

Dow and the oil cartels. It was not where I wanted to stand. I had been infected by that concept of ecology that was in the air.

My memory of that Earth Day celebration is not entirely clear, however. Maybe it's because Earth Day was really Earth Week in Philadelphia. In preparing this book I found out that Ralph Nader made a speech in Philadelphia that I must have attended. But I may not have. The speech given by Ian McHarg was in Fairmont Park, rather than on the Penn campus where my memory placed it, I now know, and was given on the 17th, not 22nd of April. My recollection of Ira Einhorn is assisted by his accusation, trial, and conviction in absentia of murdering one of those attractive young ladies that had flocked to him. A fugitive for years, he was only much later extradited from France and convicted in person.

In my memory, young Penn ecology professor Robert Ricklefs gave a speech at that Earth Day celebration, but I now doubt it. I may, instead, have been remembering a class I had taken from him. Troubled by a mind too restless for only one subject, I had double majored at Penn, taking ecology from Ricklefs as coursework for a BA degree in biology. In his classroom, instead of Bernoulli's equation and the Laplace transforms of my engineering courses, there were stories about animals and plants fitting into their environments.

That ecology course I had taken was strictly academic, filled with ideas and observations about nature, not a course on environmental issues and problems. Ecologists as saviors of the environment was something the media had been creating, especially around Earth Day. They convinced me.

The issue of *Time* magazine with Barry Commoner, "Paul Revere of Ecology," on its cover, while inside it was a picture of the Statue of Liberty seen through garbage at Fresh Kills landfill on Staten Island, brought people out to Earth Day activities who had never heard of Gaylord Nelson, I am certain. *Time* called ecology the "science of survival." It noted that ecology's lack of narrow specialization appealed to generalists, "including people with a religious sense." It also identified ecology as a systems approach that was more of an "attitude rather than a discipline."

The thumbnail sketch given in the paragraph above perfectly captures ecology at that point except for one glaring omission. Ecology was a science undergoing revolution. Whether it truly was the science of survival for humans, whether it was mainly ecosystem science, whether it was general enough to encompass all areas of the environment, and whether it was an attitude or religion were all parts of the revolution. Also part, but less visible, was a revolution on the methods and meaning of the science that is just now playing itself out. That revolution is the subject of this book. To understand modern ecology one must understand what it was on Earth Day, what it was not, and what and how it became what it is now.

It was Earth Day, rightly or wrongly, that let me think of ecology as a tool that could save the whale, put a stop to pollution, save the rainforest. And it was more interesting than engineering. Why was that? Why couldn't engineering save the whale and stop pollution? Furthermore, Laplace transforms also had their place in ecology, I was to learn. Why did they seem less onerous there than in engineering?

A childhood memory may have given me an answer to that question. Like Proust's taste of Madeleine, my trying in vain to come up with a snappy but appropriate title for an environmental science text suddenly shook loose that memory. It was of the very first science textbook I had ever held in my hands. It was back fourth grade, I believe. The book had been passed out with little ceremony except for the mention that it was to be our "science book." Yet this had been enough to make me approach the book with some childish awe, for no previous book had been announced to be "about" something. They had been "reading books," meant to teach reading. Their titles had been irrelevant. Only years later did I recognize them as being "Dick and Jane readers." The science book, however, was different. The teacher had said it was "our" book, and its title was *Our Environment*. I remember staring at the cover, puzzling things out. The easy four syllables of "environment" may have impressed me with the length of the word they created, for I was only a few years past those Dick and Jane readers; but I think not, for in my recollection is the impression that it was a word I already knew. There was a photograph on the cover which has unfortunately escaped my memory, a mountain or some other such natural vista. It was not, however, the photograph or the word, but the idea of the title that had caught my imagination. I knew that environment was synonymous with surroundings, but this science text promised to extend my surroundings out to places I had never been and probably could never reach, of the existence of which I had, in fact, been totally unaware. And this was juxtaposed with the idea of ownership. Our environment. It was mine, that environment.

Edward O Wilson, expert on ants, founder of the science of sociobiology, and champion of global biodiversity, writes in *Biophilia* that the mechanism for the feeling of pleasure that people can obtain from studying, working, or just being in nature is permanently rooted in the human mind. To Wilson it must be genetically programmed, much like other human drives. The evidence he gives is not entirely convincing, but my reactions to learning ecology from Ricklefs or picking up that fourth-grade science text are entirely consistent with his thesis. What I felt was something like religious rapture.

Analogies between biophilia and religious rapture are troublesome to ecology as a science; yet they exist and should not be ignored. Whether biophilia is real or not, it cannot be denied that many people are drawn to naturalistic pursuits for the sheer joy they give. It should be no surprise that these people might become ecologists. Is joy in nature driving their science? Have they changed the science for the better?

Unlike others more actively involved in that Earth Day and unlike the impression I may have just given, I can't say that my life was immediately and directly altered. Earth Day was mainly a political event. Biophilia's effect on me was slower and more subtle. Having already had classes in ecology as a basic, rather than practical science and possessing the structured mindset of a trained engineer, I saw the environmental movement as being only loosely related to its ecological driving force. Ecology was a science, not a goal. When I did choose to pursue ecology, it was as a scientist, not activist. My intention to use ecology for pragmatic purposes seemed no different than using my engineering skills to clean up, rather than befoul our waters. Ecology and environmentalism were to be separate for me, no less so than, say, Judaism and chemistry—or, to give a more historically conflicted example: Roman Catholicism and biology. I now have to reconsider how separate ecology really is from environmentalism.

A month after Earth Day, I married a girl who often summered at her grandparents' house on Southport Island, a short walk past some salt ponds from the Maine refuge of a little old lady who poked among sea wrack in her tennis shoes and wrote nature books. I read and enjoyed her books. Biophilia took stronger root.

Earth Day receded from my mind, however. I ignored Earth Day 1980. My struggle to solve the ecology-environment dualism problem had been nothing at all like the struggle of philosophy to solve the mind-body dualism problem. Or so I thought. Cracks did begin to appear in that mental wall I had constructed between my science of ecology and my philosophy of environmentalism. I usually did not recognize them.

An example: although, unfortunately, my path was never to cross that of Ian McHarg again, I did at one time find myself sitting next to one of his students in a graduate seminar. Rather than city planning or landscape design, the subject of the seminar was ecology, heavily biased toward genetic and evolutionary studies. Yet here was this McHarg student reaching into an academic area way out on the periphery of his own. Not only must he have been underprepared by training for the seminar, but most of it had to have been irrelevant to his profession. Still, he pressed on to its end, trying to make sense of the high levels of heterozygocity being found in populations of wild mice. What made his presence even more impressive is that my description of him as a student is misleading. He was an older man, probably in his forties, a McHarg graduate, who owned his own firm. In contrast to him, graduate students I knew in engineering or biochemistry would never be found in any course or seminar not directly required for their degrees. Their strategy was to complete their coursework as soon as possible, then have nothing more to do with it, often not even from the instructor's side. As senior scientists, new techniques would interest them more than ideas. Yet there was that planner next to me. Ecology has to be endowed with ideas having an intellectual attraction that transcends their utility. The evidence for that was right there in that seminar room.

Later, I would come across more McHarg graduates in engineering and consulting firms and in the planning and landscape departments of universities. They surprised me, every one of them, by seeing themselves as practitioners of ecology. Typically they characterized McHarg's contribution to "ecology" as "enormous."

"He got much of the world to view the earth through ecological glasses," one proclaimed.

One ecology course, apparently, even if poorly understood, seemed to be sufficient to become a practitioner of the science. Meanwhile I had spent four years of graduate study to get a PhD in the subject, and I still felt that my grasp of it was incomplete.

A one-time colleague of mine at UCLA, a University of Wisconsin-trained ecologist who went into the planning field, once shocked me by using Jeremy Rifkin as an ecological authority in his newly-written environmental planning textbook. It was given to me to review. Rifkin, a radical environmentalist who later advocated against genetic engineering and beef consumption, and had also been a fellow undergraduate student with me at Penn, had not been in the audience with me on that first Earth Day; he had to have been up on the podium. Environmental activist yes, but at no time until my friend's planning text had anyone ever accused Jeremy Rifkin of having scientific credentials.

"Ecology" had begun to creep into the popular lexicon more as a term for a point of view—sometimes as almost a religion—than as a discipline of science, probably even before that first Earth Day. At least one social analyst has concluded, "Ecology is now a political category, like socialism or conservatism."

No it is not. That may have been one view of it on Earth Day, but it was and remains a science, even if one that is evolving.

Expectations for ecology at present are great, perhaps the reason for the infrequent, but annoyingly persistent notes of backlash that keep on troubling ecologists. A recent book by a Danish statistician, *The Skeptical Environmentalist*, has raised the hackles of ecologists far out of proportion to the book's significance or the heresies and misrepresentations in it. Why? Why not damn it with silence?

The answer to that question becomes clear on learning what was hidden from the public on that Earth Day. It had been hidden from me also. Only by reflecting on what I have learned from trying to make sense of that "Decade of the Environment" or "Decade of Ecology," as the 1970s have been called, could I learn what I needed to understand the science that is "more 'useless,' yet more stimulating and challenging" than any other. That remark I borrow from ecologist and environmental maverick Frank Egler who attributed it to one of his mentors. He repeated it in an article arguing for a code of ethics for ecologists because society was placing important demands on that useless science.

Other remarks of which I should have become aware much sooner are even more telling. "It's just natural history, isn't it?" a colleague at my first place of employment as an ecologist said to me.

No! No! There is much more to ecology! How could she not have know that?

Actually, it turned out, she was just repeating what had been a rather firmly entrenched impression among other biologists. Before events suddenly brought ecology to the fore on environmental issues, it had been a valid impression.

"What is ecology?" is a refrain that has been repeated in every decade from the 1930s to the present. Strangely enough, it is a refrain sounded as often by ecologists as those outside of the field.

"Lacking in rigor" was an accusation about ecology that was difficult to defend prior to Earth Day. "Unscientific" became the accusation about at least one of the areas of ecology that had sprung up in defense of rigor in the field. "Sick" was one ecologist's description of his science. "Barroom brawl" was the description for some of the scientific discourse being carried on by ecologists well after Earth Day.

"Why doesn't anyone listen to ecologists?" might not seem to be an unusual question for the theme of an address by the president of the closest thing that ecologists had to a professional society. However, the date of the address was 1982 and people certainly thought that they were listening to ecologists by that time.

What gives?

As great as expectations for it still are, the ecology of today was not the ecology presented to the public on Earth Day. Ecology was undergoing a number of revolutions, some unrelated to environmentalism, some directly caused by it, some perhaps in reaction to environmentalism. The ecosystem concept, the fragility of which was at the heart of so much popular ecology, died and was reborn. Ecologists warred over their favorite methods of reasoning and found political justifications in some of their pet theories. In the 1970s, there were complaints—not heard by the public—of intellectual censorship by the revolutionary vanguards of a new ecology. In the 1980s, ecologists debated the legitimacy of the very foundations of their science, all quite out of public view. Political considerations crept into debates over the design of refuges, with scientists being admonished to abandon scientific rigor due to the urgency of conservation needs. We have the theory we need, the admonitions, in effect, went. Try not to confuse the issue with facts. In the 1990s, a senior ecologist warned others to turn the heat down on debates over their data and its interpretations, lest the public obtain the wrong impression about the science and turn against its efforts to preserve biodiversity.

During those revolutions, many academic ecologists decided that their professional responsibilities were in working out the patterns created by evolution in nature, rather than the more practical tasks of tracing out the pathways of energy or nutrient flows through ecosystems. Some ecologists even stopped claiming expertise in the economic and pollution issues they had championed on Earth Day, while others found useful expertise in exactly those areas.

These changes make up the subject of this book. They also lead to the question—never properly addressed on Earth Day—of what the science of ecology really is and what it can do to improve human affairs.

Loren Eiseley, in *Darwin's Century*, provided an elegant description of the birth of Darwinian theory. I borrow and reproduce it here on the justification that it is quite apt as a description for the birth of the environmental movement a century later.

Eiseley wrote: "Seemingly unrelated events, diverse scientific discoveries, industrial trends, and religious outlook can all, in historical perspective, be observed to revolve in a moment of seeming heterogeneity before they crystallize into a new pattern with Darwinism at the center. It is like looking into a chemical retort which is about to produce some rare and many-sided crystal. One moment everything is in solution; there is a potentiality, no more—and yet in the next instant a shape has appeared out of nowhere."

That shape is Darwin's theory of evolution through natural selection. Something similar was happening in the years around the first Earth Day. A number of influences and ideas as varied as the anthropology of Margaret Mead, the humanism of René Dubos, the science fiction of Frank Herbert, and the poetry of Allen Ginsburg came together with science to take shape from a shapeless mass. The shape that seemed to crystallize on Earth Day was the modern concept of ecology.

However, from my perspective, I see a science that crystallized like Darwin's theory did in Eiseley's elegant prose, but—instead of becoming a foundation pillar upon which much of our understanding of the biological world is based—continued to swirl around in eddies that dispersed farther and farther from its original core, the very picture of chaotic turbulence. There is no longer any reason, I believe, to put forth ecology as the integrative environmental science. Nor should ecology be the first science to which we need to turn for even strictly biological questions. There is much basic and applied biology available—or that is possible to do given the need—to solve most "ecological" issues without invoking ecological principles and their entrained baggage, conceptual, political, and personal.

There is a valued place for ecology at the scientific table. There always has been. It should return to and stay there, however. This is also the story I will tell.

Like others who have looked at the recent history of ecology, I see a story with threads so knotted and tangled that one often does not know where to begin or leave off. Like a famously misapplied "ecological" principle, everything seems to be connected to everything else. This causes difficulties in writing. Victorian naturalists managed to write very readable prose using only full and semi-colons and commas. I find I cannot deal with some of the complex relationships of ecological ideas without parentheses—and many of them.

Like JD Salinger, also once accused of overuse of the parenthesis, let me offer the reader as apology a bouquet: ((()). Unlike Salinger, my background in mathematics temps me to use the exactitude of logical punctuation. I may resort to that, too, so here is another bouquet: {[()]}. On occasion, I may actually resort to that powerful language that is mathematics. The reader may find it easier to follow than some of the tangled ideas it is used to represent.

There are also people involved in this story. It starts with Rachel Carson and the other writers and scientists whose words caught the attention of the public on Earth Day. At about the same time, the Odum brothers came from the corn pone South to champion the ecosystem idea. There is Robert MacArthur, the "James Dean" of ecology, and kindly G Evelyn Hutchinson, not just MacArthur's scientific godfather, but ecology's "Godfather." There is Jared Diamond, who tried to be his successor and in the effort set off a war in ecology. His antagonists were led by Dan Simberloff, who rebelled against the science inspired by his own mentors, EO Wilson and "James Dean" MacArthur. Simberloff thought that science should be carried out, well, scientifically. Seen as something of a spoilsport by those having more fun with their science, he paid for it. There are Paul Ehrlich and David Pimentel, for whom no environmental issue was beyond their expertise, attracting criticism to them like flies to paper. Then there is Gene Likens. No James Dean, he looked and acted more like an insurance salesman. Yet he found a way through the controversies swirling in his science to put it to good practical use. There are, of course, many others, each trying to find their own personal way in the broad, important science that is ecology. Their connected threads are the most difficult to untangle, but the most rewarding in understanding the story of ecology since Earth Day.

The Art of Earth Day

My standing in the crowds of Earth Day put me right in the middle of ecology's story. Being in the middle of what I set out to tell has its advantages, but it has drawbacks, too. How to tell a story from its middle may be easy for fiction writers. Start with a dramatic event involving most of the story's main characters and you are off. Not so fast with a story about science, though. My cast of characters is a bother. It has too many changes. Characters important to the beginning are missing at its end. And there are too many minor characters. And instead of a dead body, I had a dead lake. Who can kill a lake? Who could want to?

All hope is not lost, however. Earth Day did give me a special story-telling gift. Ecology on Earth Day was presented in its simplest form. And its most elegant. Those voices that were raised on Earth Day in support of ecology were most often literary voices—and good ones.

Ecology really reached the American consciousness through a series of literary and artistic events. The publication in 1962 of *Silent Spring*, Rachel Carson's famous book was the biggest. Carson had been a government-employed writer and biologist when the success of her nature book, *The Sea Around Us*, let her give up her government duties and become a full time writer. That book spent 32 weeks atop the best-seller lists and was followed in 1955 by the similar success of *The Edge of the Sea*. Both depended on her writing talent for their success. Although *Silent Spring* was an entirely different sort of book, its author was someone who had honed her writing talents to appeal to mass audiences. In her first book, *Under the Sea Wind*, for example, her story of Scomber the mackerel is a tale worthy of Dickens. She knew the tricks of popular literature.

Carson put those tricks to effective use in *Silent Spring*. The title is a metaphor. Its brief first chapter, "A Fable for Tomorrow," is a parable, meant to warn. A silent spring comes to a fictitious unnamed community. In her words, the community suffers "a strange blight," "some evil spell," and "mysterious maladies" that end with the "shadow of death." The silence in Carson's spring is that of birds whose choruses had once filled spring mornings. The "grim specter" that crept unnoticed through the community, Carson made clear, is a metaphor for the threats to countless American towns posed by the indiscriminate use of chemical pesticides, DDT in particular.

Her fable captured the attention of an admiring—and alarmed—public while giving ammunition of sorts to her detractors. Books have already been written about the public and corporate reactions to the book, but more important is the reaction it brought about in a small segment of the scientific community. As an antidote to that silent spring, Carson offered the science of ecology, simply and persuasively sketched. She had not been the first to argue that ecology should guide our use of the environment, but it was her voice that most sped the avalanche that peaked on Earth Day.

Close on the heels of *Silent Spring*, Stewart Udall, then Secretary of the Interior, gave the imprimatur of an official in a popular administration—John Fitzgerald Kennedy's Camelot—to the conservation of wilderness resources. His book, *The Quiet Crisis*, helped re-introduce the ideas of George Perkins Marsh, Henry David Thoreau, and Aldo Leopold to the public. Leopold's *A Sand County Almanac* soon became required reading at Earth Day teach-ins. By raising concerns over the direction that conservation of our natural resources had taken, Udall was in synergy with Rachel Carson's raising concerns over chemical pesticide use. His title even echoed hers. *Silent Spring* and *The Quiet Crisis* were like a one-two punch.

A large part of the environmental revolution was the new attitude toward wilderness evident in Udall's book. It was avidly picked up by young (and old) activists in general rebellion against the status quo, and it was abetted by ecologists. The war in Viet Nam, birth control, civil rights, consumerism, established religion—all of these targets of protest had an ecological dimension to them, but wilderness values were distinctly ecological.

"The ecologists' need for untouched areas was an effective argument in this battle and was one of the reasons that the Wilderness Bill was recently passed" is how, a little later, Udall summed up the process of creating the first federal bill to specifically protect wilderness that he oversaw. This special attitude ecologists have toward wilderness affects their profession and their science in a significant way. But I am getting ahead of myself.

Also in the early 1960s, David Brower had the Sierra Club sell beautiful coffee-table books of nature photography. They were in the style of the great American photographer Ansell Adams, who did with a camera what the Hudson River School of artists did with easels in helping to create a respect for the natural environment—as did Henry David Thoreau and other Transcendentalist writers. The vistas of the American West in paintings, still photos, and Hollywood cinema helped to define an important part of America for us. Europe had its history; America its wilderness, our writers and artists seemed to say.

The Sierra Club books took advantage of our new love of wilderness, especially western. I don't have the large-format version of *Time and the River Flowing*. It was first published in 1964 as a high-priced art book intended to make money for the organization. Its price made it something an undergraduate was not likely to have unless it had been given to him as a gift. It made a good Christmas or birthday present for someone in the camping-out set who already had everything he needed. I must not have had everything I needed, for no one gave me that gift. What I have is the less expensive paperback edition. This was put out in 1968 and had considerably more impact,

I can dare say, than the large format one. It stopped further damming of the Colorado River. Honest! My copy is still a beauty. True, at 6½ by 9½ it would be lost amid the clutter of most coffee tables and the $3.95 price on it would puzzle more than impress, but within its covers is the same spectacular photography of Grand Canyon scenery, magnificent even in reduced size. The photos all seem to catch the natural light when it does its greatest magic on the landforms, the kind of effect for which one finds people with cameras, amateur and professional, staking out vantage points in predawn hours along the canyon's rim.

Leafing through it now I find one beautifully composed photo of some river-running boats tied up along a bank in the foreground, towers of rock in the background. The caption tells us that a proposed dam would raise the water level there 300 feet. Outrage! Tragedy! Criminality! I can still fume in my best David Brower style of reaction. Yet from the perspective of the photograph, it is difficult to assess just how high up the cliffs in the background 300 feet might be or whether there might not be a similar tying-up place at that height. I try not to remember if there were rubber rafts running down the canyon in those days or only those beautiful, fragile-looking wooden boats shown in the photo.

An earlier book in the series brought together the photography of Elliot Porter and the prose of Henry David Thoreau. I have the $3.95 version. Its title, *In Wildness Is the Preservation of the World*, comes straight out of Thoreau's *Walden*. The photographs in it, and the excerpts from his journals used to accompany the photography, drove people outdoors with cameras instead of hunting rifles. Porter, like Thoreau, found beauty in a temporary island in flood, the bark of a tree, even a nest of tent caterpillars. The book helped to promote wilderness into some sort of American ideal—and Thoreau to the status of ecologist.

Earth Day influences were definitely not without their art, literary, visual, and musical. Joni Mitchell's "Big Yellow Taxi" became as much a pop protest anthem as Bob Dylan's "The Times They Are A-changin.'" Visual images abounded on Earth Day. One was the already-mentioned unflattering shot of the Statue of Liberty. Another was a poster prominent in California that had a photo of an obviously nursing mother with a caution imprinted on her naked breast. The DDT concentration in mother's milk had been found to exceed standards that would not allow it to cross state lines except in its original container.

The artwork that may have most strongly mobilized Earth Day efforts, however, was the simple photograph of the earth taken by Apollo astronauts. Our planet turned out to be incredibly beautiful. In that beauty was something fragile, like youth, too soon to age and be worn down by life.

Earth Day 1970 was a convergence of art and literature, of anti-consumer and anti-government social activism, anti-war protest, romantic attitudes toward nature as wilderness, and back-to-the-land organic farming sentiments merging with uncomplicated public health and quality-of-life concerns—and with the science of ecology. There was no lack of literary devices driving that merger. It left the meaning of the word ecology as somewhat of a metaphor—if not actually a myth. That's because, with a few notable exceptions, ecologists had not learned how to speak to the public. Most still can't.

There were strictly scientific voices, though, that did speak up eloquently for ecology in the years leading up to Earth Day. One of them, Paul Ehrlich, happened to be an ecologist, one specializing in butterflies. By then, he was already a leading expert on population ecology and had created, along with botanist Peter H Raven, a concept for an entire field of ecological studies and a word, "coevolution," to describe it. Their paper proposed that plants evolved in response to their butterfly predators, who in turn evolved in response to their plant prey, resulting in much of the diversity seen in nature. It became one of the most cited publications in their field. It helped to explain one particular puzzle of nature: why there are so many species.

More than just a writer, Ehrlich became a media star through his 25 visits to Johnny Carson's late-night television show. His literary device was that of a "Population Bomb," the explosive growth of the human population that would lead us to Armageddon as surely as any nuclear conflagration. I have in hand the revised version of the original 1968 book commissioned by the Sierra Club and shepherded into print by David Brower (again!). It is a remarkably cheap product, not only in price ($0.95), but in quality of paper, typeset, cover art, whatever. Let me note that it is "Dr Paul R Ehrlich" in the title. I suspect that, like Dr Spock's baby book, it was meant for use and disposal. Both books have babies on their covers. Coincidentally, both of the doctors were outspoken critics of nuclear weapons and advocates of population control.

Ehrlich chose to define the world's neo-Malthusian problem in the following way, based on a trip to Delhi: "The streets seemed alive with people. People eating, people washing, people sleeping. People visiting, arguing, and screaming. People thrusting their hands through the taxi window, begging. People clinging to buses. People herding animals. People, people, people, people."

I am jealous. Not of the trip. I would have stayed home. Perhaps Ehrlich should have, too. At the time, he and his wife and daughter probably wished they had. What makes me jealous is what incredibly good writing that passage is.

I am not the only one impressed with Ehrlich's nimble pen. (These were the days before word processors.) Bjørn Lomborg, the

Danish statistician who has taken aim at some of the numbers that environmentalists use, quoted the very same passage. Somehow the fact that the quality of Ehrlich's prose helped sales of *The Population Bomb* to exceed 3 million was cause enough for Lomborg to suspect Ehrlich's motives as an environmental scientist.

The passage's purpose was certainly literary, rather than scientific. What Ehrlich did with it was capture the attention of the reader the way Rachel Carson did with her little fable. After a statement that powerful, the rest is just detail.

And the details in *The Population Bomb* were rock solid. True, there was some silly fiction that crept into the book. More than likely that was influenced by *Silent Spring's* fable and was also purportedly illustration. Ehrlich's fable, "The Ends of The Road," is really two separate scenarios of doomsday and another of redemption that all seem profoundly out of place. Peopled by over a dozen fictitious characters, all given fictitious names, who interact in less than a dozen pages, the first scenario makes it clear where Ehrlich's talents do not lie.

However, you could do the numbers in the rest of the book yourself and get the same answers as Ehrlich, which is more important. I have done so.

You might disagree, however, with the validity of the particular numbers chosen and their interpretation. Many did. And still do. Ehrlich's interpretation of his numbers was that unless births were curbed immediately, instead of the affluent society of the 1960s, our children would "inherit a totally different world, a world in which the standards, politics, and economics of the past decade are dead."

How did Ehrlich get to this conclusion? And, since my children are living in a more affluent society than I did as a young adult in the 1960s, how did he go astray? These are not easy questions, because the argument may have to be made that it was not Ehrlich who went astray, but his science.

Getting briefly to that science. Keep in mind that Ehrlich managed to reach out to the public in an era when many ecologists chose not to stray very far intellectually from what they might learn from their butterflies, or whatever. Was he qualified to speak as an expert on the scientific issues that he tackled? Overpopulation? Pollution? War? Absolutely! The straightforward calculation of doubling time for a population is a method right out of introductory ecology. Its actuarially precise equations were developed by a mathematician working with British ecologist Charles Elton.

Another issue that concerned Ehrlich was the "subtle changes in the complex web of life and in delicately balanced natural chemical cycles." This is ecosystem ecology as it was known then. It is also what Rachel Carson referred to as the "balance of nature" in *Silent Spring*. This is what Barry Commoner's First Law of Ecology purported to be about. Ehrlich's statement "that the simpler an ecosystem is, the more unstable it is" was then embedded in ecological thought, even though it soon became a matter of controversy.

As to war—who among us is not qualified in some way to offer an opinion?

Ehrlich, ecologist though he was, became associated more with the population policies that he stressed than with the ecology that Rachel Carson had championed. The ecology of populations was, after all, his expertise. Of ecologists interested in food webs, nutrient flows through ecosystems, all of those useful things that were capturing the interests of ecologists at that time, none reached the attention of the public as well as Paul Ehrlich. Into the breach that was left jumped Barry Commoner.

Barry was a plant physiologist who became involved with issues having to do with nuclear fallout, weapons testing, and, ultimately, chemical pollution of the environment. Barry is how almost everyone who got to know him, even his most junior subordinates, soon came to call him. I never heard him ask to be called that, but he did seem pleased with its egalitarianism.

Commoner's metaphor was of a "closing circle," which he dramatically held off from defining until the last pages of his popular book by the same title. What it represented was his ideal for a society that recycled renewable goods, instead of using up its necessary resources and drowning in its own filth. It was also a device for summing up what Barry saw to be at the heart of ecology.

I happen to have been fortunate enough to be a part of Barry Commoner's Center for the Biology of Natural Systems when it was at Washington University in St Louis. It had been an exciting time to be there. Teams of scientists and economists who were taking measurements and tabulating statistics on some of the best-managed farms in the Midwest were demonstrating that American agriculture could sustain its yields using only a fraction, if any, of the chemicals with which it was befouling the environment. The results were published in the prestigious journal, *Science*. Another team, this one of laboratory scientists, discovered the potential that Big Macs (and other hamburgers cooked in an identical manner) had for causing cancer. That also was published in *Science*. And Barry was putting the finishing touches on his next influential book, *The Poverty of Power*. Exciting, but also disappointing. I am disappointed not to have worked more closely than I had with Barry.

It was not until I began to reflect on events for this book, however, that the reason for my working at a distance from him became clear. I had been suffering under two misconceptions about it. One was that Commoner had a strong interest in the science of ecology. That misimpression is the reason for the brief digression that follows.

Barry, I had been told by people who seemed to have known him better than I did, made his opinion about someone on first meeting and never changed it. Furthermore, if Barry found you in agreement on one of his pet ideas, he would forever hold you high in his esteem. By disagreement, however, you only demonstrated your stupidity. I did not find myself in agreement with him on our first meeting. (It had something to do with what role economics could play in ecology.)

On reflection, though, I realized that it might not have been any overt stupidity on my part that closed Barry's inner circle away from me. In fact, later in my tenure at his Center, he very generously gave me time from his over-scheduled working days to teach me what he knew of the fine points of grantsmanship. Finding and obtaining funding is the foremost skill necessary for survival in academia. Barry shared much of what he knew about getting funding (which was substantial) with me. To this end we would meet almost weekly in his office, where a psychiatrist's couch (what probably would have been called a chaise longue by those who know such things) tempted me. Several dozen copies of the issue of *Time* magazine with him on the cover, stacked at a convenient eye level, in an otherwise cluttered bookshelf, always caught my eye.

"Polish it and polish it and polish it some more," he would repeat over each new proposal draft I would bring to him, until that most exacting of editors—time—put a stop to our revisions.

Barry would not have gone out of his way to help had he no respect for my intellect. Why I had felt ignored at first was, instead, quite simply that I was an ecologist—the only one for most of my stay—in a nest of environmentalists. My direct superior, for example, was a PhD physicist. Others at the Center were economists, agriculturists, and biologists of various sorts. None, however, had the ecological training to understand me professionally. My ecological expertise was of little use to Barry. I had been hired to crunch numbers.

Perhaps seeing himself stepping into a void, lack of knowledge of ecology notwithstanding, Barry did set about to educate the public in what he called the four laws of ecology. Ecology, Barry defined as the "science of planetary housekeeping." By this definition he implied that anyone could understand and practice ecology, just as anyone can learn to do housekeeping chores. This was very in keeping with Earth Day tradition: the environment belongs to everyone, as do its problems.

"The First Law of Ecology: Everything Is Connected to Everything Else," is a simplification of an ecological food web. By Earth Day, that organisms were enmeshed in complex feeding interactions in a biological community was a bedrock of the science. In *The Closing Circle*, it is the hypothetical consequence, only, of the connectedness in his statement of law that is given attention, not the many details that ecologists have learned about food webs.

"A small perturbation in one place may have large, distant, long-delayed effects," he wrote in explanation.

If overstressed, an ecosystem can indeed come to a dramatic collapse. Eutrophication, instabilities in population cycles, and the biomagnification of DDT, all discussed in *The Closing Circle*, are textbook examples of food web interactions gone awry.

Eutrophication is actually a natural process by which crystal-clear lakes have nutrients and silt added to them until some may eventually fill in to become bogs, then terra firma. Eutrophication just is, and bogs happen to be as natural as pristine mountain lakes. The process—when it happens—is a slow one, on the order of centuries. Problems happen, as is true in so many cases, when humans speed it up. Excess sewage unnaturally speeds the progression of clear lake to murky bog, leading to unwanted consequences such as the death of Lake Erie, as an example used by Commoner. (Actually, the pronouncement of Lake Erie's death was unfairly attributed to him. Barry never declared Lake Erie dead in print. "If the lake is not yet 'dead,'" he wrote, "it certainly appears to be in the grip of a fatal disease." Another lovely metaphor.)

Commoner, even with his simplifications, is not far from mainstream ecological science here, but what has come down from him to the public mind is that the "First Law of Ecology" says that we will wreak havoc by any little disturbance to the environment, especially its natural parts. This is not supported by the examples Barry used. Lake Erie did not come to its precarious situation on Earth Day due to a minor disturbance. It was assaulted by the efflux of the large population centers and heavy industries surrounding it. The wolves in Yellowstone did not mysteriously disappear. They were intentionally extirpated by hunters and trappers. Neither had they been some obscure, insignificant part of the Yellowstone ecosystem when their systematic elimination was undertaken.

Still, in reaching the public, a good literary device is worth tomes of weighty scientific reports. A corollary of Commoner's First Law became "The Third Law of Ecology: Nature Knows Best." By logical necessity, to be true, this law requires that the First Law also be true, which it is not. It is also beyond the purview of science, even if we brush aside objections to his First Law. His Third Law is a value judgments, no less so than Ehrlich's opposition to war.

"The Second Law of Ecology: Everything Must Go Somewhere," Commoner claimed to be a simple restatement of a basic law of physics. I think he meant the Law of Conservation of Matter-Energy. Why, if there is a perfectly good law in physics, there was need to rename it for ecology was left unexplained.

Finally, why did he have more than three laws? Newton had three laws. Kepler had three laws. Einstein was satisfied with a single, pithy equation. Nevertheless, "The Fourth Law of Ecology: There Is No Such Thing as a Free Lunch" was an interesting statement that Commoner claimed to have borrowed from economics. Again, it is a law that has no ecological principles implicit in it. I must admit I do find it a remarkably useful law, if that is what it is. Pollution cleanup and control, in particular, seem to fit it well. To remove a pollutant from water, you must add pollutants to the air or soil, or both. In many cases, there is no real pollution removal, only transfer.

Come to think of it, Barry Commoner's Laws of Ecology would make marvelous Laws of Pollution. He just had the name wrong. He called them ecological laws no doubt because it was attractive to be identified as an ecologist. Or perhaps it was because the use of the word ecology evoked strong feelings in the listener or reader at that time that an author could put to his use. It still does.

This was the situation of ecology as of Earth Day. As propelled to public attention by the literary skills of Rachel Carson, among others, it was a concept stripped of tools and quantitative principles. Meanwhile, most ecologists were too perplexed about their field to enter the public arena—or did so at the risk of stretching well past credulity what little knowledge they actually had about the environment.

Kenneth Watt, a University of California ecologist noted for his use of computers to model insect population growth, provides a notable example. Watt offered the following conclusion in a national magazine on the eve of Earth Week: "At the present rate of nitrogen buildup, it's only a matter of time before light will be filtered out of the atmosphere and none of our land will be usable."

It fell to Commoner and others to speak on those subjects that ecologists could not or would not address publicly. I don't think he advanced the public perception of the science much beyond that created by Rachel Carson.

"Perhaps I Have Said Too Much Already"

Gaylord Nelson may have been responsible for gathering the forces needed to have an Earth Day, but it was Rachel Carson's *Silent Spring* that nurtured the views that came to fruition on that day. Unwittingly, she also created a crisis of sorts in the science of ecology.

Here are but a few of the accolades the book received. Anthropologist Loren Eiseley suggested that "Silent Spring should be read by every American who does not want it to be the epitaph of a world not very far beyond us in time." Pesticide scientist Robert Rudd wrote that "*Silent Spring* is biological warning, social commentary and moral reminder." Placing Rachel Carson even higher on the scale of scientific worth, conservation biologist Paul Sears wrote that in *Silent Spring*, Rachel Carson "has also produced an important contribution to ecology."

That contribution was one of changing ecology forever. Ecologists themselves concluded in a 1965 report on the future of ecology that "Rachel Carson's book *Silent Spring* created a tide of opinion which will never again allow professional ecologists to remain comfortably aloof from public responsibility."

"There was a major sea change in the ESA from the Applied Ecology Committee having little respect before the publication of *Silent Spring* to becoming very respectable," Gene E Likens, who helped to make that change, recalls of the shift from an essentially academic orientation in ecology to one of practical applications. The ESA is the Ecological Society of America. It is the professional society for ecologists in the US. Purists then running the Society did not think ecologists should be involved in applied problems. Indeed, a common criterion in the choice of study habitats was their relative lack of influence by man's activities. The Applied Ecology Section went on to be the biggest membership of any section in the ESA, the sea change implied by Likens.

Rachel Carson, therefore, has presented me with a logical device to explain, rather than a literary one. She has given me a paradox. Although trained as a scientist, when she wrote *Silent Spring* she was no longer practicing her science. She had long before abandoned any connection with formal scientific research. In fact, it could be argued that she never had been a practicing scientist. Early critics from the chemical industry pounced on that weakness in her background. According to them, Carson did not have the credentials necessary to speak publicly as a professional scientist: she had earned only a master's degree, had little actual research experience, held no academic appointment, and had never published in a peer-reviewed journal.

Rachel Carson had anticipated the attacks. Prior to the publication of *Silent Spring*, she had taken pains to put her scientific credentials in the best light possible. She insisted that the book jacket stress that she had studied genetics at Johns Hopkins University under Herbert S Jennings and Raymond Pearl, two highly regarded scientists, and had taken an early interest in the effects of synthetic pesticides.

Later industry attacks, especially from those who admitted to not having read the book, were less circumspect about Carson's perceived shortcomings. They accused her of more than just amateurism. They called her a nature-loving faddist.

Carson's career is traced most simply on the Signet editions of her nature books. Published after her death, their covers all carry the same biography. She had done graduate work at Johns Hopkins University and at the Woods Hole Marine Biological Laboratory, the blurbs tell us. She had taught at the University of Maryland and Johns Hopkins and had worked for several years as a marine biologist. The latter position had presumably been with the Fish and Wildlife Service, although the syntax of the blurb on the book covers, identical in each, does not make it clear. Testing the syntax of any blurb writer, the Fish and Wildlife Service had been formed during Carson's early tenure there by combining the Bureau of Fisheries with the US Biological Survey. Testing my own ability to weave a simple story from the complexities that my subject matter has presented me, a historian suggested that I look, instead, at the hardcover jackets. A used bookstore in St Johnsbury, Vermont, that had just recently uncovered a medieval illuminated manuscript that sold for six figures, managed to provide a hardcover copy—with jacket—of *The Sea Around Us* for three figures, including pennies. In this case, the author's biography emphasized her writing ability. Did the publication of *Silent Spring*, her final work, require the change in credentials, I wonder?

An answer to that question can only be found in the haunts peculiar to professional historians: the special collections rooms of the nation's finest libraries. Those professionals will no doubt be amused that I found not the answer to that question in those rooms, but another, more interesting question. It had to do with how Rachel Carson, until then almost paralyzed by timidity in putting forth her own scientific conclusions, found the courage to place her point of view on the issues of pesticides before the public in *Silent Spring*.

Carson always did well in her science courses. Meticulousness and a talent with language let her star in course work. However, research seemed to be more problematic for her. Yet research is what separates the science major from the practicing scientist, independent research in particular. Independent research requires a scientist to rely on his or her own instincts, rather than being guided by others. It is success at this that becomes the mark of an expert and leader in a field. And it is success that we equate with authority.

Many brilliant students fall during their climb up the hierarchy of science when they come face-to-face with the need to do independent research. Usually this comes to them in the form of thesis research.

I have seen budding scientists stymied by the challenge of research in a number of ways. Some just could not find a research problem, while others could not narrow the problem they chose, blundering through good ideas without somehow converting them to actual research. Some stumble on or—on too many occasions—are given a problem that is amenable to research and the resources necessary for its investigation, but somehow never bring their researches to a successful conclusion. They seem always to be troubled by too many doubts, too many loose ends. Or what they do manage to glean from their efforts turns out not to be worthy of having set off in search for it in the first place. Outfitted to trap an elephant, they bring home a mouse.

Others simply do not see or believe the significance of their findings and give up and abandon the scientific gold they might have mined. Or they may be incapable of convincing others that the base metal they have discovered is actually gold. Modesty, not ability, becomes their undoing.

It is not an unusual situation in the academic world for a capable—even brilliant—student to fail at independent research. There is even a term for the condition: the ABD (All But Doctorate). It is not limited to scientists. Many ABDs do eventually find the way through the thesis maze after years of deprivation for themselves and their loved ones, but more often they do not. Those that stay ABDs usually either leave research or become valuable, although not independent members of the scientific research establishment. They can be found in the background at famous research laboratories. They keep the labs running. They keep the research effort moving. They maintain productivity. They advance the careers of the senior scientists (almost exclusively men during Carson's life, it should be noted). They are the nurses of the operating room, necessary, but not as valued as the surgeons.

Making no progress in her initial master's research on the cranial nerves of reptiles, Rachel Carson eventually wrote a thesis on the pronephros (a kind of kidney) of catfish. Why she chose cranial nerves as a topic for investigation is not clear. I can't imagine it being from childhood enthusiasm for the subject. Her attraction to nature was to be in it, not to dissect it. Neither is the pronephros an organ she could have developed a particular enthusiasm for during her science studies. I assume its study had been suggested to her, since she is known to have worked on the topic while still enrolled at Pennsylvania College for Women. Possibly it had been suggested by Rheinhart P Cowles, with whom she studied during her first summer session at the Marine Biological Laboratory, an idyllic scientific Valhalla along the southern Cape Cod seashore at Woods Hole, Massachusetts. Cowles had Rachel spending a considerable amount of time in the library trying to narrow her research.

Whatever lack of progress in research she experienced at Johns Hopkins, Rachel bitterly blamed on the inadequate training she had received in her last year of college. The psychological makeup of young investigators has changed little over the years, apparently. Externalizing blame for lack of progress is an affliction that is common to them. It must confer needed psychological benefits. I myself have succumbed to such fallacious thinking more than once.

Abandoning her investigation of cranial nerves, Carson briefly considered studying the function of the "pit" of a pit viper, then traded the snake for a squirrel. The topic she finally settled on was suggested by Cowles, who became her advisor at Johns Hopkins, also. That a senior scientist suggested a suitable research topic to a graduate student was hardly unusual. It is even more common now. The choice of research topic for a graduate student is now determined almost exclusively by what a student's advisor can support with his (or her) research grants. Usually, the topic is on some part of the advisor's research.

Rachel Carson's lack of progress toward a degree was not remarkably unusual, but it was atypical to a degree that did raise concern among the faculty at Hopkins. Carson herself confessed to feelings of being overwhelmed in her research. "I feel as though I'm not getting anywhere as far as the degree is concerned," she wrote to a friend. The load she was feeling was just too great, she said, "at least unless you're an Amazon." Her slow progress and need for extra financial support marked her as a difficult student for the faculty.

Finally, however, she did manage to complete a Master's thesis on the pronephros, which she submitted to an examining committee for approval. Cowles and EA Andrews, who taught marine ecology, a course not listed as part of Carson's curriculum, were the members of the committee. Usually a thesis committee will critically read a thesis then have the student defend it orally before their questioning. There are three possible outcomes. On very rare occasions, a thesis will be rejected outright, the student crushed emotionally, and a career ruined. More commonly at a place like Johns Hopkins, the thesis will be approved, a Master of Science degree granted, and a recommendation made for further doctoral work. The other outcome is that the degree is awarded, but is terminal. The student never rises in credentials above the MS level. Sometimes that had been the intent of the student all along, and the occasion is one of triumph. At other times, emotionally crushing, the

MS is a sort of booby prize. No, the committee decision conveys, you are not of a quality to go on for a PhD, but, here, have this nice Master's Degree, instead.

Whatever the examining committee may have thought of it when they approved her master's thesis, none of its research ever made it into print. In addition, neither Jennings nor Pearl, whose names she used to bolster her scientific credibility on publishing *Silent Spring*, could express any enthusiasm for her research or scientific promise in letters they submitted to her credentials file at Johns Hopkins. Jennings evaluation of her on March 29, 1935, was "not brilliant."

This same thesis work also made up the bulk of the research Carson did during summer sessions at the Marine Biological Laboratory at Wood's Hole. It became the last of her research activities. Although it was clear that scientific research had been a goal for Carson, once she left academia, much of the "research" she did was more on the order of either library research or the simple observations of a naturalist. Mary Scott Skinker, meanwhile, Rachel Carson's mentor at Pennsylvania College for Women, had recommended her to John's Hopkins as a someone who desired to contribute to science even with her undergraduate work.

"She hopes to carry the work to the point of publishing on some of the forms of reptiles on which no figures have been published," Skinker had written in her recommendation.

On leaving Hopkins, Carson listed her profession as "Scientific research."

The "field research" that Carson did in later life for her books was little more than just observing and taking notes along a beach. Here is an example paragraph from notes Carson made while researching for *The Edge of the Sea*: "Saw tracks of a shore bird—probably a sanderling, and followed them a little, then they turned toward the water and were soon obliterated by the sea. How much it washes away, and makes as though it had never been. Time itself is like the sea, containing all that came before us, sooner or later sweeping us away on its flood and washing over and obliterating the traces of our presence, as the sea this morning erased the footprints of the bird." The passage did not make it into *The Edge of the Sea*. Perhaps it was too poetic for what Carson saw as her most scientific work to date. The paragraph is typical of her notebooks, however. They are more literary journal than research notebook.

At the Beineke Library at Yale University, shorn of all manner of material having pockets into which documents might fit (including standard folding manila envelopes) and all writing implements but pencils having soft lead (provided), I was allowed to examine Rachel Carson's notebooks. It was not quite the somber experience of the reporter going through personal papers in Orson Welles's famous film "Citizen Kane." That experience was more like looking at Rachel Carson's student folder at Johns Hopkins, except that the huge empty room of the movie, with its massive table, became in my experience a pleasant, book-filled room with a number of unimposing tables, some pleasingly cluttered. That room and the Hopkins staff invited, rather than forbade my examination of the documents provided, one set of which I had not asked for, but had been shown to me by the librarian in his wisdom. Instead of the dour guard and librarian in Welles's masterpiece, I was watched over by a pleasant undergraduate. At Beineke, I looked at the more extensive papers found there having to do with Rachel Carson in a bright room in the company of other scholars searching for their own historical nuggets. We were watched in stealth, I imagine, as much by some sort of electronic surveillance system as by the pleasant people staffing the reference desk some distance away.

Her notes for her first book, *Under the Sea Wind*, published in 1941, are those of a naturalist having had scientific training. Locations are relatively precise and observations are quantitative. "Shore Road near Falmouth," she wrote in a notebook. "Pond on left of road. 12 herring gulls (only 1 in full plumage)." There is mention of plankton distributions in zones having 15-foot intervals. Notes on intended revisions suggest a need to include more science. "Science explains—" she jotted down on one page, "normal range—When pop. pressure great, many spill into new territory." There are also the items that could only have the reader in mind: drawings of marine organisms, a list of plankton that eat "fishes," the latter doubly underlined, and then the question "What age child do editors prefer to attract?"

And then, in a spiral notebook, there is: "Ripples in sand in shallow water as though shadows of surface ripples had dug deep & become permanent."

It is the voice of a poet that is uncovered. That poet may have come to the fore in Rachel Carson's makeup as the years went on, but the poet never did eliminate the scientist from her persona. *Silent Spring* is evidence of that.

In evaluating her scientific development—or lack thereof—it should be kept in mind that Carson had the misfortune of completing her dissertation for a Master's Degree in the face of a lack of enthusiasm by the faculty examining it and a severe world-wide economic depression. Given her penurious financial situation and the expectation that graduate students at Hopkins would go on for their PhDs, Rachel Carson found herself in the classic situation of a for-certain ABD. Sometimes, of course, fortunate circumstances can take the ABD up the next rung of the science ladder, even if lacking ability to do independent research. For example, some ABDs, especially during that era, would obtain faculty positions. Certain of these fortunates not only would obtain faculty positions, but, with the help of generous guidance, would even be able to complete their doctoral requirements. The latter, those who eventually obtained their PhD when well into a teaching career, may then have been in possession of all the external signs of practicing authority, professors of their specialized subject matter.

All but one sign, that is. The scientific enterprise has established one more hurdle to overcome before one can speak authoritatively as a scientist. To claim expertise in a subject area, a scientist has to publish on the subject and she has to publish in a peer-reviewed publication. A dissertation on file at a university library has little more chance of its ideas being circulated within the scientific community than the copy its author keeps on a shelf at home.

Peer review is a process intended to ensure that what a scientist publishes is free from error, contradiction, or bias. Scientists are just as capable today of publishing their ideas, as did Newton and Darwin, on their own, in any form they wish. Scientists with an existing track record will often marshal together their ideas and take them to a publisher. (Publishers also send manuscripts, in parts, out for expert review before agreeing to publish.) Less common is a scientist who eschews peer review entirely. There have, however, been notable examples of publications circumventing the review process, some quite recent.

Evolutionary theorist Leigh Van Valen, for example, created his own scientific journal in 1973, appropriately entitled *Evolutionary Theory*, in order to publish his perfectly legitimate ideas. His act of "publication" also required distribution of the journal to suitable libraries, the Library of Congress, in particular. I seem to recall that he delivered some in person, handing copies of his journal over to library catalogers. Van Valen was an intellectual descendant of Tom Paine. Both were rebels and pamphleteers.

More recently, physicist Stephen Wolfram decided to do what Darwin did. Wolfram sprang his ideas on science all at once in book form. Like Darwin, he had guarded them carefully before publication. One reviewer in the *New York Times Book Review* marveled at Wolfram's ability in circumventing having "to dispatch regular reports from the field—unreadable papers published in fashionable zines like *Physical Review Letters* or *Physica A*." Physicists are still at work making sense of the ideas Wolfram published in his 1263-page tome, which presumably is more readable, at least, than the technical papers he would first have had to produce to get the same amount of attention. (Maybe less attention, actually.)

These notable exceptions not withstanding, the vast majority of scientists prefer to publish in peer-reviewed publications. These are also known as "refereed" publications, because not only do reviewers read and assess papers, they also pass judgment on them. In baseball terms, they make the "safe or out" calls. As can be expected in all things in our society, there is a pecking order in their prestige. It is based on the rigor of the review a manuscript receives before being accepted for publication. The most prestigious journals are national publications with the most rigorous review, followed by journals published by reputable professional societies. "Reputable" is often synonymous with old and stuffy. The journals of commercial publishers follow far behind. Their review process is less rigorous. They sometimes print articles based on a quick read and decision by their editor, rather than formal review processes.

Skip ahead another fifteen years and the "Dodge City" of the Internet provides multiple on-line outlets that are available essentially for hire. Publication without review or rigor is now possible, especially in certain fields. The current situation is, however, best left to some future historian of science to analyze, so we will return to the transitional era in ecology that is our subject.

Rigor in review is synonymous with rejection rate. As has become true with Ivy League colleges, a high rejection rate of applicants becomes a de facto indicator of quality. For our purposes, *Science*, in the US, and *Nature*, in the UK, represent the Ivies in selectivity. As with selective Ivy League colleges, less than 10% of submissions to *Science* are accepted for publication. The *American Naturalist* and *Ecology*, both published by professional societies, represent the level of prestige that might be accorded to great public research universities, such as the University of Michigan and the University of California at Berkeley. They have lower rejection rates. Strangely, like vanity presses, they charge authors for inclusion between the prestigious covers of the societies' low-circulation journals. These are known as "page charges" and can be a hardship to a scientist on the outside of the well-funded scientific establishment.

Commercial presses, as opposed to those of professional societies, turn a profit on their journals by charging libraries outrageous subscription fees. Some also have advertisements. They do not have page charges. The journals published by commercial presses, *Oecologia*, *Journal of Environmental Management*, and *Ecosystems* are examples in which ecologists might publish, are accorded a lower level of prestige. They are not quite the "degree-granting, but non-learning institutions" that one acquaintance of mine has used to describe the vast majority of American institutions of higher learning—this great unwashed is represented by various journals with the word education in them, such as *The Journal of Environmental Education*. Scientists shun these unless they have the misfortune to be studying education in their research. However, it must be said that sometimes a very good paper will appear in a low-prestige journal exactly because of the journal's kinder review and publication policies. Scientists, like all of us, can get frustrated by having to jump through what appear to be arbitrary hoops and rebel against them. More often, though, distinguished scientists will declare a symposium on a subject of interest to them and will get their ideas into print in a handsomely bound volume they may themselves have edited.

On-line and open-access publications were still but gleams in the eyes of futuristic-thinking scientists when the events in this book headed to a resolution. For those with an interest, more will be said of them in the new chapter that concludes this book.

Peer review, let me repeat, has the single goal of maintaining rigor in science. It is science's way of preventing crackpots and charlatans—accusations that were leveled at Rachel Carson—from foisting their ideas on an unwary public. To that end, a scientist's submission is sent out for review to one or more other scientists knowledgeable on the publication's subject matter. They should have no ties to the scientist who submitted the paper or her institution. The preference is for more than one referee, for scientists are not known to always agree. The requirement of no affiliation between author and referees makes the review an "external" one, supposedly free from the biases that a scientist's colleagues might have toward her work. These might make a referee more inclined to accept a paper, although petty jealousy and turf battles with colleagues can sometimes have the opposite effect. In either case, the bias of a colleague is avoided. Nothing, of course, can guarantee unbiased assessment, as a consideration of human nature will quickly reveal. Scientists have all sorts of biases, none more common than the perceived threat a submission might represent to their own research status. The scientific establishment is huge, but the groups of specialists in a particular research area capable of refereeing a submission are small. Commonly they all know each other. Publications or research proposals that contradict a reviewer's own research will quickly be faulted, although not for that reason. Errors in interpretation or methodology will be uncovered, instead. (The methodology might be at odds with that used in the referee's own research—clearly grounds for rejection.)

When a referee turns down a paper, the journal's editor can either suffer the scorn of the reviewers, who are often prestigious scientists, or return the paper and suffer the scorn of the author. This the editor can handily deflect back upon the anonymous reviewers. On the whole, the process works, but it is far from a perfect one. Truly innovative work has been known to be misunderstood or disbelieved and rejected for publication. Outright fraud slips past the most rigorous review. Nevertheless, even with the above vagaries, publication in a peer-reviewed journal is still the most agreed-upon imprimatur of legitimacy for a scientific work and for its author as a scientist.

A scientist without publications is usually called a teacher. That is the ultimate result of the intricate process I have just detailed. Without a refereed publication, that is the category in which we must put Rachel Carson.

A paper entitled "The Edge of the Sea," presented at an American Association for the Advancement of Science (AAAS) symposium held in Boston, was the only purely scientific paper Rachel Carson ever gave to a professional academic organization. In it, she pursued questions such as "Why does an animal live where it does?" and "What is the nature of the ties that bind it to its world?" Although the ideas in answer to those questions were not her own, still, she balked at presenting the paper, although the correspondence over it suggests that copyright issues and their ensuing monetary considerations were as much an issue as Carson's timidity in taking on the role of scientist, rather than popularizer.

The AAAS is a venerable and respected scientific organization. It is the publisher of the journal *Science*. Membership is egalitarian, however, open to all who subscribe to the advancement of science and pay the required dues. The keynote speakers are not always the giants of a science. The 2002 AAAS meetings in Boston, for example, listed Colin Powell as an invited lecturer. (Duties of State after September 11 presumably made him decline the invitation.)

Carson's only paper thus was not refereed in the manner of a submission to a research journal, if at all. Having no new research to be evaluated, her talk had no need of peer review. Neither did it find its way into print form in any official AAAS publication.

There had been another scientific paper written by Carson, however. It was written while she was still with the US Fish and Wildlife Service and was based on some of her own observations. She gave it the title, "Report on Amber Tide Studies on the Florida West Coast During July 1947." It is part of something called the "gray" literature in science, in this case, an internally circulated report. Although gray literature sometimes can make legitimate contributions to science, this report stayed forever gray and it did so out of scientific timidity.

"I have attempted to set forth the general observations made in the field together with what information of a more detailed nature it was possible to do in the field," she concluded about her study. "At the same time I have set forth the theories and ideas that appear reasonable to me on the basis of these observations. Until Dr. Galtsoff and the taxocologist [sic] are able to report on their findings from the material furnished them, it appears futile for me to attempt to say more. Perhaps I have said too much already."

It is a timidity that echoes that in her master's dissertation. After some 100 pages of bland description, finally, there comes a chance in it for some original thought. This has to do with the contrasting development of the organ under study in trout and salmon as compared to catfish. Forgive me, but Carson did not rise to the bait.

"It is not the intention of the author to offer at this time any theory of the origin of the pseudolymphoid tissue in the catfish," was her conclusion.

Rachel Carson had shied away from developing the credentials of a credible, practicing scientific authority. What, then, had given her the courage to step forth with the case histories in *Silent Spring* that she used to illustrate the hazards of pesticides to humans and the conclusions she made from them? The chemical manufacturers had gotten one thing right: Rachel Carson's talents and accomplishments were in popularizing science, rather than in making scientific judgments. And she had been warned to expect the attacks that came.

But without answering my question, two others came to my mind in poring through old documents in historical collections. What did she know of the science of ecology that she helped to revolutionize, and when did she know it?

Great AEPPS

I am fortunate that Rachel Carson inspired a biography as meticulously detailed as Linda Lear's *Rachel Carson: Witness for Nature*. It helps me establish that Carson never had a formal course in ecology. That may be one reason that questions about Rachel Carson's scientific knowledge base arose not just from detractors in the pesticide industry, but also from those in the field she adopted late in her popular writings and used to good effect in *Silent Spring*—ecology.

Ecology courses were not readily available on US campuses circa 1930. There were none at Johns Hopkins University, a leader in graduate education in America, during the years Carson was enrolled there. Neither was ecology mentioned in the description of the content of any of the courses offered there, although at least one, "The Life Histories and Classification of Plants," sounds like it could have used the term in its description. Botanists in the United States at that time were quite active in what they specifically labeled as ecological studies. Few American colleges and universities, however, offered courses in ecology even as late as the 1950s, leading a number of publishers to turn down Eugene P Odum's original version of his classic textbook. Like Macmillan, which first encouraged then declined Odum's popular ecology text, those publishers concluded from surveying American college courses that there would be no demand for the book.

Complicating things for me, ecology in the middle of the 20th century was a science in transformation. Its transitional and polymorphous nature during Rachel Carson's life makes it difficult to use specific targets as indicators of ecological expertise on her part. I could not, for example, scan her resume the same way I might a current scientist—or job applicant fresh out of college—to see if she was trained in ecology. I could not exclude the possibility that Rachel Carson picked up her knowledge of ecology in her course work, even without courses labeled specifically "ecology" on her transcript. Without actual lecture notes, who can know what was taught in a course the better part of a century in the past? A few specific ecology courses have been identified in graduate curricula in the US as far back as 1901 (notably at the University of Chicago), but in many cases the material was taught as part of other courses, botany and zoology, for example. In fact, ecology was seen from its beginnings to bw a point of view, rather than a separate discipline. Practitioners of that point of view were found in a variety of academic departments. Plant ecology, animal ecology, oceanography and limnology have been identified as courses offered through the 1920s in which an ecological point of view could be found. In addition, practical courses in agriculture, game management, and fisheries science might also have had some ecological content—or not, depending on the vagaries of the instructor.

The identity crisis in ecology was then so bad that, like others, I have been forced to distinguish between "self-conscious" ecologists—those who called themselves ecologists—and those who presumably contributed to the science of ecology unwittingly. When I looked at the Hopkins faculty, I found that, although none were charter members of the ESA, three of the six-member faculty of 1929 were listed in its first membership directory, published in 1917. Now on the surface, membership in that society should be an indication of expertise in that new science, but real life is never so easy as logic might dictate. To this day, the society has tried to be all-inclusive in its membership requirements. There is no testing, no degree or curricular requirements to meet in gaining membership. Nothing more is needed than a desire to belong that is spurred (apparently) by a love of nature and natural history studies.

None of the three Hopkins "ecologists" listed specifically ecological topics of interest in response to questions posed to them by the ESA, unlike some of the more famous "self-conscious" ecologists listed, in addition to a number of lesser known, but apparently self-aware ecologists. (EA Andrews, who was on Carson's examining committee, dropped out of the society before she enrolled there.) Yet all three published in the ESA's journals. Articles on the distribution of vegetation and on techniques for measuring humidity were published just prior to and during Rachel Carson's time there. She therefore had the opportunity at Hopkins to be exposed to two venerable parts of ecology: animal and plant physiology. That and the natural history studies that make up the part of ecology most accessible to the general public—there is no need for a professional degree to read a nature book—provided sufficient background for her books about the sea. It was not a sufficient background for *Silent Spring*, however.

Other influences on Carson are similarly ambiguous. Raymond Pearl, who developed the logistic growth equation that is so fundamental to ecology and in whose laboratory Carson finished out her education at Hopkins, could reasonably be called a population ecologist, but he was in fact not an ecologist at all. His specialty was human biology. He pioneered the science of demography. That Pearl's work could in turn exert the influence it did on ecology is indicative of what an amorphous field of science it was at that time. Similarly, Herbert Jennings, who had judged her not to be brilliant, was not recognizable as an ecologist when Rachel Carson was enrolled in his classes, yet a 1965 compendium of ecological literature includes a 1904 paper on methodology by him.

Given the density of ecologists—avowedly so or otherwise—at Hopkins during Rachel Carson's stay there, she could not have avoided exposure to ideas of that science. But which ideas? And to what effect? Pearl's logistic equation, a foundation for ecological and environmental thought, never finds its way into Carson's works, even by inference. His influence may have been more through his holism than his population studies. It was and still is a popular way to look at science. In Pearl's holistic view, one that looks at the totality of interactions, biological studies should serve to promote a better understanding of the human condition. Pearl's point of view—with man a part of, not apart from biology—was shared also by Jennings. Man being a part of, rather than apart from nature was exactly what Carson urged in *Silent Spring*.

Carson had to have been exposed to ecology at Hopkins, but it was not the ecology that was to burst out in America after World War II. It was not what would in essence simultaneously be proclaimed as the science of the environment. The ecology to which she was exposed was the animal physiology and descriptive plant ecology of her professors at Hopkins. The ideas of Frederick E Clements, Victor E Shelford, Charles S Elton, and Giorgii Gause (and Pearl), which would soon be stirred together with those of others to create a new ecology, left no tracks leading through her graduate experience at Hopkins. We can tell that by reading her works and examining her notes.

First, there is that timid Master's dissertation. One hundred and one pages and no hint anywhere of those ecological ideas that had to have been in the air at Hopkins.

Her experience at the Woods Hole Marine Biology Laboratory might have been more important in her training than her classwork at Hopkins. When Carson wrote "research scientist" as her career goal on coming out of Hopkins, it must have been those scientists at Woods Hole who were her model. The lifelong influence of youthful experiences and environments should never be underestimated. The "Grand Tours" on which Victorian youth were sent accomplished much more than an appreciation for scenery. Our exchange programs for youths and the opportunities for studies abroad prevalent on many college campuses are democratized versions of those "Grand Tours." Rachel Carson was twenty-two when she found herself at Woods Hole in the setting of a picture-book village situated by the sea next to wharves from which fascinatingly equipped research vessels set out. It was one of those "quaint little villages here and there" romanticized in the song by Patti Page a generation later, making of Cape Cod, sight unseen, a favorite American place. I was in my fifties when I first set foot there and I was not immune to the pull of the place on one's career choice. It convinced me that her books on the sea would not have been written without her experience at Woods Hole. It must have initiated a lifelong passion for the seashore that culminated in her cottage on the Maine shore. It also provided her with lifelong contacts. Until *Silent Spring* caused her to broaden her contacts, the scientific advice she sought was most often from people with a Woods Hole background.

Yet as impossible as it is to establish what she might have read while at Hopkins or heard in lectures, Woods Hole presents an even greater mystery. We cannot even establish what the subject of her research was her second summer at Woods Hole, when her view of science must have matured considerably from the undergraduate one of her first summer there. Modern ideas of ecology were at least in the air at Hopkins, but the quality of air was different at Woods Hole. One pioneering animal ecologist of the day, for example, called "that Woods Hole establishment" anti-ecological. In the photographs that have come down to us of Woods Hole scientists of those days, they are almost invariably posed with microscopes. A laboratory emphasis is prominent in the topics for seminars and lectures during Carson's stays there. However, there was in residence at Woods Hole in the 1920s someone who was helping to shape what ecology was to become at mid-century. Warder Clyde Allee would go on to head what became known as the "Chicago School" of ecology. He also co-authored the highly influential text, *Principles of Animal Ecology*. It came to be known among ecologists as "great AEPPS," after the initials of the five authors. (If you don't get it, move the "E" to the other side of the first "P" and delete the second.)

Just as important as formal presentations to the development of its scientists and students at Woods Hole, there were informal interactions, such as in the "mess," the laboratory's highly informal dining room. Human nature has not really changed that much since Rachel Carson's youth. A spirited discussion could have had as much impact on Carson's thinking as a lecture. And there were many of those at the "mess."

Besides the "mess," the Woods Hole library could have been a source of information to Carson. Then as now, it was first rate. On its shelves, all of the latest in ecology was there at her disposal. Unfortunately, there was no old library card waiting for me there with lists of the materials taken out by her. Both the discussions and subjects of her reading are lost to us.

Therefore, I have gone through Carson's writings and examined the works she attributed. Presumably these were the sources of her knowledge. The revised edition of *The Sea Around Us* is found to list *Ecological Animal Geography* (Allee was one of its author-translators) as further reading. It is not known what Carson absorbed from it. "About a fourth of the book is concerned with marine animals," she made note of it for her readers.

The marine environment was, after all, her love. Although there are no notes in her research notebooks from *Ecological Animal Geography*, she probably would have been interested in specific species, their distribution, and their life histories that can be found in it. Her research notebooks, preserved at Yale University, give me confidence in that conclusion. In a folder of material labeled "Ecology" that was used in preparing *The Edge of the Sea*, there are 23 pages of notes on a paper having to do with the species present in a tidal inlet and their distribution. In the same box is a request for an issue of *Ecological Monographs*. The object of that request managed to be preserved in the Yale collection.

Scientists then—as now, supposedly—published in order to communicate the facts they uncovered and their ideas in interpreting those facts. Technology today pretty much allows for instant access to such work. Personal research files the size of some libraries can now be kept on disks smaller than kitchen plates or even an ignition key. In my scientific youth, copying machines let me bring knowledge back with me from a library when it was found in a non-circulating journal. In Rachel Carson's day, she would have needed to take notes on works in the library. Remember those stacks of 3x5 index cards? (If not, have you wondered what their purpose might have once been?) Or she could request a reprint from the author of the piece. By arrangement with the journals—and at a price—an author could have his article printed by itself. He could then distribute these "reprints" to others out of vanity or upon their request. In this case, for reasons unknown, Rachel Carson purchased an entire issue from its publisher. They often printed a larger number than were subscribed to and would store and sell off the extras as long as there was a supply.

What had interested her in that volume was a report on the ecology of sand beaches in Beaufort, North Carolina. Parts of it are heavily annotated with underlines and the occasional parenthetical remark. The annotations provide an open window to what interested her. She was particularly drawn to the descriptive sections. The section titled "Adaptations of Sand Beach Animals" is heavily annotated. It is a who-is-who and who-does-what-to-whom of that seashore. Unmarked by Carson is the main data table. Unmarked also is a section entitled "Seasonal Progression on Sand Beaches." That section, at least, attempted to bring the research more into line with 20th century ecology, something that strived to be more than description and simple natural history. Neither did Rachel Carson seem to care much for what was written about the plants in the paper. Marine organisms, what they eat and what eats them, appear to have been Carson's overriding interest in the ecology of sand beaches.

Another monograph she requested was on a tidal inlet at Cape Ann, Massachusetts. I tried to map, in effect, the rather dry information in the monograph to Carson's lively prose. There is a section in *The Edge of the Sea* that just happens to be on that very inlet. Here was an opportunity to see what science she used from the monograph and how she used it. I was totally without success.

What had first captured my attention in that paper was a portion of its title. It proclaimed itself to be "A Study in Bio-ecology." Bio-ecology was the term used by Shelford and Clements for their attempt to combine animal and plant ecology through the use of what was called the "community concept." Their idea was that an organism's characteristics had to be examined not just in relation to non-living factors, but also against the framework of its interactions with suites of other organisms. Recognizing the amorphous nature of ecology, as spread through various academic departments, they also saw in the term a way to escape the ambiguous meaning attached to "ecology." Bio-ecology was very much the science of communities. Clements had also championed the "superorganism" concept, about which more will come later. Shelford, who turned his attention to the role of animals in plant communities while at the University of Chicago, managed to help build the Chicago School, even though, having trained Allee, he then moved to the University of Illinois, where he finished out his long career. Allee stayed on at Chicago.

Unfortunately, whether Carson bothered to follow up these ideas through the sources cited in the monograph on Cape Ann is unknown. The paper, composed mostly of species lists and descriptive material, was not a good example of bio-ecology. Along with lists of species, there were also lists of terms, another sore point among ecologists of the day. Most likely, Carson's interests in it were similar to the interests she showed through her annotation and underlining of the study of sand beaches in Beaufort.

In spite of the anti-ecology establishment at Woods Hole a generation before, Howard T and Eugene Odum did offer pioneering advanced course work there in ecosystem ecology, the next development of the community idea, in the summers of 1957 to 1961. Had she been there, Rachel Carson may have learned some modern ecology while working on *Silent Spring*. Her seashores of preference by then were the rocky coasts of Maine, however, rather than the beaches of Cape Cod. She had stopped going to Woods Hole.

Rachel Carson's very use of the word "ecology" in *Silent Spring* offers clues that her concept of ecology was far from that of mainstream ecologists such as the Odum brothers. Carson's use is quaint in contrast to theirs, antiquated even for her era. She uses it mainly as "the ecology," letting it stand for an abstraction about nature. "The ecology" appears to represent the benevolent workings of nature to her. I know that "the ecology" was a concern to Marvin Gaye, but a serious ecologist would never formally use the term in the same way as in Gaye's song, "What's Going On?" The Odums studied the ecology of ecosystems. "The ecology," independent of a specific system, was not one of their concepts, not even for Howard T Odum, who went so far as to promote both a "world system" and an energy basis for religion, both indicative of a willingness to expand the ideas of ecology beyond its normal confines within science. "The ecology" was a concept I am sure he might have supported, but it was not a term that he used in his formal writing. "The ecology" was, however, a concept that Rachel Carson and Aldo Leopold posthumously brought to the public by way of Barry Commoner and the speakers on Earth Day.

Prior to *Silent Spring*, in promoting what was to become *The Edge of the Sea*, Rachel Carson emphasized that it would be a creative work in which the "ecological concept" would dominate and that she knew of "no existing book on the Atlantic coast that stresses the ecological approach." There is no need to reassess that statement —whether there were such books and whether she really used any ecological science—now some than 65 years later, but it is testimony that Rachel Carson had a concept of what ecology was quite some time before embarking on her work for *Silent Spring*. Yet it also leaves me wondering about the intellectual genesis of that concept, Was it only from other works on the Atlantic coast? I know she studied and read much about that coastline, which so stirred her intellectual and creative passions. But where did she find that "ecological concept?"

In her review of the work of others, some current, some from the past, in the paper she presented to AAAS, Carson almost parenthetically quotes, without citation, words of WC Allee, whom she identifies as an animal ecologist at Woods Hole in the early 1930s. What contact Carson had with Allee—or any of the *AEPPS* authors— is difficult to establish. Allee's long career had been almost entirely at the University of Chicago, but he had actively pursued research at Woods Hole, as noted already. Thomas Park, his student and one of the "Ps" in "AEPPS," arrived in Pearl's lab as a post-doc in 1933, but Carson's assistantship there had ended in 1931. Woods Hole's *Biological Bulletins* have Allee as an independent investigator in 1931, 1934, and 1936, but Carson was there in 1929 and 1932. However, Allee was a member of the corporation in 1932 and might have made a brief appearance. Nonetheless, I can find no smoking gun linking Carson to Allee. Whether they met or not, however, he managed somehow to exert an influence on her through the Woods Hole experience.

Judging from her research materials, Carson learned what she needed as she needed it when preparing her books. She did so by selective reading and personal observation. There is no evidence of her having selected "the ecological concept" she was so eager to pursue in *The Edge of the Sea*. There are no notes of her having sat down to read *AEPPS*, or Charles Elton, or Eugene Odum before pursuing the "ecological concept." With the exception of correspondence around *The Edge of the Sea*, ecology is not seen in her correspondence until she was hot in pursuit of the abuse of pesticides for *Silent Spring*. Neither is correspondence with the leading ecologists of her day evident until then. She learned her ecology on the fly.

It is in researching *Silent Spring* that the name Charles Elton appears in her notes. Elton's work represents one of the paths to the ecosystem concept promoted by Eugene Odum that shared the organicism (synonymous for our purposes with holism) that has been seen to underlie Carson's "ecological concept" in *Silent Spring*. This is the path through G Evelyn Hutchinson and Raymond Lindeman. The emphasis on energy and nutrient dynamics in the systems approach of Lindeman, in combination with the radioactive tracer studies of the Odums and others, is what eventually allowed a fundamental ecological explanation for the decline of raptors due to DDT use.

Yet it is not Elton's classic text, *Animal Ecology*, but his popular work, *The Ecology of Invasions*, that Carson made note of. It is a work intended for a non-technical audience, although it is nonetheless a useful work even for the specialist. Carson, however, seemed to have somehow missed an opportunity for entrée into modern ideas in ecology. "Elton recalls the youth of ecology as a science by saying that only 25 years ago it was in its Neolithic age," Carson had carefully typed. What ecology was at the time she wrote those words was left unnoted. She had gone on to conclude: "One has only to look about to see that, in terms of its philosophy, applied entomology is still in its Stone Age." What she had taken from his text was not so much its science, but a literary device.

Carson depended in great part on correspondence to get information she needed. We know that Carson first became aware of Elton's popular—it grew out of three BBC radio broadcasts—book when she was introduced to it by Edward O Wilson while she was already working on the manuscript that was to become *Silent Spring*. It suggests a haphazard way into modern ecology for her. And some of her guides were not those to lead someone unsure about her way. For advice on wildlife and pesticide issues for *Silent Spring*, she relied on personal friends.

Then there was Frank E Egler, a scientific maverick with an ax to grind and a decidedly not dispassionate approach to the science of his choice. A prolific letter writer—they were truly missives in his case—he waged war against herbicide use, enlisting any and all who might help as allies in his cause and writing off as enemies all those who disagreed with him in any way. "I was once an Assoc. Prof. of Physics," he announced in their correspondence on a page that could be instantly recognized as his from across the room. (His writing style could only be described as early Tom! Wolfish, abetted by a recalcitrant typewriter; it let his cantankerous personality show through.) He had his own view of ecology and ecologists.

"I would sooner trust an intelligent and self-educated layman, than a PhD in ecology who is an 'expert' in one of the fashionable specialties of the day," Egler once had announced to the ESA membership in the midst of its controversy over professional certification for ecologists.

It was an opinion Egler had earlier shared with Carson in their correspondence. It is perhaps his influence that resulted in Carson's quaint use of the word "ecology" in *Silent Spring*.

In *Silent Spring*, Rachel Carson equates ecology with "interrelationships" and "interdependence." The indirect poisoning of robins by insecticides reflects "the web of life—or death—that scientists know as ecology," she wrote. This poetic, but careless use of the term is stretched even more in her next sentence, with which she begins a discussion of "an ecology of the world within our bodies."

It must be fair to say that the ecologists who reviewed her book were squarely on the side of protecting the environment—much like ecologists today. Had that not been so, there would have been additional ammunition for the pesticide manufacturers for use against Carson's professionalism.

In truth, though, her ecological arguments had concerned the chemical industry much less than the many more pages that argued that chemical pesticides cause cancer in humans. Perhaps they saw her use of ecology as simply metaphorical, something of a literary device only. And in that there is a sobering thought for me. Perhaps they had investigated the nature writer, like General Motors soon would do with consumer activist Ralph Nader, and had found nothing more in her professional arsenal than the poetic ideas of *A Sand County Almanac*.

An Ecologist Reconsiders Silent Spring after 17 Years

An interest in, but a lack of training in ecology could have been the reason Rachel Carson was so pleased with LaMont Cole's 1962 review of *Silent Spring* in *Scientific American*. It was important in a number of ways. Cole was then one of the nation's leading ecologists, soon to hold the highest offices in the ESA and be its spokesman around the time of the first Earth Day. The issue of *Time* magazine having Barry Commoner on its cover identified Cole as one of the nation's leading ecologists. In *Scientific American*, he would be communicating his science to a very broad scientific audience. (The journal had not yet changed ownership and adopted the popular science format it has today. It was then an outlet for scientists to communicate their latest and most important results in a manner accessible to scientists from all disciplines, as well as to the public.) More importantly, not having assisted in the preparation of Carson's work and having a position at Cornell University, then and now the site of one of the world's leading agricultural schools, Cole could be considered impartial. His ecological credentials were beyond reproach, having produced studies bringing mathematical precision to animal distribution and abundance, population cycles, and life history phenomenon. In each subject, his papers were and still are classics in population ecology. In addition, he was among the first ecologists to touch on the practical environmental applications of general ecological principles. His review was important enough to be revisited by Paul Ehrlich 17 years later.

"As an ecologist," Cole wrote in the review, "I am glad that this provocative book was written." His criticism of it was mainly with the "highly partisan selection of examples and interpretations" chosen by Carson. He found errors of fact to be "infrequent, trivial and irrelevant."

On a negative note, Cole criticized her use of the idea of a "balance of nature," calling it "an obsolete concept among ecologists." This is "the ecology" concept that Cole takes issue with. To Cole, its use was a sign of a lack of training in ecology. It pops up again and again in this story. It is, after all, what the public still thinks ecology is.

Cole especially took Rachel Carson to task for what he saw was her misunderstanding of the evolution of insect resistance to pesticides, claiming "not for a moment" to believe "that the chemicals are producing superinsects." Actually, Cole had it all wrong. Carson had it right. In hindsight, Cole displayed a primitive concept of the genetic and physiological means of insect resistance.

The basis for Cole's criticism of Carson's knowledge of evolution is an idea much older than the science of ecology: that selection—artificial or natural—must be a compromise of sorts. A jack of all trades can be master of none, the old saying goes, and it has found its way into evolutionary theory. Cole uses the example of the sickle-cell trait in humans, in which the mechanism for resistance to malaria also causes anemia. The anemia can be fatal to those known as homozygous for the trait, although decreasingly so in our era of modern medical magic. Heterozygotes inherit both a sickle-cell gene and a normal gene from their parents. Homozygotes are of two types. Both inherit only a single form of the gene from their parents. Either both genes are normal or both genes have the sickle-cell trait. The sickle-cell trait causes hemoglobin to deform, causing red blood cells to take on sickle-like shapes that prevent malaria-causing organisms to multiply, but that also interferes with hemoglobin's function of carrying oxygen to cells. (To those unfamiliar with antique farm implements, a sickle is a curved blade used to harvest wheat. It was part of the hammer and sickle symbol of international communism. Normal red blood cells are as pleasingly round as Frisbees.) A heterozygous individual has normal hemoglobin along with sickled hemoglobin. It is the best and worst of both possibilities. The normal homozygote is healthier than the other two types, except for resistance to malaria, an important exception in many parts of the world.

The trade-off is clear for a population: either anemia or malaria; there can be no freedom from both, for both types must be present by the laws of inheritance. The sickle cell example and that of industrial melanism are the two best examples there are of natural selection at work. In industrial melanism, light colored moths were replaced through natural selection in polluted industrial areas of Great Britain by darker "melanic" forms. Light-colored moths resting in the daytime on soot-darkened tree trunks have been shown to be easier prey for their bird predators than the more cryptically colored melanic forms. Both phenomena provide strong support for the modern theory of evolution. They are taught in grade schools that have not succumbed to the pressure of religious Neanderthals.

Within the sickle-cell example (and industrial melanism), though, there is hidden a mostly philosophical point of view that is based on observations of artificial selection. Animals and plants that are bred for traits valued by humans tend not to survive well in competition in the wild with their progenitors. Selection for one set of traits necessarily deselected a set of other traits. Organisms, as found, are compromises, the idea goes. However, evolutionary trade-off need not occur in every case. Where the total amount of energy available to an organism is limited, using it for one goal must reduce the amount available to others. This cannot be argued, but the energy at issue can become fairly abstract, in a sense representing the energy available for evolution. There is no system of accounting yet developed for evolutionary energy. Furthermore, energy need not be in limited supply in all cases for all traits. Years ago, although well after *Silent Spring*, I found that the trade-off can be an informational one, rather than energetic, and be no trade-off at all. More of one enzyme variant than another can be produced, for example, with no energy difference.

There **were** "superbugs" out there. They were resistant not only to DDT, but also to insecticides to which they had not yet been exposed, with no loss of fitness in other respects. Insects in Australia were found to have resistance to organochlorine insecticides, such as DDT, persisting 15 years without any selection pressure for resistance. And they were no less healthy in other ways than insects lacking resistance. A simple change in cuticle properties was all that the trick took for certain insects. Rachel Carson had this information in front of her when writing *Silent Spring*, but it never made it into the final version of the book. In material she dropped from "Chapter 16, The Rumblings of an Avalanche," she did consider changes in cuticle, behavior, and metabolism in populations as possible mechanisms of resistance. Her training in genetics was showing. Why that material was dropped may never be known.

In fairness to Cole, the research results on DDT and insects in Australia, among other similar findings, were not yet available for him in writing his review. Years later, when media superstar and ecologist Paul Ehrlich reexamined *Silent Spring*, however, he failed to correct this part of Cole's criticism, even though I found numerous reports of insect cross-resistance in research publications that should have been readily accessible to him. One would have thought he would have looked into the matter.

Cole's delving into evolutionary questions begs another question that may already be in the reader's mind. How did we go from ecology, the interrelationships of organisms, to evolution, the change in characteristics of organisms over (usually) long periods of time? That was exactly the focus of an intellectual battle in ecology. Was ecology one or the other? It was already underway when *Silent Spring* was published and it was to intensify after its publication. *Silent Spring*, tying ecology more tightly to issues of the environment rather than evolution, was instrumental in that intensification. It is a can of worms I have been avoiding. For now, let's just take a quick peek inside it.

There needs to be great care in defining the science of ecology. Maverick ecologist Frank Egler once defined it "by the contents of ECOLOGY, ECOLOGICAL MONOGRAPHS, and the E.S.A. BULLETIN," then the official publications of the ESA. He was being facetious. It is an easy way out of a difficult question. Yet ecology in his time was a grab bag of topics. It still is.

Ecology has had a symbiotic relationship with evolutionary biology. Ecology provided the facts to be interpreted using evolutionary processes. Where an animal or plant is found (ecology) is as much an adaptation (evolution) as any physiological function of an organ under study in isolation. That the adaptations could be interspecific was recognized by Darwin.

It was as studies of physiological adaptation by botanists that ecology started out in the United States near the end of the 19th century. German biologist Ernst Haeckel had given ecology a name, extending physiological studies beyond organs to organisms. According to Haeckel, "ecology is the study of all those complex interrelations referred to by Darwin as the conditions of the struggle for existence."

Biogeographers, those intrepid voyagers, among whom were Wallace and Darwin, then looked at adaptations even more broadly, seeing them in the various associations, assemblages, and communities of plants that they identified. Species making up desert communities, for example, seemed to be adapted to their environment in concert, rather than individually. It did not take long before ecologists specializing in both animals and plants on both sides of the Atlantic began to call for ecological studies to break away from strictly laboratory investigations of physiology and from assembling the lists of species and vegetation units that ecologists then produced when studying nature in the field and turn, instead, to investigations on adaptations of species to each other, which could be just as important as adaptations to climatic conditions. This brings us to Clements and the Chicago School and takes us through the ecosystem concept and on to modern ecology.

This strand following the study of adaptations of aggregations of organisms is not the only one that can be drawn through that history, I must admit. And even it is tangled with loose ends. Ideas, methodologies, and fads come in and out of the picture that is ecology in its historical development. Historians want ecology to be a combination of diverse ideas that let them link the science to Gilbert White's pastoralism, Thoreau's transcendentalism, the British organic farming movement, and Nazi vegetarianism, as some have done, yet ecology sprang forth in the nineteenth century as a science having, as all sciences do, unifying ideas. Identifying a mainstream through 20th century ecology may be like tracing the past, current, and future flows of the Mississippi River through Louisiana to the sea. Flowing waters join and commingle then separate out again. Which is the main stem and which a tributary? Which is but a minor channel flowing to the sea and which the navigable river? There is, however, a single progression of ideas that can be traced through the science. Hopefully, I have not forced an unnatural structure to ecological history in the way the Army Corps of Engineers has kept the Mississippi from flowing into the Atchafalaya, but that single strand from plant ecology to ecosystem biology does suitably identify a *mainstream* to ecology that is relevant to *Silent Spring*. How the "loose strands" interwove with the "mainstream" of ecology led to an identity crisis for ecology that has continued to the present.

Just as Eugene Odum and others were trying to make ecology legitimate through the ecosystem approach and another character soon to appear in our story, Robert MacArthur, was trying to make it rigorous and elegant with his mathematics, Carson made it relevant, rather than arcane or esoteric. It suffered additional exacerbation through confusion between ecology and environmentalism, which was also fostered in the public in part by *Silent Spring*.

Ecology today as a specific scientific discipline is the biological study of groups at different levels of organization: populations, species, communities, and ecosystems. In the 1950s, it was being pulled toward energy- and nutrient-oriented ecosystem studies, new population studies, and the new community studies that make up a great part of this story. Thermodynamically based studies of ecosystems were something new and unpleasant to many ecologists. Population and community studies represented a return to the more familiar working out of evolutionary adaptations that had represented a major part of ecology. And then, of course, there was the pull to make it all relevant to the environment. Any way you looked at it, ecologists were being stretched in different directions.

Cole was not stepping out of a recognized area of expertise in ecology with his criticism. Evolutionary studies are inextricably woven into the fabric that is ecology.

Cole also took issue with something Rachel Carson never said, perhaps associating her silence on it with ignorance of the topic. This had to do with views that were then current of what kept animal numbers in check. One view was that mortality was density dependent. A simple equation for population growth promoted by and sometimes named after Raymond Pearl was at the heart of this view. According to it, populations suffer higher death rates when numbers are too high to support unlimited growth. Think of thickly sown seedlings competing for moisture or a herd of wildebeest wallowing in the mud of too little water to sustain the entire herd. The other view was that mortality resulted from factors that were independent of density. Organisms consumed by fire or animals that freeze to death perish independently of their densities. The argument came to be known as one over "density-dependent mortality" or, more simply, "density dependence." As far as *Silent Spring* is concerned, the issue was a red herring. Cole's bringing it up puts ecology in a bad light. Already wrong on the possibilities for evolution in pesticide resistance, why was he taking Rachel Carson to task on an issue over which there is still controversy?

I find the following paragraph of his to be particularly telling. "Some ecologists contend that poisons lack the feedback qualities necessary to regulate population size," Cole wrote, "except when man creates the feedback by poisoning more intensively when populations increase. In this view poisoning should be carefully integrated with continuous study of population trends. I prefer not to commit myself as being for or against this generalization, but once again it illustrates the complexity of the problem of pest control. Most users and distributors of insecticides have probably never even heard this argument."

What Cole was saying, although not committing himself to it, was that mortality had to be density dependent in order for resistance to pesticides to evolve. The reference to "feedback" qualities is to negative feedback controls borrowed from the then infectiously popular science of cybernetics. It had particularly virulent activity in ecologists, as we will see. Ignoring cybernetics as not germane, what

Cole showed in the paragraph above was that he had an ax to grind. Carson's ecological transgression was really no more than having crossed his turf. He is the ecologist and must be deferred to. What he finds to be important must be important. This argument repeats itself in the ecology-environment dichotomy almost as a syllogism. Because it is an issue in ecology, the syllogism goes, and because ecology is important to the environment, the issue must also be important to the environment. It is used as much by non-ecologists as by those actually trained in the science.

Cole's concern in the quoted paragraph was over an unresolved issue in ecology that he did not adequately explain to the reader. In order to have insects evolve any trait, the idea went, it must be selected for in a density-dependent manner. That is the implication of the "feedback" that regulates population size in his paragraph. Yet density-dependent mortality had not been firmly established as fact in ecology when Cole wrote those words. Cole was rather cryptically taking a stance that is still currently under dispute.

I have trouble finding a suitable description for the attitude evident in his final sentence from that paragraph, other than sheer snobbishness. "We, the ecologists," he seemed to say, "are privy to vital knowledge to which others are not."

That Rachel Carson had never heard of an argument that is untrue or irrelevant is hardly an indictment of her science. That an ecologist could belittle others for not knowing an important, but bogus ecological issue says something about ecology and ecologists at that time—and perhaps now.

Finally, there are issues of human nature to consider. Cole was a true curmudgeon. He may also have been, as many of Carson's critics (almost exclusively male) revealed in their criticisms, a misogynist. I can offer personal experience on his curmudgeonliness, having been once one of his targets.

As a young graduate student at Cornell, I had had the good fortune of shepherding a number of other graduate students through a little independent study that turned out to warrant publication in *Science*. Our paper soon found itself on the suggested reading list of an invited lecturer for a course required of PhD ecology students at Cornell. That was heady stuff to this unknown graduate student.

The week of its publication, I happened to go to a seminar at Cornell's Langmuir Laboratories. The home of the Ecology and Systematics Department, Langmuir was not on the Cornell campus, but located several miles away, next to Ithaca's airport. Having to drive there, it was not a place I frequented often, even though it was technically where my home department was housed. However, for a good seminar speaker, as on this day (I think it was Ehrlich), I would find transportation to it from the basement labs and offices of Comstock Hall in which I was a denizen on the main campus. Arriving close to the scheduled starting time, I looked around for a good seat when I noted Dick Root, a well-loved Cornell ecologist, pointing LaMont Cole my way. "There's one of the authors right there," Root said to him in my direction.

It was not difficult to guess what their discussion had been and I gratefully extended my hand to Cole in anticipation of a handshake and maybe even a few words of praise. Graduate students —at least this graduate student, at that time with a wife and new baby at home waiting for me to start supporting them—need all of the positive stroking they can get. A complement from someone like Cole would have gone a long way toward making worthwhile all of the sacrifices of graduate school.

Instead of a handshake, though, I got the kiss-off. Apparently having trouble finding the right words as he kept his hand to himself, Cole blurted out "Terrible" in response to my question of what he thought of our study. He shook his head instead of my hand. He might even have repeated himself. I was suddenly at a loss for words myself. Before I could ask for explanation, Cole and I both noticed that the invited speaker was already at the podium, with whoever was about to introduce him making throat-clearing noises. Awkwardly, we returned to our separate places.

Although I had been looking forward to that day's speaker, after the contretemps with Cole, I could no less have followed what he was saying had he been speaking Chinese. So I left. I slunk out the back, abandoning the seminar and any chance to argue its points with potential colleagues after it. Later my advisor suggested I go talk to Cole and find out what his issues were with the paper. He knew Cole well and probably assumed that he would soften on conversation, but I never took the advice, using some such utterance to myself as "Never truck with fools and drunkards" to cover my humiliation. That Cole would have been very much different a decade earlier in 1962 is highly unlikely.

The point of my story is that Cole seemed constitutionally unable to resist exposing the soft underbelly of some idea that seemed hide-bound on its surface. He admitted it, even taking some pride in the fuss that was raised over his sweeping away of the basis for studies on the causes of periodic population outbreaks. There were no causes because there were no periodic outbreaks. Outbreaks, yes, but not with any regular cycles. It must have been his curmudgeonliness coming out when he faulted Carson for being anti-insect and anti-pesticide. He pointed out that honeybees faced a more difficult threat to survival from the old, non-synthetic pesticides than from DDT. He attributed her "bias and oversimplification" to "what it takes to write a best seller." We now know, however, that synthetic pesticides do not exactly lead to peace of mind in the honeybee hive.

In the end, Cole's review, although obviously meant to be critical, gave Carson's science a stamp of approval. Ehrlich's review in the *Bulletin of the Atomic Scientists* left it there. Both left Carson's interpretation of the ecological aspects of pesticide use essentially untouched by criticism. These are the side effects, as we might call them, of pesticide use, such as destruction of natural enemies, development of resistance to pesticides, and food chain effects (think eggshell thinning in eagles). That Carson's treatment of them escaped criticism should be attributed to the fact that she had it all correctly. Cole could find nothing to criticize beyond what has been already covered, what was, on hindsight, criticized in error.

The revisions to *Silent Spring* that Paul Ehrlich suggested given the passage of 17 years was to place less emphasis on the "balance of nature," a criticism Cole had already leveled, and to add that "plants and herbivores are in a 'coevolutionary race.'" This latter point is a bit of promotion for one of Ehrlich's pet theories. He had coined the term "coevolution." Ehrlich should have spent a little more time in the library examining the literature on resistance to pesticides by insects before writing the review, rather than beating his own drum. He could then have even more emphatically answered in the negative the question he posed to himself: "Could it be, as one physician critic wrote, that the book as science is so much hogwash?" It was not!

And both Cole and, 17 years later, Ehrlich must be taken to task for their insistence that Carson represented pyrethrins as simple molecules. Here is the offending passage: "The result has been a seemingly endless stream of synthetic insecticides. In being man-made—by ingenious laboratory manipulation of the molecules, substituting atoms, altering their arrangement—they differ sharply from the simpler inorganic insecticides of prewar days. These were derived from naturally occurring minerals and plant products—compounds of arsenic, copper, lead, manganese, zinc, and other materials, pyrethrum from the dried flowers of chrysanthemums, nicotine sulfate from the relatives of tobacco, and rotenone from the leguminous plants of the East Indies." Readers should make up their own minds. Was Rachel Carson implying that pyrethrum, nicotine, and rotenone are simple inorganic insecticides? Isn't it more likely that her point was that these toxic chemicals did not require synthesis of new molecules? My conclusion is that the paragraph reflected careless writing, rather than careless chemistry. That Ehrlich and Cole saw it differently suggests that they were looking for things to fault. Why?

Even though Carson had failed to use the right ecological words and possibly (given the science of the time) have even misapplied—although not fatally so—ecological theory at least once, she clearly presented all of the necessary principles, without the technical jargon, in her chapter, "Nature Fights Back." In it she gave a solid, clear, and convincing account of the workings of nature as they apply to or are disrupted by the use of pesticides.

Why then is there not found in *Silent Spring* a discussion posed in those ecological words? Why are words such as niche, habitat, community, ecosystem, trophic level, diversity—any of the words ecologists use to make what they say more clear and specific —not in her text? Why was Carson's lovely synopsis of ecology organized along the "balance of nature" thesis? Certainly the ecological terms cannot be said to have been too technical for her readers, not when chemical names such as triorthocresyl phosphate or cholinesterase did not warrant editing out.

The grudging acceptance of Carson's ecological expertise had to do with the science, I think, rather than the personalities. A science insecure about its status suddenly found itself promoted to a highly visible role in solving environmental problems before it had developed the body of knowledge and tools to fill that role. Indeed, it was before it had come to terms with what it actually was as a science. Ecologists were still trying to sort out whether ecology was a general point of view, a specific predictive science, or an arcane set of descriptive terms and data. And if it was on the verge of becoming a predictive science, there was little agreement of what that science would be like. Would it be physiological? Ecosystem oriented? Or would it return to its roots in the working out of evolutionary adaptations? It was not a time to have what appeared to be an outsider communicating to the public what ecology was.

How could Carson, the amateur ecologist seen in a photograph in tennis shoes poking under the bladder wrack along the Maine shore in search of crabs, have the apparent deficiency in ecological knowledge that Cole pointed out, and yet have hit the nail right on the head? And how is the extent of her influence on the science to be explained?

Maybe her knowledge of ecology was not at all what the evidence I have uncovered putatively suggests. She may, indeed, have had more than the rudimentary knowledge relating to the Chicago School of ecology that must be granted her based on her writings. Maybe she knew the terms all along, sere, ecosystem, and geotome, but took the advice of Frank Egler, a larger-than-life person in the story of the environmental movement in ecology. His enthusiasm toward Rachel Carson, whose cause in reducing needless pesticide use paralleled his own campaign against herbicide use on rights-of-way, took the form of a flood of advice, comments, names of contacts, appeals for appearances, and his own idiosyncratic observations about ecology and ecologists. She had come to know him in the dual personas of Frank Egler and Warren Kenfield, a pseudonym he adopted in writing an eccentric book on landscaping. Always one who despised the needless terminology of his field of "vegetation science," as he called it, Egler seemed to have devoted his career to being ecology's conscience. In praising a draft version of *Silent Spring*, he confessed disillusionment "because of the continuing weakness, failure, and sabotaging of my own colleagues, with their adherence to textbook academicism" over his own 15-year struggle.

In his previous letter he had written: "I am happy that nowhere in your book—even tho this is a book strictly in 'ecology'—did you feel obliged to 'explain' things in terms of plant succession, of climax, of biomes, of all that fantastic terminology and esoteric theory against which, also, I have been fighting for years. Your book, tho it was not intended as such, is the sharpest possible rebuke to the academic ecological world that their peculiar jargon which forms their stock-in-trade is quite superfluous, you have handled the subject matter with no use of it."

Autecologists, Synecologists, and Genecologists

Here is the paradox: little old lady in tennis shoes scoops experts in ecology. It is clear that Rachel Carson was not a practicing ecologist in the sense of the science as it is today. Somehow she managed to produce a significant ecological work. How can that be?

The Ecological Society of America's Board of Professional Certification relatively recently stressed a need "to define standards and formally identify the profession of ecologist" in their society's *Bulletin*. It's a worthy goal. It was a matter on which Frank Egler raised his distinctive voice almost a half century ago. As much as he hated pompous ecologists who would not deign to dirty their hands on practical environmental issues, he came to hate amateurs even more. When at one time he could praise "a remarkable assemblage of individuals" that included Rachel Carson (of course!), Rene Dubos, Barry Commoner, Loren Eiseley, Ian McHarg, and LaMont Cole (surprise!), among others, who "out-ecologize the ecologists," he would later rave in print about ecological consultants that he termed "Instant Ecologists" reaping "the harvest of their Greenbacked Revolution, be they litter-pickers, or specialists in irrelevant fields."

A more recent urgency for certification was driven by the following pronouncements: "Ecology is poised to be one of the lead sciences of the 21st century. Some have called this period 'a century of ecological repair of the planet.'" These are interesting statements. Was it—and is it—ecologists *only* at work to repair the planet? Is all of ecology dedicated to planetary repair? Just what is ecology?

These are not easy questions to answer. I punted this question in the last chapter by following a single set of ideas through the science, my "mainstream" of ecology. Defining a science is always ticklishly elusive. A high school textbook definition of physics, for example, will turn out to be similar—if not identical—to a high school textbook definition of chemistry. Neither is incorrect, but neither is very useful.

Sometimes a science is what its practitioners say it is through what they do, as Egler suggested. Fads, influential ideas (and personalities), the availability of instruments, all of these drive the focus of scientists here and there like leaves before the winds of history. As in my analogy to flowing waters (which may well have been influenced by first-generation-ecologist William F Cooper's "braided stream" analogy for vegetational succession), specific disciplines appear and disappear, coalesce with or break off from others, or extend or contract their boundaries. What had once been heresy may surface as dogma, often by the simple expedient of the heretics outliving their scientific opponents. Ecology suffered these difficulties more than other sciences.

I could also just simply let ecology be defined by what an ecologist does, as ESA's Board of Professional Certification appears to have done in agreement with Egler's facetious definition. This is a serious trap, though, both to the science and to those outside of it trying to understand it, for unlike other sciences, ecology exists simultaneously as a science and as a philosophy of life. I saw how easy it was for people to step into the role of ecologist on Earth Day. I can't say it was all for the good.

Let's go to a definition, after all. Let's go to authorities on the subject and see how they define it. Unfortunately what we get is not one definition, but a multitude. In his famous textbook on ecology, Eugene Odum failed to offer up a single, clear definition of it, opting out with "in the long run the best definition for a broad subject field is probably the shortest and least technical one, as, for example, 'environmental biology.'" There certainly are environmental biologists. I have met some and discussed with them what it was that they were doing. They would have been quite concerned, as would I, to find out that what they were doing constituted all of ecology.

(By the way, his famous textbook was, for the longest time, actually two textbooks. One was a yellow version (3rd Edition) published just after Earth Day; the other, a green version that came out a few years before *Silent Spring*. People tended (and still do) to quote the yellow one while claiming that the earlier, green version, was much better.)

In his book, Odum first offered the simplest definition given by Haeckel. that "ecology is the study of organisms 'at home.'" He then followed that with the usual definition of ecology "as the study of the relation of organisms or groups of organisms to their environment." Another definition Odum gave was "the study of the structure and function of nature."

Odum even offered up a dictionary definition: *"the totality of patterns of relations between organisms and their environment."* No doubt, this is the sense in which Rachel Carson studied ecology. This dictionary definition seems to be the one that has stuck. Robert Ricklefs also used it ("the study of the natural environment, particularly the interrelationships of organisms and their surroundings") in a textbook he was preparing concurrently with the first Earth Day.

This broad definition that has come down to us today serves us poorly, however, in understanding how ecology is a science, rather than just an idea. It is too inclusive, too much like definitions of physics in high school textbooks. For physics, there are historical excuses for a broad definition such as "the study of matter and energy," not the least of which is that the development of science is pretty much synonymous with the development of physics. The development of ecology is not synonymous with the development of environmental thought or of biology, however.

Another way to gain an understanding of a particular science is through its subdisciplines. Often a scientific discipline—physics, chemistry, biology—is broken up into smaller morsels of subject matter—astrophysics, solid state physics, particle physics, for example. Ecology was a subdiscipline of biology. Most would still place it there. Subdisciplines can also have subdisciplines. The smaller the morsel of subject matter, the more intellectually palatable it becomes, apparently. Looking for the commonalities of the subdisciplines of a science can sometimes lead to a working definition for it. Alas, for ecology, these are no more orderly or limited than is the overriding idea of "ecology."

Autecologists, Synecologists, and Genecologists

By its very nature, as any science marches toward the accumulation of more and more knowledge and, one hopes, the solution of more and more problems, its discoveries often lead to new and more interesting or more important questions to answer. The multiplication of knowledge that results from this march of progress leads to fewer of its practitioners being able to master all aspects of their science. Specialization has been the inevitable result, with multi-syllabic names for each specialty as a consequence. We will need to go through those multi-syllabic names for ecology, but I first offer a quote as incentive to do so. I discovered it in Loren Eiseley's *Darwin's Century*. It is by Linnaeus, the great categorizer and namer.

"The first step of science is to know one thing from another," Linnaeus wrote over two centuries ago. "This knowledge consists in their specific distinctions; but in order that it may be fixed and permanent distinct names must be given to different things, and those names must be recorded and remembered."

He offered his binomials, genus and species, *Homo sapiens*, for example, to assist with naming. They were actually simplifications of the system, or lack thereof, then in use. In order to know the science of ecology, according to Linnaeus's advice, one must note and remember the names of its parts. Be forewarned, however, that—as with taxonomic categories—some of the categories of ecologists might not be as distinct as the good Linnaeus might have wished. That had been another of his simplifications.

This tendency toward Balkanization of scientific study areas has not been universally true throughout history. Along with the exponential growth of the human population in the last two centuries there has been an exponential growth in its knowledge. This is what has led to the myriad of subspecialties in science. One cannot now know everything. Michael Lomonosov, to take one historical example, may have been able to reform the language of his Asiatic 18th century country by producing the first effective Russian grammar while at the same time making significant contributions to science and publishing a few not-insignificant literary works. Nicholas Riasanovsky, in his *The History of Russia*, describes Lomonosov "as a pioneering grammarian, an important literary scholar, and a gifted poet, [who] was also a chemist, a physicist, an astronomer, a meteorologist, a geologist, a mineralogist, a metallurgist, a specialist in navigation, a geographer, an economist, and a historian, as well as a master of various crafts and a tireless inventor." It's all true. This is not just typical Russian exaggeration. I dare say that Lomonosov did know quite a bit about each. But his impressive resume was accomplished during the century of the "Renaissance Man." It was a time when a learned man could contribute to knowledge in a number of diverse fields, not because he was a kind of intellectual giant who does not exist now, but because the part of the exponential curve that represents the accumulating new knowledge that Renaissance men helped to set off was much less steep then and had less area under it, that area being the total knowledge. Whereas Lomonosov and his contemporaries (Newton and Franklin come readily to mind) could study "science" in general, the scientists of today must restrict their researches to subject matter such as quantum solid state physics or microbiogeochemistry. In many sciences, the biological sciences, ecology in particular, it was not until the 20th century that specialization became a requirement of scientific study. In certain ways, ecologists are still being asked to perform like Lomonosovs. Ecology as the integrative environmental science demands it.

    Those subdisciplines of a science are as often defined by the tools that a given group of scientists are using as by the subject to which they turn those tools. Nuclear physicists might just as easily be categorized in a taxonomy of science as scientists that study nature by using particle accelerators, rather than as those who study the atomic nucleus. Microbiologists were first defined as studying those living things that could only be seen using a microscope. The instruments used by ecologists range from highly sophisticated particle accelerators and scintillation counters down to humble field glasses and notebooks. There can be no ordering of the science based on the instruments of choice of ecologists.

    The subdisciplines of ecology are more often defined by the taxonomic group (a collection of related species, closely so, such as

Autecologists, Synecologists, and Genecologists

finches, or broadly so, such as fish) that is under study. Insect ecologists, a large group, apply ecological principals to the study of insects. Other ecologists seem to be defined according to the habitat that they study. Forest ecologists are an example. Why are they not simply entomologists and foresters?

Good question.

If ecologists are identified primarily by their biological subdiscipline, the science of ecology vanishes. Ecology becomes more a set of methods or a point of view than a separate science. Ecology then is nothing more than an instrument for others. An insect ecologist might just as well be declared an entomologist who happens to use ecology in the study of insects. Some see themselves just that way. That is the reason for the need to differentiate between "self-conscious" ecologists, those who called themselves ecologists, and others who contributed to the science without seemingly realizing they were doing ecology—or even, in some cases, that such a science existed. This is a curious situation for any science.

This identity disorder troubled me enough in graduate school to make me deliberately avoid a specialization on any one habitat or group. I thought of myself as an ecologist with a "big E." Ecologist. No adjective. What did that leave me for a subject of study? The ideas that are the subject of this book.

Rachel Carson thought of herself as a marine ecologist. Today, she might have collaborated on *Silent Spring* with a chemical ecologist, one who studies how chemicals, both natural and artificial, mediate ecological interactions. So there must be ecologists who are not tied to a taxonomic group. There are. Theoretical ecologists, as another example, have as their domain the strictly mathematical description of ecological interactions.

Then there are paleoecologists and microecologists; soil ecologists, terrestrial ecologists, and aquatic ecologists; plant ecologists and animal ecologists; bird ecologists and butterfly ecologists; autecologists, synecologists, and genecologists. Paul Ehrlich might variously describe himself as a population ecologist, insect ecologist, or butterfly ecologist; perhaps he is a butterfly population ecologist or, maybe, a butterfly evolutionary ecologist. There are lizard ecologists and conservation ecologists, tropical ecologists and algal ecologists, restoration ecologists and desert ecologists. Some distinctions begin to blur. Some seem to be drawn too finely. More and more of them seem to appear as time goes on.

Sometimes, too, the ecological label is unnecessary. Population biology and population ecology are synonymous, as are a number of other matched pairs.

There is an explanation. Ecology has spread through the body of biology like a malignant cancer. It has found its way into every aspect of biology almost as much as evolutionary theory has. And there is the explanation. It is a truism today that one cannot study any part of biology without first having a knowledge of Darwin's theory of evolution through natural selection. The root of ecology is firmly grounded in that theory. Evolution, after all, is an explanation for how adaptations through which an organism interacts with its environment come about. Without a preexisting concept of ecological relationships, there can be no evolutionary theory.

Historically, then, ecology is the study of evolutionary adaptations. Quite as much as the useful knowledge ecology can provide to help manage the environment sensibly, it is the working out of natural selection's designs in nature that attracted many to the field. The attraction is that of a puzzle that when solved rewards with a feeling that all is well with the world, that things do make sense, after all. Monks studying variations as evidence of the workings of God's designs studied ecology no less than any past or present atheist ecologist—given the substitution of natural selection for God's will—and with much the same emotions. Biophilia runs rampant both in cloisters and in laboratories.

To be fair, there is another way of looking at the development of ecology. This is from its major origins: management of species of economic importance, investigation of nature's patterns, and examination of the balance of nature. These points of origin would be located in the decades during which modern ecology was being synthesized, from the 1930s to its supposed crystallization on Earth

Day. However, all actually had origins farther back in time. Except for its practical aspects—game and fisheries management and pest control—all of ecology is rooted in the ideas of Darwin's *Origin of Species*. The patterns in species interrelationships that ecologists study are explained as resulting from natural selection. The balance of nature, such as it is, would be a direct outcome of those adaptations. Even game management, post-Leopold, and pest control, post-*Silent Spring*, are rooted in the idea of adaptation through natural selection.

Yet if ecology is the study of evolutionary adaptations, then its time frame of action exceeds that of the actions of man. With the exception of very short-lived species, such as microbes and most insects, nothing that any of us can do in our lifetimes will alter the adaptations produced by evolution over millennia. Ecology can't be a practical science. Ecology tells us that animals and plants will not likely adapt to rapid man-induced environmental changes. Period. It can not tell us how to manage a landscape "ecologically," no more than an expert on the evolution of Monarch Butterflies—or any species or group of species—can predict the future evolutionary direction of the creatures in any but the most general of ways. Extinction, of course, is a rather safe prediction; any other prediction must be hedged. If climate change continues, for example, and if drought and fire destroy the butterflies' wintering grounds in Mexico and California, and if there is present in the population genetic variation for wintering grounds, then we might predict that Monarchs will shift to wintering in, as a complete supposition, Batsto, New Jersey. Or they just might all go extinct before evolution can act. Or, maybe, the habitat destruction will not be severe enough and the genetic variation insufficient, so the butterflies will still find places in Central Mexico and coastal California to roost over the winters. That is about the kind of predictive theory that can be expected from the evolutionary theory that is at the heart of even "modern" ecology.

But this can't be all there is to ecology. This is not the ecology to which we turned for salvation on Earth Day. There must be more.

There is. There is ecosystem ecology. On Earth Day, 1970, it was ecosystem ecology that was poised to become the tool of environmental management. It was a part of ecology that really was being born in the decades between the 1930s and Earth Day. It was so new that some were not sure that the science really existed in the way its proponents said it did. Ecosystem ecology was suspect for the simple reason that there was doubt over whether ecosystems really existed.

Eugene Odum was the greatest proponent of what has come to be called the ecosystem concept. As have others, Odum found it useful to also divide the discipline of ecology according to levels of organization, starting with populations, then aggregating through communities and ecosystems, until eventually reaching the level of the biosphere.

Although this division by level of organization is complicated by the specific tools used by individual ecologists (ie, pure mathematics, chemistry, or Have-a-heart traps) and the organism toward which the tools are aimed (ie, birds, insects, plankton, tropical trees), much important ecological research does fit into categories of single organism, population, community, and ecosystem. The level above ecosystem, the entire earth, or the biosphere, is just a larger and more complex ecosystem. The level below that of the individual organism (however that may be defined) is a fruitful one for an ecologist only if it reflects back upon some adaptation by the organism to its environment, specifically an adaptation that has evolved through natural selection.

Not surprisingly, ecologists have found that processes that go on at one level can take on new meanings when seen from a different level. A fish dying from a toxin is an interaction between an organism and its environment. Concentration of the toxin through food webs, however, is an ecosystem phenomenon, while the loss of fecundity (the ability to reproduce) in the top predator so that its extinction is threatened is a population phenomenon. *Niche*, one of those ecological terms I have accused Rachel Carson of slighting, is used to define an animal's role in a community. The empty niche that loss of the top predator presents is best studied at the community level. The boundaries between levels have never been sharp, but the divisions have remained essentially unchanged since Earth Day.

The organism level has a somewhat arcane term associated with it: autecology, defined as the relationship of an individual organism to its environment. It deals mostly with physiological interactions. Eggshell thinning in birds in response to the presence of DDT in their diets is an autecological example used in *Silent Spring*. The birds of prey were maladapted to DDT, not having evolved in its presence. Evolution clearly resulted in the particular physiology of eggshell production in raptors. Evolution is clearly germane to complete understanding of the menace of DDT in the environment, as it is for full understanding of any biological phenomenon. It is not, however, as useful to identification and solution of the problem.

Autecology is synonymous with physiological ecology, which is synonymous with a number of other terms, all professedly referring to ecologists. Actually, that physiological ecology still exists as a separate discipline is a mystery. It truly does just represent a point of view: that physiological adaptations should be studied in the field. The question of what is physiological ecology drew 14 responses from leaders of its practice in the 1980s to an issue of the ESA's *Bulletin*. One did not "believe that there is a lizard's eyelash of difference between 'physiological ecology,' 'environmental physiology,' 'ecophysiology,' or (these days) 'comparative physiology.'" Other physiologists, well aware that lizards have no eyelashes, tried to come up with suitable definitions, but one suggested it had more to do with modes of transportation and types of wardrobe favored by ecologists versus physiologists.

"Somewhat in jest," Boyd R Strain, of Duke University, wrote, "I tell my students that physiological ecologists drive four-wheelers, wear desert boots, and use Schollander bombs, while environmental physiologists drive sedans, wear dress shoes, and use Waring blenders."

"Both approaches are valid, of course," he made certain to add.

Although it can be incredibly sophisticated in theory and methodology, autecology is the ecological specialty that is the closest to ecology as simply scientific natural history. On this basis it is by far the most attractive and accessible to the layperson. My wife's late grandmother, sitting with field glasses on her porch on the Maine coast close to Rachel Carson's former Maine home, concerned herself with making exactly the type of observations an autecologist might have made on the birds about her. There is a caricature in ecology of unknown origin, that I have already used, with much due respect, I wish the reader to know, rather than scorn, in description of Rachel Carson, that of the "Little Old Lady in Tennis Shoes." Today Birkenstocks, rather than tennis shoes, keep her feet comfortable and she has many male companions, some rather strapping, sharing her interests (as she had also in her tennis shoes stage), but the essential category of the interested amateur remains unchanged. When the Little Old Lady in Tennis Shoes watches the behavior of a plover nesting, she is gathering data not much different from what an autecologist might refer to as behavioral ecology, except that he could cloak his observations with technical language, measurements, and mathematics.

On the next level of organization above the individual, population ecology, groups of organisms of the same species are studied as they interact and respond to their environment, essentially *en masse*, or by some average aggregate property. Mass deaths of Western Grebes following DDT application in Clear Lake, California, resurgence of agricultural pests, and development of resistance to pesticides are all phenomena for study by population ecologists. Wolves eating moose (or not being there to eat them) on Isle Royal in Michigan represent a subject for study by population ecologists. Being mathematical, it is not as easily absorbed by backyard naturalists.

Take into consideration all of the species in an area (as well as one can) and you are studying community ecology. The community ecologist sees how species interact with other species in assemblages called communities that in some ways mimic human communities. From the community perspective, each species is seen to have a particular community, or part thereof, in which it is found (its habitat) and a specific role (its niche) in that community. It is the upset in community properties that leads to insect pest resurgence, to give one practical example of a community interaction. It is the community feeding relationships in aquatic communities that result in biomagnification of DDT that caused the eggshell thinning in fish-eating birds.

The ecosystem level, defined as a biological community along with its non-living environment, is the largest scale at which ecologists view the world. It is distinguished from community ecology by its concern with the flow and storage of energy and chemicals through biotic communities. The biomagnification that led to eggshell thinning and ultimate extinction of the American Peregrin Falcon was an ecosystem process. (Do not be deterred by my having just identified it as a community process a paragraph above. It is both. The differences between ecological communities and ecosystems are more historical than real, but the difference between community ecologists and ecosystem ecologists is substantial and important. It will all be explained in subsequent chapters.)

The actual size and the boundaries of an ecosystem, as in other systems analysis, are operationally defined. That means there are no real boundaries; there is no maximum size or minimum size. An ecosystem can vary from a microcosm, as little as a drop of rainwater with algae, to the whole earth. The microbiome of organisms that live in an individual human digestive system is a currently popular ecosystem to study, mainly by nutritionists, however.

The community concept, however, does not extend to the entire earth. The earth is made up of a large number of communities that do not sum the way ecosystems do. This is one of the differences between the community and the ecosystem. Both are similar logical constructs, but they came about differently. Communities were first seen to be biological entities. They were thought to actually exist in nature. Ecosystems were created by a method of analysis, that of cybernetics, or systems science.

I will be using systems science and cybernetics interchangeably out of convenience and out of tradition in ecology. The terms are not quite synonymous. Cybernetics, which deals with self-correcting mechanisms, is a part, albeit a large one, of systems science. A systems approach need not be cybernetic—that all depends on whether the system or its parts are self correcting. Whether ecosystems are actually cybernetic was a matter of some contention. That does not, however, preclude a systems analysis of an ecosystem.

Although that may cover the taxonomy of ecology, even if far from perfectly, I fear, there is one final issue that also must be examined before turning to the successes and failures of the science of ecology. This has to do with what actually constitutes the science. All legitimate sciences must have a set of principles, a body of knowledge, and a set of techniques that its practitioners use and agree upon. The reason I know that a legitimate science must have these attributes is because I was called upon to deliver the principles, body of knowledge, and techniques that constituted a science called environmental science and engineering while at UCLA. Silly me, I managed to do so and the current recipients of UCLA's Doctor of Environmental Science and Engineering degree (it's a DEnv instead of PhD) are, I suppose, grateful to me. The bit of prestidigitation involved was easy, actually. UCLA's interdisciplinary program did possess all three attributes, but not *separately*. It possessed them only in the collective sense that each of the actual scientific disciplines that make up environmental science and engineering possessed them. In performing this particular slight of hand, however, I was impressed with the truth that those three areas, principles, body of knowledge, and techniques, do define a science.

The necessary body of knowledge for ecology can be found on the pages of journals such as *Ecology* and summarized by a variety of textbooks that appeared soon after that first Earth Day. The techniques employed by ecology: census, modeling, measurement in the field and in the laboratory, may not be strictly specific to ecology, but they are integral to the science by the need to measure the environment along with the organism, as required by definition.

Ecological principles are easy to find in Eugene Odum's *Fundamentals of Ecology*. They turn out to be less numerous and much easier to make sense out of than all of these subdisciplines that have occupied our attention for the last too many pages. Actually, in her later writings, Rachel Carson showed as good a grasp of these as any ecology textbook author.

First, there is the idea that organisms are adapted to their environment in fascinating ways that are the result of evolution through natural selection. In one of her letters, Carson revealed that she had chosen the seashore as the subject for *Edge of the Sea* because "it is a place that demands every bit of adaptability living things can muster. And by adapting themselves to the shore, sea animals have taken a long step toward adapting themselves to the land. So this is a place where the dramatic process of evolution can actually be observed." Her words describe the Darwinian concept of ecology that is at the heart of the science.

Carson's address to the AAAS began with a definition of ecology. "In recent years I have been dealing with the ecology of the seashore: with the animal and plant communities of the rocky coasts, the beach sands, the marshes and mud flats, the coral reefs and mangrove swamps," she wrote. "I have been thinking about the relations of one animal to other animals, of animals to plants, and of the animal or plant to the physical world about it. Always in such reflections one is made aware of the complex pattern of life. No thread is found to be complete in itself, nor does it have meaning alone. Each is but a small part of the intricately woven whole, for the living organism is bound to its world by many ties, some of them relating to biology, others to chemistry, geology, or physics."

In another of her writings, Carson placed the science of ecology squarely at the heart of conservation efforts. "Awareness of ecological relationships is—or should be—the basis for modern conservation programs, for it is useless to attempt to preserve a living species unless the kind of land or water it requires is also preserved," she wrote. It's unfortunate that she followed it with, "So delicately interwoven are the relationships that when we disturb one thread of the community fabric we alter it all—perhaps almost imperceptibly, perhaps so drastically that destruction follows." This is the concept of balance of nature seen also in Barry Commoner's First Law. Attractive as it seemed, evidence just did not support it. Still, it creeps unnoticed into ecological thought even today.

In the chapter entitled "Nature Fights Back" in *Silent Spring*, Carson gave an adequate overview of the main principles of ecology. There on page 248, for example, stripped of all mathematical obfuscation, are the basic principles of predator-prey dynamics. Earlier in the book Carson wrote: "The key to a healthy plant or animal community lies in what the British ecologist Charles Elton calls 'the conservation of variety.'" This is the premise behind controversy over the relation between diversity and stability (or function), a topic still under investigation today. A few pages later is a nutshell explanation of how biomagnification of DDT led to reproductive failure in the Bald Eagle.

Perhaps educated by her work on *Silent Spring*, a more modern view of ecology was given in one of her final communications. "There is nothing static about an ecosystem; something is always happening," her perfect prose tells us. "Energy and materials are being received, transformed, given off. The living community maintains itself in a dynamic rather than static balance." She goes on to use numbers, in parts per million, to illustrate biomagnification. "The contaminant does not remain in the place deposited," her words continue, "or in its original concentration, but rather becomes involved in biological activities of an intensive nature."

Odum organized his first chapters around the following principles and concepts: the ecosystem, energy in ecosystems, biogeochemical cycles, limiting factors, communities, and populations. Robert E Ricklefs's *Ecology*, used again because of its proximate publication to Earth Day and its differing viewpoint on the science from that of Odum, is organized around natural selection and evolution as played out in organisms, populations, and communities. There are no simply stated principles to be found in Ricklefs. The organizing principles seem to be given in answer to questions about nature and are descriptive, rather than predictive. F does not necessarily equal ma in ecology, but we have some interesting ways of measuring both m and a, the text seems to say. (Actually, other than those principles relating to ecosystems, Odum's other principles are no less descriptive than Ricklefs's.) There is a reason for the difference in emphasis between Odum and Ricklefs, Odum focusing on ecosystem interactions between organisms, Ricklefs, on natural selection and evolution. The reason is that, unbeknownst to Earth Day audiences, there was that revolution under way in the science of ecology with which I have been teasing the reader; Odum representing one set of forces, Ricklefs, not exactly a leader of the others, but a stand-in for their deceased champion, Robert MacArthur, his mentor.

The dichotomy between an ecology based on evolution and natural selection and one based on systems science, in which natural selection merely provides the materials of study, has not been unnoticed, especially by those looking at the development of ecology from an outside perspective. Some saw the dichotomy as more than just historical. One suggested there were pecuniary aspects to the schism. Alston Chase, for example, an outsider looking at the science, has written in *In a Dark Wood: The Fight Over Forests and the Rising Tyranny of Ecology*: "Ecosystems ecologists had a political agenda, and evolutionary ecologists did not." It is a comment that at once demands inspection. It is also a bit simplistic. Being an outsider, Chase, along with others, has not appreciated the scientific depth behind the split.

Where are we? Have we finally come to the subject of this book? Let's look more closely at some of the characters in our story.

A Tale of Two Brothers and an Uncle

Where are we? Have we finally come to the subject of this book? Let's look more closely at some of the characters in our story.

They were tall (at least Howard was) and lanky sons of the corn pone south, Eugene Odum and his brother Howard. In photographs from their youth they had that lean, hungry look of southern sharecroppers, but without the mien of poverty defacing it. They were, instead of sharecroppers, perhaps, more like Andy Griffith's sheriff of Mayberry, a few years and pounds lighter, their aquiline noses more suited to Army privates than Hollywood hunks. Another stereotyped impression the Odums suggested was that of the Southern preacher. That found its way even into staid scientific literature. One comment on a scientific paper of his sums up its tone as: "Eugene Odum, with some of the fervor of a Baptist preacher, is heard in the land." The preacher is heard calling for money: government funding for ecosystems studies in order to provide salvation for all of society's long range environmental problems.

Years later, in a documentary about Gene Odum produced by the University of Georgia, Gene Likens, an ecologist whose worldwide stature at present is equal to Odum's at its peak, says of him, "He is clearly a giant in the field with wonderfully creative ideas. He never stops teaching, whether he is with a farmer in the field or a student on the sidewalk. He is an evangelist for the values of ecology." Born-again former President and fellow Georgian, Jimmy Carter, had similar words of praise for Odum.

One time I had the good fortune to squire Howard T (the "T" is necessary to distinguish him from his father, Howard W) around Los Angeles in order to get him to the speeches and seminars he was to give that were my responsibility. On greeting him at LAX as he came off a cross-country flight, I did come to wonder whether I had invited a noted ecologist or a Baptist preacher. No, make that a deacon. Both Odums had grown up in the south, Georgia and North Carolina, but their father was sociologist Howard W Odum, developer of the concept of regionalism, rather than a Baptist preacher. Howard W was, however, in all respects a southerner, born in Georgia and educated in Mississippi (with brief forays for doctorates from Clark and Columbia Universities). Although the Odum family was Methodist, not Baptist, Eugene Odum's biographer Betty Jean Craige has him remembering his father's "folksy speech, dress, and manners" as being out of character with his academic accomplishments. Howard T was clearly Howard W's son. (Within the family, the son was called Tom. Those who worked closely with him reveled in calling him Tom, rather than HT or Howard, as those who knew him only through his publications might have called him.) The folksy dress in the younger Howard's case extended to having loosened his trousers for comfort during the four-hour-plus flight to LA then forgetting what it was that was making him comfortable. He continued to forget all the way off the plane. His wide tie did not quite reach to cover, but pointed to his mistake. That and the rumpled suit made him look more ready to sit out on a front porch scratching a hound dog's ear than to give a lecture on the cybernetics of ecosystems.

Fathers, like certain teachers, can have influence in totally unexpected ways on their progeny. That both Odums were staunch advocates of the ecosystem may well have been influenced by the wide perspective from which their father approached his subject, sociology. He found principles of sociology based on the study of regions, the South, for example, rather than individuals or small groups. To view an ecosystem, one must broaden, rather than narrow one's perspective. Howard T called it putting on a "macroscope." Eugene, putting on his macroscope, saw that the ecosystem was the "basic functional unit in ecology." He also saw it as of "extreme importance in human affairs generally." The words echo the holism of Rachel Carson's mentors at Johns Hopkins.

"Conservation of natural resources," Odum wrote, "a most practical application of ecology, must be built around these viewpoints." The ecosystem was not only the heart of the science—in his view—but also what gave ecology its practical significance. Ecology on that Earth Day and many subsequent ones was ecosystem ecology, thanks mainly to Eugene Odum.

Odum defined an ecosystem as: "Any unit that includes all of the organisms (ie, the 'community') in a given area interacting with the physical environment so that a flow of energy leads to clearly defined trophic structure, biotic diversity, and material cycles (ie, exchange of materials between living and nonliving parts) within the system is an ecological system or *ecosystem*." This is a mouthful—and a mindful—that pretty much embeds the history of the concept within its definition. It is also pregnant with ideas that need explication—and testing.

"Ecosystem" was first used by the British ecologist Arthur G Tansley for what he saw as the fundamental ecological unit that included living and nonliving things. Tansley's idea was equivalent in many ways to a category then in use called the "biome" or "biotic community," a group of organisms having characteristics peculiar to their climate. (Biome has a slightly different meaning now. It is no longer synonymous with ecosystem, but the difference is a technical one that need not concern us here.) Biomes are named according to the climate and type of vegetation of a region. Thus Arctic tundra is the name of a biome, as is tropical rainforest. Tansley needed to coin the new term in order to distinguish his ideas from those of Frederick Clements, a plant ecologist working mostly in Midwestern grasslands in the first half of the 20th century, who saw specific groupings of plants as being like organisms in their interactions and changes. Clements was probably the first to use the term *community* for them. He likened the stages in going from bare soil through forb and grassland to scrub forest and finally to mature oak-and-hickory-dominated forest, what Clements called a "climax" community, to the developmental stages of an organism. From a seed through embryonic stages to maturity, perhaps even to senescence, the community grew and progressed. In the example of the hardwood forest, change beyond that of the oak-hickory climax, according to Clements came about only through regression to an earlier stage, as caused by some catastrophe, such as fire, storm, or disease. His 1916 work expounding his ideas, *Plant Succession*, immediately became the definitive work on the subject, shaping the efforts of many ecologists for the rest of the century.

Clements had called for ecology to be a rigorous experimental science that would be a "rational field physiology," physiology being a part of biology then making great progress. His concept of the plant community as a developing organism gave him fertile ground for analogy to physiological processes, but he did not stop at analogy. "The unit of vegetation, the climax formation, is an organic entity," Clements wrote. "Furthermore," he went on, "each climax formation is able to reproduce itself, repeating with essential fidelity the stages of its development." Clements' communities **were** organisms in every sense of the word.

According to this view, plant communities could evolve through competition with other plant communities, replacing each other through some sort of Darwinian survival of the fittest. However, a mechanism for evolution of large numbers of species in concert, as required for communities to evolve as a whole, has not been established to this day (although what roots of various species of trees are doing by interlacing and sharing with each other has just been opened to further investigation). It troubled Clements not at all, however. As did many of his contemporaries, and as, for some time, did Darwin, Clements believed in a Lamarckian type of inheritance in which acquired characteristics could be passed on. This provided both Clements and Darwin with a muddled, but ready explanation for the phenomena they were so ably observing.

Clements's plant community was a growing, evolving creature. As such, according to him, it showed the quality of homeostasis that all living things show: the ability to self-stabilize from internal and external disturbances (up to a point, of course.) Change a living organism's state variables, such as by lowering body temperature, and the organism will make adjustments, such as (for us) shivering, to return to the more optimal state (98.6° F, measured rectally). Shivering is an adaptation acquired through the process of evolution to keep us from wild downward swings in temperature.

(More important for us than the shivering may be the urge for covering one's self that it induces.)

Tansley's ecosystem kept the analogy but denied the actuality. Raymond L Lindeman, another American ecologist, gave us a simpler and generally accepted definition of ecosystem, that of "the biotic community *plus* its abiotic environment." Odum's longer definition brings a certain exactitude to the simpler statements of Lindeman and Tansley.

The concept of an ecosystem, living things in the nonliving environment, points ecologists in the direction of the physical sciences, for in order to understand an organism's interactions with other organisms, the nonliving factors that mediate those interactions must also be addressed. For example, one plant may outcompete another in one area, but have the contest's outcome reversed in a second area. Factors such as sunlight and shade, soil chemistry, and moisture can shift competitive outcomes. Similarly, a plant can so change its environment as to make it more suitable for some other plant to grow. So far, this was nothing new to biology. Ecologists had long been aware of the effect of physical factors on the distribution and abundance of species.

What was new in ecosystem ecology was that the ecosystem concept brought to ecology the convergence of two strands of scientific thought. One had to do with the scale at which ecologists looked upon their subject; the other, a method. The first is the study of landscape-sized pieces of nature, such as sections of forests, watersheds, lakes, or entire islands. Plant ecologists were particularly instrumental in taking natural history studies beyond those of individual species, and to include interactions between species to explain patterns seen on the scale of landscapes. Competition—for space, water, sunlight, various minerals and other nutrients—defined plant communities by its outcome. Competition, mediated by changes made by plants to their environment—to soil in particular— was also seen as resulting in the process of change called "succession." Picture asters and goldenrod sprouting in an untended New England yard, followed over the years by shrubs, then trees, and you are watching an ecological succession. The shrubs, pin cherry, perhaps, and trees, such as oak, would have gotten there as seeds in bird droppings or through the absent-minded work of squirrels with acorns. So throw in that animals, by way of fruit dispersal and grazing, for instance, can also alter plant distributions, and a level of complexity is quickly reached in which the student of scientific natural history is faced with a perplexing number of possible interactions. Yes, all those factors contribute to explanations why there are kangaroos in the Australian outback and lions on the African veldt, but science calls for more than "just so" stories to fit the evidence. Science seeks predictive principles. Yet in community ecology, merely identifying which variables need to be studied—or, indeed, can be studied—becomes a difficult task. Competition, edaphic (related to soil) factors, history of disturbance, trophic interactions—where does one start?

What Raymond Lindeman added to the plant ecologists' concept of ecological communities is a method of study—that of systems science—and a focus for it: the energetic relationships within an ecosystem. This is the second strand of thought that, together with the community concept of plant ecologists, converged to produce modern ecosystem ecology. It provided a starting point from which to launch that search for predictive principles. The part of systems science that studies controls, cybernetics, was particularly attractive to the young ecologists who grappled with somehow trying to make scientific sense of the myriad interactions that are a grassland or lake. What more logical place is there to begin investigation than the control of a system, its heart and mind, in a sense?

Systems science focuses on studying the functions of parts of complex systems. In general, its method is to describe, using mathematical equations, rates of changes and exchanges of materials in a system and the physical conditions of the system. In systems terms, a process (the system) is analyzed to determine how various parameters affect its performance. (Parameter is a systems word for a very old concept in science, that of the variable, but a parameter can also be a variable that is not the specific subject of focus. In that case, parameter might as well be thought of as a constant in a mathematical equation.) A system itself is defined as a set of interacting parts. Often, it is conveniently seen to consist of a "black box" of unknown properties having inputs and outputs that can be measured and varied. In addition, it has what physical scientists call properties of state that also can be measured and varied. Temperature is a common "state" variable. (It can also be a parameter.) In theory, the black box could be a factory, interconnected phone lines (a communication system), a bank, a house suspected of holding terrorists, or a forest. Respectively, inputs could be raw materials, messages, money from depositors, unsavory visitors, or sunlight and water. Outputs could be finished products, messages, money for mortgages, exiting unsavory characters, or additional growth of existing spruce. You can imagine for yourself how quantifying the amounts of inputs and outputs can lead to useful conclusions in each of the above cases. Even the flow of messages can, when looked at by an expert in information theory; recent changes in levels of alertness to terrorist actions resulted from precisely such analysis.

The beauty of systems science was that it allowed ecologists to jump right into an ecosystem and start collecting data on its workings, either as a whole, or in part. It gave them a mathematical framework within which to place data and a set of analytical techniques to make sense of the data. The plant community and its successional changes through time, particularly after a disturbance, provided promising subject matter for analysis. The flow of materials resulting from the interactions of various species in a lake, because of the clear boundaries of a lake, was another attractive subject for systems studies. Of course, ultimately, a systems science study of the entire world was a reward in the waiting. Most importantly, systems science provided the tools to identify variables that could and should be studied at the higher levels of biological organization which until then were treated descriptively.

Raymond Lindeman had been inspired to put what turned out to be a systems science framework over the analysis of lake ecology by a curious figure in the history of ecology, Yale's avuncular George Evelyn Hutchinson. Hutchinson, although not physically involved in the ecological *Sturm und Drang* of the 1970s that it is my goal to recount and understand, was nonetheless very much involved in it through his ideas. He can unambiguously be shown to have started the revolution to come and, more curious still, have started what turned out to be both of the competing sides of it. Hutchinson, by his inspiration, had lit ecology's explosive candle at both ends. If not the father, he was the godfather of the superorganism, while at the same time being the sire of the champion who rallied science to kill the creature.

Personally, I find Hutchinson's influence difficult to fathom, even though I know it to be quite real. It is the why that I do not understand. Guest lecturers who came through Cornell in the 1970s seldom failed to point out a connection to Hutchinson, if there was one, usually through the pretext of some amusing personal story about him. It conferred, I imagine, extra status on them. I rarely found the anecdotes amusing. Instead, I was more perplexed by their point. A common Hutchinson story (in which the speakers, biologists all, could have had no actual part) went as follows. A physics major wanting to know the difference between space dust and the composition of our more earthly form is sent by the chair of the physics department at Yale to go see Hutchinson. "Ask him," the physics chairman says about Hutchinson, according to the anecdote. "He knows something about everything." Another of the stories recounted Hutchinson's habit of greeting a student or post-doc coming to see him by handing him a specimen of some aquatic creature or other and dryly suggesting, "You may find this rather interesting," with the best British diction the teller could evoke.

EO Wilson in his autobiography, *Naturalist*, also puzzled over Hutchinson's effect on others. "I asked several after they became my friends what 'Hutch' did to inspire such enterprise in his disciples. The answer was always the same: nothing. He did nothing, except welcome into his office every graduate student who wished to see him, praise everything they did, and with insight and marginal scholarly digressions, find at least some merit in the most inchoate of research proposals."

In his later years, Wilson recounted, Hutchinson took on the appearance of a guru, "seated in his office with wispy white hair and basset eyes" beside a stuffed Galapagos tortoise. "Head bobbing slightly between hunched shoulders, a wise human Galapagos tortoise, he would murmur, Wonderful, Wilson, well done, very interesting.

"It would have been pleasant to stay near him," Wilson recalled of his meetings with Hutchinson, "the kindly academic father I never knew."

Yoda of *Star Wars* could fit Wilson's description of "Hutch."

When I first began to work on this book, I regretted never having met or seen in person some of the principals of this story. High among these were Robert MacArthur, Eugene Odum, and GE Hutchinson. MacArthur having died before I came on the scene, I could not fault myself for having never met (although we must have roamed the same hallways at Penn on concurrent occasions), but Hutchinson I particularly faulted myself for having missed.

It turns out, however, that I had not missed him. Research on this book led me to an obscure little volume entitled *The Changing Scenes in the Natural Sciences, 1776-1976*. It was a symposium called together in Philadelphia to celebrate science in America on its 200th anniversary. With its theme of how science contributed to our understanding of the natural world, it became more a celebration of Darwinian biology, evolution and ecology, in particular, than of our nation's birthday. I had been in attendance there when GE Hutchinson opened the symposium to a large audience in a cavernous lecture hall on the Penn campus. It was the same room in which I had once sat with 800 others, trying to master introductory biology. The crowd Hutchinson had drawn was large, but a bit short of the enrollment for that biology class. (Memory has it that I scored the highest of all 800 enrollees on the course's first test, but memory is sometimes flawed. It had been a damned good performance, though, whatever the exact rank was, and, instead of an elective course, biology soon became that elusive second major I had been searching for. Years later, my son also received an "A" in the course, with the same professor, possibly held in the same auditorium. He did not, however, become a biologist. Identical stimuli can elicit different responses, as you might well know.)

The memory of Hutchinson in real life, refreshed now from having perused the text of his address, is of a gnomish figure giving what would have been a monotonous talk had it not been liberally seasoned with the distraction of slides. As is the case with many scientists at the ends of their careers, Hutchinson had taken to looking back instead of forward at his science. Instead of taking on the revolution he had helped to set in motion, Hutchinson's lecture had bogged down in and never escaped from the 18th century. It was an erudite talk, full of historical asides, Latin proverbs, and works whose titles were referred to in their original languages. The talk had left me with no impression at all of the man nor his intellect. Much more vivid is my recollection of ecologist Dan Janzen from that same day. His literature review of much more recent works morphed into a plea to save the tropics. Had he asked, I would have that day followed him to his research site in Costa Rica. Admittedly, Janzen was some two generations younger than Hutchinson, but it seems in no way possible that the staid, elderly British scholar (still British even though most of his life had been spent in America) could have metamorphosed from anything but a staid, young British scholar. I would have followed him only to where there might have been brandy and cigars.

But then my research led me to stumble onto another view of Hutchinson. It was a portrait of Hutchinson printed in the Ecological Society of America's *Bulletin* on the occasion of his having been awarded the society's Eminent Ecologist Award for 1962. It is a formal portrait of a man in late middle age—possibly around fifty— before his hair grayed and thinned and his face took on the lines that gave him a gnomish appearance in old age. He is looking off to the side, his jaw held proudly upward. His eyes take on the lazy squint that moviegoers from that era would have seen as typical of certain

Englishmen. His eyes are those of James Mason. There is a charm to that gaze that is both overtly playful and full of hidden possibilities.

Pedigree may have meant much more in ecology than I—semi-revolutionary immigrant that I fancied myself—had thought at the time. In Hutchinson's case it was ecological pedigree, not his own, but his progeny's. On the occasion of his retirement from Yale a year after Earth Day, *Limnology and Oceanography*, the journal most appropriate to Hutchinson's expertise, devoted an issue in his honor, in which Alan Kohn, a former student, penned an intellectual phylogenetic tree of Hutchinson's descendants. Hutchinson's students made up the branches and their students, the twigs. Some of the names on the twigs will pop up on occasion in the story of ecology after Earth Day. Two of its branches are HT Odum and RH MacArthur. Lindeman, a post-doctoral associate, rather than student, is a dragonfly at rest on the ground before Kohn's fanciful tree.

It is interesting how conferral of an honor makes the conferee worthy of it. Frank L Baum understood it clearly in his *Wizard of Oz*. I am thinking of the scarecrow. Kohn's act of producing a tree for Hutchinson, both the sketch and the research behind it, helped to create a legend surrounding the kindly Englishman. It also boosted the professional stock of those—Kohn among them—on that intellectual family tree. I wonder if an equally impressive genealogical tree might not be created for Victor Shelford, a pioneer animal ecologist and contemporary of Hutchinson's. Shelford's tree would have two major branches: that of WC Allee and Samuel Charles Kendeigh. The latter's branches at the University of Illinois include Eugene Odum and Robert H Whittaker, of whom there will be more later. Also interesting is that five out of the twenty students successfully obtaining PhDs under Shelford, a full quarter, were women. Kohn's tree used the then-current convention of using only first initials, but of the names on it that I recognized, none were female.

Hutchinson's actual pedigree may be just as fanciful as Kohn's tree. His father was on the science faculty at Cambridge. His grandfather appropriately made his fortune "as an East Indian broker dealing in tea and silks," according to Hutchinson's autobiography, *The Kindly Fruits of the Earth*. His mother descended from Italian nobility, at least based on the amateur genealogy Hutchinson found it necessary to include in the autobiography. Noble or not, it was admittedly a privileged existence. He delighted in being the "only living person to have incarcerated one of Charles Darwin's sons" at a family dinner party, an anecdote that he found "often provided a useful link with the past when giving a class on evolution." Hutchinson subtly reveals in his autobiography the view shared by many members of the British upper class that the upbringing of an English country gentleman, even if the "country" is slight, is a valuable privilege.

"What I have written will I hope indicate the extraordinary richness of the environment in which I grew up," he tells us. "The role of social structure in providing a really rich mental and artistic environment might well be a significant subject of study for an intellectual historian," he continues. "I suspect that quite unexpected parts of society may be involved."

Indeed. Although he does not explain what those "unexpected parts of society" might be, from my point of view—that of an impoverished immigrant, rather than privileged member of an extant social structure—what Hutchinson was describing was a set of connections that would take him through the best of the British educational system and guarantee suitable placement afterward.

Like many "country gentlemen," he was a butterfly collector before the age of eight. He later switched to water boatmen, an unusual passion for a collector, but one that suited his physiological bent. It was as an expert on these and the pond and lake environments in which they are found that Hutchinson's indisputable reputation was built. I suspect that his career rested on the foundation of his upbringing as much as on the ideas spawned during it. In anecdotes about him related by his acolytes, there is rarely a *bon mot* or a spark of brilliance. In appearance, he was an unremarkable British gentleman. In action he was described as helpful, kindly, and—most importantly—reserved. Without exception, anecdotes about him touched upon the British brand of reserve that Americans came to expect from English speakers from across the Atlantic. It may have been all there was—or all he needed—to exert his influence.

His academic career prior to Yale was not distinguished. Even at Yale, there were not the breathtakingly spectacular breakthroughs we expect early in the careers of men of genius. His came quite late in his career. He was dismissed from his first academic post at the University of Witwatersrand in South Africa, where a former student of his father was his protector. Hutchinson argued that it was an unjust dismissal, but admitted about his teaching that, "My course at first was not good, but it would have been better if it had not been attended by the professor." (In the European model of those days a department had only one professor, who was all-powerful.) "I gradually came to realize I was not wanted," he added in understatement, for he was suspended from all teaching duties in the second year of his two-year appointment on the grounds of incompetence and had to face formal dismissal proceedings. Yet he had connections who could have saved him from that fate. One of them was at the University of Cape Town. A trial arrangement was set up for Hutchinson to work in his lab over the Southern Hemisphere's summer break (our winter months) as a prelude to the offer of a permanent position, but it soon became "evident to both of us that I was not the kind of person he wanted," Hutchinson admitted without further explanation. Nonetheless, the Cape Town professor thought he might be suitable for Yale, where he had connections, for he wrote the Yale faculty a laudatory letter in Hutchinson's behalf. Applying for a fellowship, Hutchinson landed a permanent position by the luck of having an instructor leave Yale unexpectedly at the same time "that the sheer improbability of receiving an application by transatlantic cable from the Southern Hemisphere" could work strongly in his favor, according to his own account. Having already resigned in order to take up the Yale offer, the dismissal hearing turned out to be an unnecessary *pro forma* proceeding in order to save face for himself (and gain him some sort of monetary benefit.)

His promotion to associate professor at Yale did not come with stellar commendations. One letter solicited on his behalf concluded that he "did not stand out" and that the promotion committee had on its "hands a problem which the only accusations that can be levelled are a general tepidity." In hindsight, Hutchinson concluded "that this letter, even though accompanied by three fairly good ones, would today be enough to damn any candidate before some committee or other engaged in evaluating promotions."

Although having an inauspicious—to say the least!—start, Hutchinson's career had staying power. He spent forty-three years at Yale, an excellent base from which to build a reputation and one that guaranteed capable students and post-docs. Staying power along with a good pedigree (especially combined with British reserve and enunciation) can create a legendary career from mediocre beginnings, apparently back then, at least.

When Raymond Lindeman came to Yale in 1941 to work with Hutchinson as a post-doctoral fellow, he had already begun to write up his thesis work at the University of Minnesota. His focus was on food flows through the trophic levels of a bog. Hutchinson influenced the work by having him apply some simple mathematical equations that "Hutch" had worked up as mimeographed lecture notes for his class on limnology. They constituted only a small part of the paper Lindeman submitted to ecology, but may have had much to do with its rejection. Hutchinson, in turn, had been influenced by the mathematical descriptions of population dynamics that were just finding their way into ecology and, as was true for the rest of his career, was eager to apply the tools of math to ecology, urging those under him to do so as well. Theory without data, however, was the objection made by reviewers of Lindeman's paper. It was turned down by *Ecology*. Hutchinson immediately went to bat for Lindeman with a letter to the editor of *Ecology* having two interesting points. The first was the need to get theoretical studies into *Ecology*. Hutchinson called it an "important question of policy," words that are bound to get the attention of any editor, but especially a young one, as Thomas Park of *AEPPS* was at the time. In those days, the journal accepted data papers almost exclusively, mainly based upon the premise that ecology was still within the inductive phase of a science, when facts need to be assembled. Explanations would come when the facts were in. It was a situation that was rampant in the field of ecology, leading it to be awash in unrelated information that no amount of analysis could pull together into general principles. Hutchinson made just this point.

The other interesting part of Hutchinson's letter was that he identified a hostility in the "middle western regions," as he termed them, "for obtaining data confirming or disproving the hypotheses that have been forced on us by our little lake here." The little lake, Linsley Pond, was (and still is) a decidedly Ivy League lake. Acceptance of the paper may have been evidence of an institutional pecking order developing in ecology, in which the Ivy League and British country gentlemen had risen to the top.

In his memoirs, Hutchinson put the following postscript to the episode. He found himself coming away from "wherever a Puritan attitude was strong," such as in Wisconsin, "with two feelings of dissatisfaction. One was that it would be nice to know how to pull all their mass of data into some sort of informative scheme of general significance; the other was that it would be nice to have either tea or coffee, without seeming decadent and abnormal, for breakfast. I now suspect a connection." Yes indeed! I do, too, but the connection—or connections—were all on the tea-drinking Hutchinson's side.

Lindeman's paper initially made about as great an impact on ecology as a single water boatman did on Linsley Pond until, assisted again in great part by Hutchinson, the science of cybernetics was introduced to ecology. Part of its attraction to ecologists was the promise of complete control over nature. At the heart of cybernetics are feedback loops, and feedback loops of various sorts seem to be what comprise ecosystems.

A cybernetic system is also reducible to a set of mathematical equations, or even to simple electrical circuitry. Those equations and circuit parts, in their general forms, had already been developed, tested, and—through the enthusiasm of proponents of the method—

were made readily available to the ecologist willing to learn just a little mathematics. Not all were willing to do so. Resistance to mathematics has always been an issue in biology. It is what kept Mendel's revolutionary findings about inheritance out of mainstream biology. To many of his contemporaries, Darwin among them, perhaps, Mendel's paper seemed to be a mere mathematical diversion. It stayed that way, obviously not entirely in obscurity, for almost half a century. Lindeman's paper was first rejected on somewhat similar grounds. One of its reviewers, on a different occasion sneered at "mathematical limnologists … applying mathematical formulae used in sub-atomic physics where all the forces are presumably uniform to limnological problems"—situations in which they could not apply. Eventually, he suggested, like spiritualists, they would be able to forgo any necessity "to visit a lake to get its complete chemical, physical and biological history." The beauty of mathematics is not obvious to everyone.

      Hutchinson had taken an interest in cybernetics almost before its progenitor, Norbert Wiener, could get his ideas into print. (An advantage of being situated somewhere like Yale. New ideas get to you while they are still new. Today, the Internet may be more of a leveler on this score than even its champions might imagine.) Because he had broken up the cybernetic black box into subsystems that had obvious meaning to biologists, Lindeman's version of systems analysis became especially attractive when its time came some decade after its posthumous publication. (He had died at a young age of a rare liver ailment before his classic paper found its way into print.) Lindeman's black boxes were British animal ecologist Charles Elton's well-known trophic (feeding) relationships converted into energy terms. Elton's Pyramid of Numbers became Lindeman's Pyramid of Energy.

      With the advent of systems science, economic terms such as producers, consumers, transformers, productivity, and efficiency found easy entry into ecosystem vocabulary and theory. The study of communities, previously a disorganized mass of data and descriptive terms—leading one respected ecologist to complain that "The aimless and disoriented ecological survey is a bottomless pit for precious time, effort and enthusiasm"—now had more than enough concepts around which to organize observations. Ironically, ecosystem science became the perfect reflection of early definitions of ecology as the "economy of the home" or "nature's economy." Not surprisingly, mathematics borrowed from economics found broad applications in ecology, the invisible hand of nature substituting for that of the market in the law of supply and demand.

With the additional insights provided by cybernetics, the hope was that the complex workings of nature could now be measured, analyzed, and perhaps controlled. At the least, the science appeared to allow prediction of the consequences of a disturbance. This was at the heart of the Odums' enthusiasm for the ecosystem concept and techniques for its study.

Butterfly Collectors

It's very attractive, that idea that self-regulating mechanisms have evolved in ecosystems. It then follows that one should not carelessly mess with something that has evolved to a steady state. (Another of Barry Commoner's Laws: Nature Knows Best.) A climax community, the idea goes, like the human body, can withstand only so much disturbance before it loses its self-regulatory properties, at least in the short term. Lower body temperature too much and death, rather than shivering, ensues. The same should be true for the superorganism. An indicator of superorganism health, when identified, like body temperature in mammals, could then be closely watched in our forests and lakes.

An attractive idea it is, but more is necessary to have bright ideas incorporated into scientific thinking. Reasons have to be developed for adding the idea to a science and a champion must be available to take up its cause. Oddly enough, the champions that came forth were developed thanks to a most unecological event.

A mushroom cloud rising over white sands in New Mexico promoted the growth of Clements' unusual organism. Almost as if by radiation-induced mutation, it began to grow as the ecosystem concept of Eugene and Howard Odum.

It was at first a symbiotic relationship between the concept and those who nurtured it. There were clear benefits that could be obtained from the superorganism. The Odums had remarkable successes in applying Lindeman's ecosystem concept following World War II. It was a small bandwagon they were joining, but it was gathering momentum and it was doing so in a way that was highly visible to all—financially. The Manhattan Project, the code name given to the development of the A-bombs dropped on Hiroshima and Nagasaki, was as lavishly funded as any project deemed necessary (by Albert Einstein, no less) to win the most horrid war in all of history. In doing so, it had also created what became known as "Big Science," big in goals, big in personnel, big in government spending. Between the end of the war and the Russian launch of Sputnik causing the creation of the National Science Foundation, Big Science relied entirely on the military for funding. For ecologists of the day, it meant tracing the paths of various radioactive materials through the environment. Plant ecologist Robert H Whittaker and ecosystem scientist George M Woodwell, at Brookhaven National Laboratory on Long Island, and Stanley I Auerbach, at Oak Ridge National Laboratory in Tennessee, got their professional starts by tracing the movement of radiation through ecosystems and its effects on those ecosystems. (Whittaker less so than the others, of which, more, later.) All three played important roles in the history of ecology. At the time, they were essentially employees of the Atomic Energy Commission (AEC, now the Department of Energy).

Howard Odum came from Hutchinson's laboratory just after radioactive phosphorus, "accidentally" spilled there by the kindly Englishman, helped him trace the flow of phosphorus in "our little pond." It was as fortuitous an accident for Hutchinson's career as having Russian geoscientist VI Vernadsky's son, George, and the son of Vernadsky's closest friend both at hand at Yale to translate the father's seminal ideas on biospheric cycles. It was about at this time that Howard Odum's bent for the physical sciences led him to become a student under Hutchinson. It was just as the kindly Englishman was trying to fit the science of cybernetics to Lindeman's trophic-dynamic concept of the ecosystem. HT did not just run with the idea, he took it to the next level, taking his brother Eugene along with him. Having done his dissertation on strontium, a key element in radioactive fallout, HT was well situated to obtain military funding. Gene, with his ecological studies around the Atomic Energy Commission's Savannah River Plant, also had a place at the pipeline of funding for radiation-related research.

Big Science helped ecology overcome a somewhat tarnished reputation. In the past—in fact, right around when Rachel Carson may (or may not) have been learning it at Johns Hopkins or Woods Hole—ecology was seen as somehow unscientific and held in ill repute by other scientists. Scientists at the AEC, for example, the best and brightest of a scientific generation, saw ecology as being "soft," lacking in rigor compared to their own "hard" sciences. (Yet, in place of ecology's warnings, their science produced an unwanted bomb.)

This perception of a science being rigorous or "hard" is usually made relative to the density of the haze of mathematical symbols that cloak a science's fundamental principles. The mathematical haze around nuclear physics, for example, is well nigh impenetrable. Thus, the familiar phrase, it's "rocket science." Ecology was not rocket science.

"Classical" ecology, classical in the sense that it is what ecology was when its practitioners discovered themselves to be practicing a separate science, even when dressed up with mathematics, remains to this day mainly a point of view. That point of view was the organicism or holism seen in Rachel Carson's writings, her "balance of nature." As only a point of view, ecology is vulnerable to the accusation that it lacks substance by itself. It is an important point of view to many areas of biological research, but it does not make a science on its own. That is an attitude ecologists have struggled with for most of their existence. As late as the 1950s, ecology was "barely respectable in academic circles," according to one observer. It was a science that little old ladies in tennis shoes could master without effort, something perhaps still stuck in the 19th century. I suspect that the terminology adopted by pre-1950s ecologists may have been meant more to obfuscate than enlighten, but in doing so it separated them from the amateurs.

Way back in the 1920s, Charles Elton lamented that it would be "desirable that ecology should not be made to appear much more abstruse and difficult than it really is, and that it should not be possible to say that 'ecology consists in saying what every one knows in language that nobody can understand.'"

More recently (2002), ecologists at Purdue University may have taken an opposite tack. They bemoaned the low quality of ecology majors, a result of introductory ecology courses being taught without all of the hard math, presumably to maintain enrollment numbers.

"Too often," according to them, "we 'lower the bar' in introductory courses, watering down ecology to a level comfortably accessible to the majority of students, yet staid or trivial to the brightest of them. By doing so, we run the dual risk of rewarding mediocrity and losing outstanding students to other disciplines that are viewed as more challenging."

If ecology needed a savior to take it out of the wilderness of amateurism and into the Promised Land of hard science, it got two of them in the 1950s, and they were not both Odums. One was Robert MacArthur, who taught ecologists to frame their speculations in mathematical form and has begun to hang around the boundaries of my story like a ghost. The other, of course, was Eugene Odum, champion of the ecosystem concept. The ecosystem concept and the methods used in its analysis gave ecologists an avenue toward rigor they had been seeking since the century before—as did MacArthur's math. That gave the hope of making true the prediction made by one turn-of-the-last-century scientist that "the study of animals in relation to their environment, long the pastime of country gentlemen of leisure, will become a science." He was speaking about ecology. It was no longer to be the pastime of British butterfly collectors.

Yet it was not very long after the ecosystem concept was introduced that it came into question. The fact that it came into being as a "concept," whatever that means scientifically, rather than a law or principle was a vulnerable point. Rigorous sciences have theories, hypotheses, and laws for their principles, all testable predictions about how the world works. The more mathematical, the more precise, the more all encompassing are a science's laws and principles, the more rigorous it seems. Ecology had these, but also a litany of concepts, models, axioms, theorems, aspects, rules, ideas, and metaphors. On inspection, many of its principles turned out to be nothing more than descriptions of how nature might be organized. Many turned out to be erroneous when tested against good data. Ecology's concepts seemed to be something less than those of other sciences, something fuzzy, maybe even unscientific.

Scientists have ways to pull themselves out of situations in which their rigor is questioned. They can hide behind terminology. Technical jargon can be useful in making sure that all scientists in a field are saying what others think they are, but a spade is still a spade, even when it is called a geotome. A technical term for a shovel only serves to artificially separate the technician from the amateur. Ecology could no longer get away with that.

Another way out is to adopt the methods of another science, one that is indisputably rigorous. Chemistry, physics, mathematics, computer aided analysis, even when misapplied, can give the aura of rigor and beckon to all having low scientific self-esteem. Ecosystem analysis used all those methods.

By the 1950s, the social sciences began to offer another way out of the basement of the scientific establishment. Postmodernist critics questioned whether there was one true way of doing science. Chemists and physicists remained blissfully ignorant of these new views, but ecologists took them to heart. An ecologist with enough moxie could now continue his chase after butterflies or the birds of the New Hebrides, while fending off ridicule by claiming to be redefining his science in socially appropriate ways.

Ecologists soon took up the word "paradigm" for their ideas, mainly in defense, I think. Paradigm is a term used by historian Thomas H Kuhn in his influential work, *The Structure of Scientific Revolutions*, so that he could describe how ideas replace each other in the sciences. Kuhn saw science progressing under two different conditions. One he termed "normal" science, in which an accepted paradigm is in place and data collection proceeds under it; the other, a scientific revolution during which fundamental ideas or techniques compete for acceptance in all the ways—some quite unscientific—that humans have for one-upping each other.

Kuhn's concept of paradigm is not easily defined. The closest he came to a straight-forward definition was that "By choosing it, I mean to suggest that some accepted examples of actual scientific practice—examples which include law, theory, application, and instrumentation together—provide models from which spring particularly coherent traditions of scientific research."

When there are two competing theories to explain the same phenomena is a difficult condition for a science, although it makes for an exciting time for those going through it. According to Kuhn, progress in a science comes to a halt at such times. For science to progress normally, one of the theories must become the new paradigm, the one that is overwhelmingly favored and used by practitioners of the science. There being no current paradigm in sight means that the science must be undergoing a revolution, or it is in a very early stage of development, almost not a science. Once a paradigm is in place, scientists can put aside their bickering over the meaning of the puzzle they are all trying to solve—or the method by which it should be solved—and go about the business of filling in pieces of the puzzle. James Hutton's and Charles Lyell's Uniformitarianism, for example, is the paradigm that replaced Catastrophism. Newtonian dynamics replaced Aristotle's. The heliocentric solar system replaced the geocentric. I watched modern plate tectonics replace the orogeny of past geologists during my own adult lifetime. (It should not be surprising that a scientific revolution might complete itself in approximately a lifetime. According to Kuhn, most scientists are not convinced to abandon their old paradigm for a new. They are more likely to die off and be replaced by the proponents of the new. Aren't all revolutions the uprisings of the young against the old?)

Ecologists found many competing ideas to label as paradigms and avoid the implication of immaturity.

The behavior of the combatants in scientific revolutions in science, it turns out, is little different from those in political ones. Theories in competition with a reigning paradigm become heresies. Inquisitions are launched to drive out heresies in those times representing "paradigm shifts," when one paradigm replaces another. Whether ecology has truly undergone paradigm shifts or not, it has had its share of heresies.

It might be simpler and more in keeping with revolutionary terms to call the ecosystem a "doctrine," rather than concept—and some have—a doctrine being something that is taught as a set of actions to be followed, often unquestioningly. Doctrine, concept, or paradigm, within the ecosystem is also—guess what?—a literary device, the metaphor, that of the superorganism.

The ecosystem also had systems science behind it. A big appeal of systems science was that engineers used it. At Oak Ridge, where nuclear fuel had been produced for the Manhattan Project, ecology had sneaked in through the back door in 1954 in the person of Stanley Auerbach, who was hired to assist in health physics. An empire builder, Auerbach was a slight man with great ambitions. By Earth Day he had created one of the largest ecological research programs in the world. He did so with an eye always focused on how the ecological research carried out at Oak Ridge affected how the physical scientists who then dominated research at our national laboratories looked at ecology. The national laboratories, in turn, dominated research until the National Science Foundation began to pump money for research into the universities. Always, Auerbach's concern was that ecology not be seen as lacking in scientific rigor.

I unwittingly once found myself under his focus. Needing to get out of Los Angeles in the worst way, I had pulled some strings and managed to get invited to Oak Ridge for a visit to explore mutual suitability. I gave a seminar in order to show my stuff before the assembled research staff of what was by then called the Environmental Sciences Division. My first slide was a beautiful photo my wife had taken of the Klamath River, one of California's Wild and Scenic Rivers, an environment I would have happily spent my life studying. In contrast, my second slide was of a manure pile, an environment that was what my career would more likely have me studying. It was supposed to be a joke. (At the time, I was involved in a study of organic farming in the Midwest. Manure is very big on Midwestern farms, organic or otherwise.)

Something cool seemed to emanate toward me from Auerbach when it came time for our interview. He parried my assay of a greeting and endorsement from a former colleague of his with, "Who?" I dropped that tactic immediately. So, instead of chitchat over joint acquaintances, our conversation quickly turned to the issue of the manure pile: I had shown and still avowed, when given a chance by him to retract, a preference for scenic vistas along the

Klamath River over the useful countenance of the manure pile. That was no joke.

"We need more ecologists willing to study a manure pile," was what Auerbach had to say at the end of our interview. He ushered me not just out of his office, but out of any future at the lab.

Physical scientists, I learned Auerbach later said in an interview, saw ecologists as not much more than "a bunch of butterfly chasers." He had pegged me as a butterfly chaser. I was not offered a job.

Convinced that "only a quantitative, physicochemical perspective would find acceptance at Oak Ridge," Auerbach had chosen to focus on the ecosystem level of ecology after consulting with Gene Odum. It was the closest thing in ecology to physical science. That was what let the ecosystem approach and ecological research flourish at Oak Ridge. It had kept out the butterfly chasers. It let Auerbach build his empire.

Respectability for their science has long been important even to ecologists not building empires. I have already guessed that a reason for the condescending tone of Cole and Ehrlich toward Rachel Carson could have been insecurity for their science. Similar condescension was evident in other ecologists. It was their science that amateur Carson was using to make her case. For example, reviews of Robert Rudd's *Pesticides and the Living Landscape* contrasted Carson's "bold" and "dramatically" written popular work with Rudd's "textbook" in which personal judgment is "scrupulously" distinguished from evidence. Rudd was a university professor. One British ecologist smugly attributed the lack of surprise over Carson's revelations to Rudd having published on the topic since 1955 in the United States. That smugness was not justified. The 1956 position of the Ecological Society of America was that "on the whole, great care is being exercised by most federal and state agencies" in the use of chemical controls. "Instances in which beneficial animals and plants have been killed are surprisingly few and usually occurred where the applicators failed to follow instructions" a study committee concluded. That was a few paragraphs before the same committee reported that Rudd had sent it a letter emphasizing the need for more data that could be brought to bear on the issue. The following year, the conclusion of the ESA was that, "when applied to agricultural crops at the dosages and in the manner prescribed by federal and state authorities, they have caused little or no losses to wildlife," even though acknowledging fears of conservationists and "others concerned with the preservation of our wildlife" that "such treatments will destroy nature's balance."

Pest control was not an item on which butterfly collectors were particularly knowledgeable. In both instances there was more concern over losses of natural lands and damming of rivers than over pesticides. These were more traditional concerns for ecologists. In fact, Victor Shelford had the preservation of natural areas in mind as the goal of the ESA on its founding. His disappointment in the Society's inability to take direct action on that goal led to the formation of the Ecologist's Union as a way of doing so. That organization transformed itself into today's Nature Conservancy.

In the days before *Silent Spring*, the Applied Ecology Committee of the ESA had "very rough going" according to Gene Likens. With some exceptions, the ESA was still overpopulated by butterfly collectors. Likens kindly called them "theorists." In 1964, however, the ESA was sponsoring a well attended symposium put together by Likens in the hope of allowing ecologists to exchange information on the "increasing problem of pesticide pollution," and pumping new life into a self-study committee that had become rather moribund. *Silent Spring* was the acknowledged impetus in both cases. Also at this time, the committee on applied ecology was charged by the ESA President "to formulate an ecological context for the use and conservation of natural resources." Its findings were not reassuring. Present programs were not providing the factual material needed to avoid future disasters and students were repelled by ecology's lack of rigor. Narrow specialization, the kind that presumably missed the warnings given by Rachel Carson was "a real dilemma."

A number of ecologists in the 1950s, Paul Sears and Frank Egler among them, had been pushing for ecologists to take stands on environmental issues. By 1963, instead of just a few voices, a Committee on Public Affairs was appointed as "the most important action," by his own estimate, in that ESA President's term.

"There was a major sea change in the ESA from the Applied Ecology Committee having little respect before the publication of *Silent Spring* to becoming very respectable," Likens recalls of the shift from an essentially academic orientation in ecology to one of practical applications. Purists running the Society had not thought that ecologists should be involved in applied problems. Indeed, a common criterion in the choice of study habitats was their relative lack of influence by man's activities. The Applied Ecology Section went on to be the biggest membership of any section in the ESA.

In 1964, *BioScience* gave space to Stewart Udall to urge biologists to "spread this Gospel" that Rachel Carson had presented. That same year it devoted an issue "to cover the basic concepts and ideas of ecology." As if heeding Udall's call, Eugene Odum used it to preach a new ecology based on the ecosystem concept. In the same issue, a wildlife ecologist claimed that "'ecosystem ecology' is the ecology of the future," while Frank Blair blamed the primitive state of knowledge about ecosystem interactions "in part on the modesty of ecologists in seeking financial support for their research and in part on the failure of both ecologists and formulators of public policy to face up to the fact that knowledge of the interactions and interdependencies at the levels of organization with which ecology deals is essential to man's present and future welfare." Meanwhile, Frank Egler and others were making Rachel Carson's "ecological concept" synonymous with "Ecosystem Ecology."

Ecology was in crisis.

To add to that crisis, radical philosophies such as Deep Ecology plucked their moral justification from the ecosystem concept of Eugene Odum. It was not an association welcomed by all ecologists.

"Why is ecology so near the lunatic fringe?" one asked his colleagues in despair. "It has made all of ecology suspect, and it has weakened the credibility of every ecologist."

A drive toward professionalization in ecology was set in motion. It was not entirely compatible with the multifaceted subject that is ecology. It was complicated also by the peculiar identity crisis suffered by ecologists to this day having to do with philosophical viewpoints of nature, environmental activism, and the need for scientific detachment. All of these issues were coming to a head. Some ecologists needed a way out of that dilemma. Some thought they found refuge from the ecosystem perspective and its practical imperatives in the theoretical evolutionary ecology then being made popular by Robert MacArthur, even though that did not provide complete refuge, for the biodiversity issues of today are rooted in his brand of ecology. One could, however, taking Darwin's example, study biodiversity for its own sake, as a puzzle demanding solution. Some did.

The exact effects of *Silent Spring* on classical ecology began to become clear with the passage of the National Environmental Protection Act (NEPA) of 1969 and the developing character of the International Biological Program (IBP). Both were strongly influenced by the book and both resulted in changes to the quietly subversive science.

The IBP had been inspired by the IGY (International Geophysical Year), which was actually more of a year and a half in 1957 and 1958. The data obtained added much new knowledge to the geophysical sciences and overturned some basic precepts. It set the stage, for example, for acceptance of the Theory of Continental Drift. A similar enterprise was proposed for the biological sciences. Its original theme had been "The Biological Basis of Human Welfare," not ecology. However, its theme eventually morphed into "The Biological Basis of Productivity and Human Welfare" and IBP might as well have stood for the "International Biome Program." The "year" had to be stretched to several. The changing character of biota over time assured that important phenomena would be missed by too brief a period of study. By having its emphasis changed from human welfare to biomes between its 1961 inception and its 1970 funding authorization, the IBP established ecology as Big Science with a big budget. That ecology was ecosystem ecology. (Biome was no longer synonymous with ecosystem as it had been in its usage by Clements, but nonetheless still had an ecosystem perspective.) If butterflies were being collected, it was in order to study the flow of energy through their populations. Frank Blair no longer had to apologize for the modesty of ecologists in securing funding.

The IBP funded ecology because it had the proper image to be the basic science to solve environmental problems. That came about in large part through the efforts of Eugene Odum, Stanley Auerbach and his colleagues at Oak Ridge, and the major boost given ecology by the new environmental consciousness punctuated by Earth Day. Funding for the project was approved by Congress in August of 1970, exactly the time during which Odum and Auerbach were trying to raise the status of ecology in terms of scientific rigor by pushing for ecosystem science and systems analysis. What better way to do so than to promote ecology to the ranks of "Big Science," those endeavors, such as the Manhattan Project and the Apollo Program, that stirred the government into creating and funding research programs of unprecedented scope? What better time to do so than the moment at which the public and Congress had been primed by the metaphors of Earth Day into accepting ecology, ecosystem ecology in particular, as the one integrative science that could solve environmental problems and manage the earth?

Then there was NEPA. In the words of Oak Ridge's Stanley Auerbach, NEPA was an "ecological 'Magna Carta'" by which ecology was "rather suddenly thrust into a period of great individual and collective opportunity." Read that as "money."

A legal and policy analyst with Oak Ridge announced that: "The courts have, in effect, legitimized ecology."

That the ecology being legitimized was an applied science not necessarily being practiced by most ecologists was not lost on Auerbach. Those purists had not disappeared from the Society. In the same address in which he extolled the opportunity provided by NEPA, he also decried the "aloofness" of ecologists who insisted on staying within the confines of their "pure" research. It had been Auerbach's Presidential Address to the membership, traditionally given by the retiring, rather than newly elected President. In it he stressed his "not being a traditional President." He was an administrator, rather than researcher or professor. In addition, what he administered was applied science, outside of academia.

Auerbach failed to mention at this time any feeling of being "atypical" in other ways. "I was a somewhat atypical student," having turned to the study of ecology after military service in World War Two, he recalled for a biographer. He was doubly atypical in "coming as I did from an urban milieu and a culture that did not send students into the field of ecology." The typical "culture" was an Ivy League one.

Regardless of his cultural background, once more what an ecologist was was at issue post-Earth Day. It has been an issue in the decades before and since, but Auerbach and others were placing the question directly before ecologists in a new way. Ecology was something of a profession, they argued, not just an academic avocation, and they presented the argument from positions of financial power.

Not all ecologists wanted to follow them into that profession. In a bewildering juxtaposition of roles for a member of the ESA, one responded to Auerbach's call by describing himself as a "biologist first, ecologist second, activist third." Another sought to separate ecology from environmental management. What was needed on practical environmental matters in his view was environmental engineering. At the same time, ecological theorist Richard Levins was promoting "theoretical research in the broadest sense" above engineering systems-type ecology as being more fruitful for the solution of applied problems and more worthy of funding. In the background, nodding in favor at Levins, I can imagine the figure if Robert MacArthur.

There was also a heightened concern over who could claim to be an ecologist. The "Commentary" feature of signed editorials inaugurated in 1969 in *Ecology* with a plea by Robert Whittaker for the founding of an Institute of Ecology in order to help solve the problems of the environment soon had complaints about "overnight 'ecologists'" and "amateurs" usurping the voices (and consulting fees) of real ecologists. Unlike LaMont Cole, who left Rachel Carson's lack of ecological credentials out of his review, a reviewer of 1972 book that was subtitled *Conversations with Ecologists* felt a need to point out to the readers of *Ecology* those conversations which were with personages who were "not strictly ecologists."

Even Frank Egler got into the act. Of the authors of a symposium on power lines and the environment, he complained that there was "not one single major paper in the field by any of them. Not one!" Not being ecologists like him, they were not experts.

Simultaneously with this crisis in identity and not independently of it, I am sure, the ecosystem concept started to be battered around by whatever wave happened to sweep through ecology—and, lacking a single paradigm, there were many of them. There were ways, it turned out, to be environmentally relevant, without embracing the ecosystem. Although triumphant on Earth Day, like Rome at its height, ecosystem science could not maintain its gains. It began to be picked apart little by little until its empire began to crumble, at least, intellectually, if not factually.

To ascertain who or what set about to kill the superorganism, the clearest metaphor of ecosystem science, we must put on our Sherlock Holmes hats. Arthur Conan Doyle's detective, remember, would eliminate all possibilities that could be eliminated. The one who remained had to be the culprit. Suitable to a mystery, clues to the demise of the superorganism, seemingly so healthy in the years after *Silent Spring*'s publication, are as confusing at first as those in any drawing room in which a corpse is suddenly discovered. The reasons for the superorganism's rise and fall have to do with ideas and attitudes found in the fields of physics and engineering, as just noted, the philosophy of science, and trends in evolutionary biology. They also had to do with power, personality, and prestige—human motivations difficult to identify and thus easy to ignore, except in fiction, where they can be fabricated out of whole cloth. I will try to make my explanation of this episode clear, but I can not make it simple. Each idea must be taken up, at least in brief, before it can be dismissed—or remain as the prime suspect in the superorganism's fall, and of the ecosystem concept it took with it.

Some of those ideas arose directly from ecology's emerging status as the leading environmental science, others were perhaps in defiance of that status, while still others occurred independently of, but entangled with it. The story has threads so knotted and tangled that no single one can be followed from beginning to end without having to unravel some other thread that gets in the way. Like a famously misapplied "ecological" principle, everything seems to be connected to everything else. On one thing, though, I can reassure the reader. The changes that the science underwent were not conceptually difficult to understand. This is good fortune, for it allows us not to shy away from the principles and theories at issue. However, like a large family event at which you might know most of the attendees, sometimes keeping track of the connections between them can be tedious. Imagine a wedding in which the bride is the niece of the groom's father's first wife and you may get the idea.

One thread follows the money. Like Auerbach at Oak Ridge, the Odums readily found funding from the government. Their 1955 study of Eniwetok Atoll, pre- and post-nuclear detonation, won a prestigious award from the ESA, a sign that both the ecosystem concept and their methods of describing and analyzing it had won acceptance. Howard Odum followed up with a 1957 ecosystem study of a Florida spring that became an instant classic, used as a textbook example in graduate and undergraduate texts. Eugene Odum's *Fundamentals of Ecology*, in any of three editions from 1953 to 1971, with a chapter on energy by his brother Howard, became the standard text for a generation of ecologists.

The collaboration that the Odums profited from at Eniwetok, that between ecologists and radiation and nuclear scientists, had both theoretical and practical results. Radioactive materials, besides providing the hazard that necessitated a variety of environmental studies, also provided a tool with which to study the environment, especially from the ecosystem perspective. The conclusion that DDT was accumulating through food chains, for example, so important to the message of *Silent Spring*, would not have been available to Rachel Carson had there not first been studies of the bioaccumulation and bioconcentration of radioactive materials.

The other tool necessary for ecosystem studies, and also one made available by the government through its defense efforts, was that of the digital computer. Without the computer's ability to combine large amounts of data into simpler and more readily interpreted forms, ecosystem studies would not have succeeded in gathering the quantitative information that was their capstone.

Unlike the physical sciences, in which much work goes into designing, preparing, and calibrating apparatus of various sorts which —when working properly—accumulate data quickly, ecological data collection goes much more slowly: tree by tree, leaf by leaf, beetle by beetle. And each must be tabulated. The bookkeeping needed for an ecosystem-level study was an insuperable bar prior to the advent of high-speed computers with large memories, those large and incredibly expensive (not to mention onerous to work with) mainframe computers of early ecosystem studies. (The computer revolution is such that those old IBM 360-somethings have been replaced by slim laptops or tablets with spreadsheets. (Better still, automatic remote sensing.) Modeling is no longer an object in itself, but has become the tool for forecasting that it had been promised to become. Data collection, however, is mostly still as tedious as it was before the advent of the computer, a tree is still a tree and a leaf a leaf, but the ability to record, collate, and analyze it makes the tedium worthwhile, or even not at all tedious.)

And data is important. It is a truism in science that before one can understand a phenomenon one must be able to measure it. In the Cartesian view of science that most scientists follow, even if unknowingly, one must not only measure, but also make a quantitative prediction, then test that prediction. The prediction is usually in the form of a mathematical equation. We call such an equation a model, for we hope it to be an understandable approximation of the real world, even as we admit it is not a perfect representation. Newton's famous equation, $F=ma$, is one such testable model, as is Einstein's $E=mc^2$. Perhaps these two geniuses took all of the best letters for themselves, for models with such elegant simplicity and momentous significance are not easily found elsewhere in science, in particular not in ecology. That made another need for massive computing capabilities in order to apply systems concepts to ecosystem studies. Not only can ecological data be unmanageable without a computer, but so can the mathematics that must be used to model it. In fact, if one were to model all of the possible interactions in even a relatively simple ecosystem, computers would also fail at the task. As much as it is a science, there is also an art to modeling nature.

However, one very simple—many would say overly simple—model of the world is found in the black box of systems science, in which, as an example, a complex mixture of plant species can be considered to make up a simple process that takes in materials, changes them somehow, and passes them on as different materials. It turns out that the complete workings of the process need not be known in order to determine its controlling factors. Input material clearly will have an effect on output. Quite simply, no sunlight to the plants, no new leaves. One need know nothing about the identity of the plant species involved, nor a whit about photosynthesis, to reach that conclusion. One might also conclude that there is no need for input-output analysis or systems science to reach the same conclusion, given its simplicity. However, what systems science promised was the ability to find set amounts of outputs for given levels of parameters (input and otherwise) through the process of simulation. For example, given that rabbits eat plants, can we model the consumer-plant interaction numerically, then see what changes in outcome would happen if we tweaked some parameters, such as how hungry the rabbits might be, how nutritious the plants are, etc? If so, we might begin to develop an understanding of what controls, if not the whole, at least parts of complex entities such as ecosystems.

Alarmingly, perhaps, we need not deal with rabbits and plants at all. Lindeman gave us the framework for identifying the plants, regardless of species, as being the producer compartment, or black box, and the rabbits, along with their plant-eating ilk (actually in the most simplified of schemes, even those creatures that eat rabbits) as consumers. Neither Odum brother, for example, could identify most of the species of coral at Eniwetok, but they managed to describe the trophic dynamics of coral very ably. (Not to mention making the discovery that coral is symbiotic with the algae growing on it, much like fungus and algae make up lichens.) Having no need to explicitly identify the species under study removes much of the romance of studying ecological interactions, however. Instead of bugs and bunnies, one has boxes and arrows—stand-ins for numbers, actually—as subjects of study.

The reader may agree that systems science is not for everyone. One needs to develop an appreciation for mathematical beauty to love systems science. Love of butterflies and bunnies may well be a bar to embracing—or even accepting—systems science.

Strangely enough, Eugene Odum, came to the science of ecology as much a naturalist as Rachel Carson or EO Wilson. Like MacArthur, he was an avid bird watcher from youth. His earliest scientific publications are notes on birds, written while still in his teens, in journals such as *Bird-Lore*, *Wilson Bulletin*, and *Auk*. That he came to believe that nature could best be studied in the units of ecosystems may have been due mostly to the influence of his younger brother and the influence of GE Hutchinson, in turn, on him.

Collaboration between the two brothers early in their careers led to both becoming unabashed proponents of the ecosystem concept. Going a step farther than his older brother Gene, Howard came to believe that the equations governing ecosystem behavior were like electric circuit equations. Imagine a laboratory with a model of a tropical rainforest within it. In Howard Odum's lab it would look more like an analog computer, which it would be, than a rainforest, which it would pretend to be. It would be a rainforest made of wires, resistors, capacitors, and inductors, whose voltage and current readings represented the various states of a real rainforest. He may have originally been led to such an outlandish representation of reality by difficulties in obtaining digital computers with sufficient computing power early in his career, but he remained a proponent of his analog diagrams well into the age of the laptop computer. No doubt a number of tropical plants would be set out in some heuristically useful way in the laboratory to represent the biological reality. It was the plants, however that would have received real water daily, even though the circuitry could be watered several times a day with brief analog showers.

That Howard Odum had begun to see nature in terms of electrons flowing through circuits became clear to me as I listened to his sales pitch for their use circa 1980. The pitch included the provision of special plastic templates, free of charge, for easy reproduction of the symbols required to produce diagrams of the energy circuits that Odum believed best described ecosystems, nature, and even society. That the symbols of his circuit diagrams might stand for trees or birds was irrelevant. In the end, that insistence of his on the primacy of the diagrams and the energy transformations they represented in understanding nature may actually have led to a loss of credibility for him with ecologists, he later confided to an interviewer.

Robert MacArthur, another of Hutchinson's students, who led ecology into a totally different area of abstraction—and away from the ecosystems of the Odums—was eulogized by his former professor with the statement, "Robert MacArthur really knew his warblers."

Howard Odum knew his circuits.

The Curious Case of the Superorganism

If both Odums, Eugene at Georgia and Howard at Florida, argued with almost a religious fervor for their form of ecology, ecosystem ecology, as the basic tool for environmental management, they found converts not so much among their peers, but in the ranks of environmentalists. Howard Odum's circuitry diagrams, however, turned out to be a bit too esoteric for public consumption. It was Eugene Odum's description of an ecosystem, then, that was picked up by environmentalists as a basic principle. It was an ecosystem within which material cycled and energy flowed to produce "a self-correcting homeostasis with no outside control or set-point required," in Odum's words. Odum's ecosystem maintained itself. It was self-regulating. It was balanced. It was Clements's abandoned "complex organism" idea, the superorganism. Environmentalists helped bring it back to life.

That connection between Odum's concept of the ecosystem and pop ecology has caused concern. Philosopher and conservation writer Alston Chase has called it a "sledgehammer of an idea with which to change the world." He attributed to it the origins of everything from biocentric politics and philosophy to Deep Ecology and ecotage. Suddenly we are on very tenuous ecological grounds here. Odum's ecosystem was supposed to help bring rigor to the science, wasn't it?

Now Deep Ecology is hardly a point of view whose creation, if ecosystem ecology can be given credit for it, must be looked upon with any sense of shame. It is a gentle philosophy that Gandhi would probably have embraced. It substitutes an *anthrocentric* (anthropocentric is the accepted and longer version of the term) point of view that places man's needs above nature's with an *ecocentric* or *biocentric* one in which man is but a part of nature. It is the organicist view of *AEPPS*, again.

But there is a dark side to ecocentrism. Chase claims the ecosystem concept, as promoted by Eugene Odum, inspired advocates of Earth First! and ELF (Environmental Liberation Front, a highly secretive organization devoted, apparently, to ecoterrorism, such as burning mountainside condominiums, presumably to protect "the ecology.")

A 1977 paper by Eugene Odum, "The Emergence of Ecology as a New Integrative Discipline" in the journal *Science*, actually did call for the advancement of at least two radical environmental philosophies.

"We are abysmally ignorant of the ecosystems of which we are dependent parts," Odum wrote. "As a result, today we have only a half a science of man."

He continued, "Although 'ecology' is frequently misused as a synonym for 'environment,' popularization of the subject is having the beneficial effect of focusing attention on man as a part of, rather than apart from his natural surroundings."

This is as much a statement of the ecocentrism behind Deep Ecology as Arne Naess's concept of *ecological self*.

"We may be said to be in and of Nature from the very beginning of our selves," Naess has written.

There is also agreement and encouragement in Odum's paper for another, less radical popular movement. Almost as an aside, Odum used his paper to resurrect an idea of his father's. It had met its demise by being ahead of its time. It was "Regionalism." In Howard W's case, the region was the Southeast.

Eugene says of his father's work, "as a major theory of sociology the concept stalled because there was no appropriate linkage with natural science (applied ecology had not yet emerged to this level of thinking.)"

Ecology having by then emerged to the proper level, he called for linking national and international programs based on ecosystem ecology to damp the sectional conflicts that hamper development. *Bioregionalism*, a movement calling for ecological zones and boundaries to be given preference in understanding and organizing the earth over boundaries that are merely political, was also born of the superorganism.

There is then, in the form of that one paper by Odum, the potential genesis of bioregionalism, biocentrism, Deep Ecology, and ecotage. If it did not exactly create all of those movements, it at least gave them scientific legitimacy. Only organic farming and ecofeminism, among environmental movements, are not tied somehow to the superorganism. And all this was going on as the ecosystem concept was falling out of favor in its science. What gave sustenance to radical ecology accelerated the decline of the superorganism.

In a time when ecocentrism must have sounded like some sort of forgotten Clementsian term—maybe it was the center of a succession?—there was no question in Odum's mind that studying the functional parts and processes of ecosystems promised to be as useful as studying the functions and processes of organs. He had previously set his ideas before a wide audience in a classic paper entitled, "The Strategy of Ecosystem Development," also in *Science*. It did not take over ecology by storm, although its ideas had already gathered to it numbers of staunch adherents. By then, Odum was preaching to the choir, as far as ecologists were concerned. The paper did, however, lead to some remarkable and unpredictable consequences. The first is that it alienated a number of ecologists from ecosystem science. It split the science of ecology into two camps. At the same time, it kept the ecosystem ascendant in issues of environmental management policies and practices. It became dogma, the new view of the balance of nature, in which populations were naturally regulated at what could sometimes be a fragile equilibrium. Finally, the paper suggested that diversity, per se, was an aim in itself in environmental protection. That's a lot for one concise little paper in *Science*.

First the split. The title alone was a call to arms of sorts. "Strategy" was a word in common—but not formal—use to describe an evolutionary adaptation. At the time, it was more likely to be used informally in conversation by scientists than in their published works. By using it in his paper, Odum had fallen into a dual trap. Many ecologists then were finding that talk about evolutionary strategies or ecological strategies was "perversely teleological" as one historian put it. "Strategy," if used by ecologists and evolutionists out of carelessness or frustration was what is called an anthropomorphism. "Try to avoid that," William L Brown, Jr, once warned me. He was an ant taxonomist and evolutionary biologist who had proudly helped Rachel Carson with *Silent Spring*. He had also felt the need to sit down with me at around this time to carefully explain his use of the "s" word. I was then a teaching assistant in his evolution course. He called his use of "evolutionary strategy" "anthroponominalism," rather than anthropomorphism. His term means something as follows. We know that an organism does not have human attributes (that's the "anthropo-"part) such as a need to evolve a certain adaptation, but we use the term "need" for the sake of convenience in discourse ("-nominalism"). The species didn't really need the adaptation in the sense of, say, going about acquiring it.

That a bumblebee "wants" to gather enough pollen to make up for the energy expended in its search is an anthropomorphism. We have no knowledge of a bumblebee having wants or desires, but are projecting our own characteristics to it. We can guess that if the bumblebee's trips result in a net loss of energy, it is not likely to survive. A more insidious use of anthropomorphism would be to argue that bumblebees need to adapt a strategy, through evolution, to maximize the energy return for their foraging in order to increase fitness. It all may sound reasonable, but, again, there is no evolutionary "need," except in the heresy of Lamarckian evolution. Evolutionary change is not directed to a specific purpose. That is what is meant by "teleological." Change happens in ways that, in hindsight, can be interpreted as having led to a final result, but that result was never the goal of the changes. The evolution of a photosensitive spot into the vertebrate eye, for example, is a series of evolutionary modifications, each adaptive in its own right, rather than as a precursor for the final form of the eye. Thus, anthroponominalism was an act of desperation, a shorthand for a true explanation too cumbersome to repeat.

There is reason for the hair-splitting semantics: admitting that evolution can be directed admits also that there is something to direct it. This is a camel whose nose has to be kept well away from evolution's tent. Seeing direction in evolution is seeing the hand of God at work.

Odum's use of "strategy" had to have been deliberate, rather than careless. It accomplished his purpose of getting across the idea that there may be more than analogy between ecosystems and organisms. It has been called "a clever device," not a careless use. However, it put Odum in a bad light among some ecologists. Not only did it suggest that ecosystems might be capable of evolving—which he carefully avoided saying even while writing about selection at the ecosystem level—but that the evolution could be teleological, or goal directed, a matter on which there was almost universal opposition.

Damned if I don't have to go all the way back to Darwin again to explain why some ecologists drew up their wagons against Odum, but I do. The natural variation we see all about us, both in terms of the kinds of creatures, species, and the different characteristics that two organism of the same kind can possess, can all be explained, according to Darwin, by the process of evolution through natural selection. Just as a breeder selects for desirable traits, so does evolution choose from natural variation to create an organism that is adapted to its environment. Such a creature is more "fit" in Darwinian terms. Much of what ecologists see when they study organisms must be adaptations to the environment and, conversely, the interactions between an organism and its environment must lead to adaptation to that environment. Evolutionary adaptations are facts of life to an ecologist.

Ecology is mainly the study of evolutionary adaptations. Most ecologists, unless they were involved in practical, applied research—and most weren't—around 1970, regardless of what specialized name they used for themselves, were working out the adaptive results of natural selection. However, the evolutionary viewpoint had been minimal in the work of ecologists (mostly botanists) of the early 20th century from whom the ecosystem concept derived. They saw little point to an evolutionary perspective on what were basic descriptive studies at a time when Darwinian theory was in some ill repute. It was not until the Modern Evolutionary Synthesis combined Mendel's peas with equations for gene changes in populations in the 1930s and 1940s that the current Darwinian view came into being. In the meantime, American botanists were keeping ecological studies alive as a separate discipline. When, just after the turn of the last century, an editor of the *New York Evening Post* wrote to *Science* to complain that he was unable to find the word "ecology" that he had found in its pages in a dictionary, it was botanists who wrote back to set him straight. (They sent him back to his dictionary to look under "oecology," a spelling they were modernizing.) Those plant ecologist-botanists were mainly concerned with the distribution of various plants and associations of plants. Henry Cowles, an early ecologist and influence on the Chicago School, identified one of ecology's "great tasks" as the unraveling of "the mysteries of adaptation." He did so by studying the development of plant associations, what came to be called community—or ecosystem—succession.

They were studying what plants occurred where and when and trying to make sense of the changes they were seeing spread out in time and space, those ecologists. Evolutionary considerations were unnecessary. After all, community succession happens on time scales too short for evolution. The age of a mature oak in a climax stand might well be the same number of years that the stand had been undergoing succession from some previous stage. No chance for evolution there. Not in a single individual organism. Not in a single lifetime.

Those early ecologists also had a hazy view of what natural selection actually operated on. Cowles, who saw ecology as unraveling the mysteries of adaptation, thought Lamarckian and Darwinian theories were "not necessarily inharmonious." Jean Baptiste Pierre de Lamarck (the name was actually longer, de Monet rather than de Lamarck, also) had theorized in the 18th century that organisms with an "internal will" for an adaptation were more likely to produce offspring with the characteristics necessary for the adaptation. The environment directed evolution through the immediate needs it imposed on an organism as passed on to its progeny. In simpler terms, acquired characteristics (those needs) were transmitted to offspring. For someone like Frederick Clements, who saw communities evolving and who had not the benefit of the Modern Synthesis, Lamarckism had obvious attraction. Here was a way for collections of species to evolve entire suites of characteristics in concert.

It was not only plant ecologists who held a favorable view toward Lamarckism. Before Mendelian inheritance was rediscovered, the only choice to explain heredity had been either some sort of blending inheritance or Lamarckism. Blending inheritance combined with natural selection had a logical flaw pointed out during Darwin's lifetime by Fleeming Jenkin, an engineer. He saw that any new variation would be swamped by the vastly greater amount of the characteristics that were already present in a population. Jenkin posed the challenge to Darwin's supporters of explaining how a single Englishman marooned on an island of aborigines could pass on his "superior" genes to that population, if with the progeny of every mating those genes were reduced by half. Darwin eventually capitulated in favor of a Lamarckian transmission of acquired traits. So did many other biologists. Lack of a genetic mechanism muddled Darwinian evolutionary theory for almost a century. It did not, however, muddle the facts of evolution. The end products of evolution are available for all to see and do not change with changes in the underlying mechanisms of evolution. Natural selection, Lamarckism, or God's will, evolution's story is the same. Only the putative causes differ.

We know that it is individuals that are born and die, but to geneticists, it is the gene, a discreet, particle-like dose of hereditary material in the form of DNA (in the simplest sense, ignoring the niceties of such things as regulatory genes), possessed by an individual organism, that selection chooses or finds wanting. It is the gene that produces an adaptive trait or not. Early geneticists built up an impressive logical structure starting from the handful of genetic traits in peas for which 19th century monastic Gregor Mendel found mathematical regularities. They were not about to abandon it based on discordant observations made by ecologists.

Mendel had found that a pea plant having red flowers, for example, when crossed with one having white flowers, did not produce pink-flowered offspring. The colors did not blend as in the objections of Fleeming Jenkin. Both red and white flowers were present in the progeny in the proportions predicted by the dominance and recessiveness of the genes, as can be worked out by high school biology students. All the details are not of concern here, but even though not all genes behave this way—Mendel reported on eight traits that we now know were each present on a different one of the pea plant's eight chromosomes—variants can be explained away by geneticists, either mathematically or as oddities of development.

Ecologists, however, saw a number of phenomena that suggested selection might be more complex than dominance relationships and gene reassortment. Ernst Mayr, a major figure in producing the Modern Synthesis, has disparaged the simple models he termed "bean bag" genetics. (The models reduced heredity, if not evolution, to chance events that could be simulated using a bag of variously colored beans. The proportion of colors in the bag represented probabilities; pairs of beans randomly selected, the actual outcomes.) Odum, in his "The Strategy of Ecosystem Development" now (it was 1969) suggested something that the bean bag geneticists could never contemplate—that entire ecosystems might be subject to evolution in the Darwinian sense, with competition being replaced by mutualism and symbiosis. It was evolutionary progress of a kind.

Odum had revived the superorganism, given it measurable attributes, and argued that it should have a central place in environmental thinking.

It was disagreement over what natural selection actually acted upon, along with certain intangibles of human nature, not Eugene Odum's choice of level of study or method of analysis that made ecologists, even systems ecologists, and even some ecosystems ecologists, shy away from accepting his arguments that ecosystems could evolve. There soon began to be denial that ecosystems could even have what are called *emergent* properties, a point of great importance to Odum. Emergent properties are those that are unique at a particular level and cannot be completely explained by the study of a lower level alone.

Emergent properties were what made Odum's ecosystem concept worthy of attention. No emergent properties, no need for ecosystem-level study.

As an example of an emergent property, one would be hard pressed to predict the density or viscosity of water from a study of hydrogen and oxygen atoms in isolation. Organisms, as well, obviously have properties that cannot be predicted from their constituent parts as studied at a lower level of organization. However, populations of organisms may or may not.

George W Salt, the editor of *The American Naturalist*, a journal in which many of ecology's modern controversies were played out, cleared up whatever semantic confusion the term may have had by separating emergent properties from what he called collective properties. Emergent properties for an ecosystem were those that could not be predicted on the basis of the populations of which it was comprised. They should not be confused, in Salt's view, from properties that were only the sum of properties measured at a lower level. There were no emergent properties in ecosystems, was Salt's unambiguous conclusion. What he left unsaid was the next step in his logic. If no new properties emerge at that level, then there is no need to study ecosystems.

Eugene Odum's reply was that properties that resulted from interactions between components of a system did emerge as new properties of a system. His was a losing battle, though. Salt may never really have been clearing up a semantic confusion. What he was accomplishing instead and, perhaps deliberately, was to define away the concept of ecosystem. To rid his tidy field of evolutionary ecology of that messy creature was his more likely goal.

For example, Odum had stressed that the developmental stage and physiological condition of an ecosystem could be characterized by its Production to Respiration Ratio (P/R), a measure of new growth divided by the energy used for growth. To Salt this was merely the sum of the new growth produced by all of the producer organisms in an ecosystem, divided by the sum of all the respiratory energy losses of all the individual organisms in the ecosystem. To Odum, the particular P/R ratio of an ecosystem was not a simple sum, but the result of the interaction of the organisms in the system. If the whole matter sounds like unnecessary semantics to the reader, the reader is being very perceptive. The actual issue behind the semantics was whether an ecosystem could be a useful object for study. There were some who wished it was not.

Eugene Odum had defined a number of whole-system properties based mainly on materials and energy cycling through ecosystems. He also raised what in hindsight was a moral imperative for whole-system study. This was his perceived relationship between ecosystems and human ecology. He placed man's activities squarely into the analysis of ecosystems. But then he muddled things by adopting an essentially Clementsian view, no matter how carefully presented, of an ecosystem as a developing, evolving superorganism. This, and a number of non-ecological issues—personal preference, funding for big versus small science, the development of allegiances, hints of elitism—were more responsible for the development of opposing camps in ecology than the question of holism or reductionism (whether the whole item or its parts are more worthy of study) that raged furiously and needlessly in ecology at this time. Richard Levins, very much in the evolutionary camp of MacArthur-style researchers (these will be taken up in subsequent chapters), took delight in identifying himself with holistic thinking in a number of lectures I attended in the 1970s. In fact, holism has always been an attractive point of view to those who saw themselves as theorists, rather than data-gatherers.

Still, having ecosystems somehow capable of evolution, as Odum claimed, did present a number of problems to the science of the ecosystem. For example, part of his definition of an ecosystem was that of organisms "in a given area" and their physical environment. This is compatible with a systems approach. Arbitrary scales and interactions are allowed. Just draw the boundaries and label the box.

Odum, however, went so far as to state that an ecosystem could be "...a small pond, or a meadow or old-field. In fact, any area exposed to light, even a lawn, a window flower box, or a laboratory-cultured microcosm, can be the 'guinea pig' for the beginning study of ecosystems, provided that the physical dimensions and biotic diversity are not so great as to make observation of the whole difficult."

Flower boxes competing for survival with each other gives one pause. It would make something of a kindergarten science project to pursue, the kind NASA liked to send into space for publicity purposes. Odum actually favored the watershed "*as the minimum ecosystem unit.*" This is the biota and physical environment of an area that drains into a single body of water. However, he qualified his minimum as being for "*man's interest.*" (The Italics were in the original.) As far as theoretical ecology was concerned, though, anything went.

Correcting the issue of ecosystem size and boundaries, however, was waving a flag to try to slow down a train that was already miles beyond. Evolutionary ecologists stopped reading the papers of ecosystems ecologists. Odum conceded as much when he co-authored a paper lamenting that, "Many ecologists prefer not to think about ecosystems at all. For them, the theory of ecology is complete with the organism and its evolving population."

So, one reason given for rejecting the superorganism was that it had no boundaries. It was also rejected for non-existing emergent properties. It even once had to face the accusation of not being cybernetic, for heaven's sake. Yet none of these alleged shortcomings are serious enough to damn a theory as broadly reaching as Odum's ecosystem concept. Ecosystem boundaries, admittedly porous and indefinite, can still be set in some arbitrary, but useful ways. A forest or prairie edge, a watershed boundary, the area between high and low tides, for example, are all imperfect boundaries that have been used by ecosystem and community ecologists to gain insight into ecological goings on within them. Emergent properties might well be brushed aside as being nonexistent at the ecosystem level, as was done by some, but emergent properties are no more necessary to turn a science's attention to higher levels of organization for living creatures than they are for the inanimate objects geologists and chemists study. The properties of rocks and polymers are not at all apparent from the properties of the molecules making them up. The properties of a forest enjoyed by nature lovers are hardly the sum of the individual trees and birds in it. These could be more easily enjoyed in isolation.

Why then was the ecosystem concept deemed unworthy for study by so many leading ecologists? I have hints and Sherlockian suspicions. Could it have been its champion who was rejected, in favor of another, rather than the concept?

Personalities are not supposed to interfere with scientific progress. Published arguments must be based on facts and ideas, only. I have some circumstantial evidence, however, that they are not always. As with all circumstantial cases, this one will take some time to build. It will require analysis of issues that at first glance seem to be irrelevant. Evolution of ecosystems is one.

Whether or not an ecosystem can evolve may well have been beside the point. For example, there are ecological—and some morphological—equivalents in the faunas of the African Veldt and the Australia Outback, even though the mammals are (were) exclusively marsupial in Australia. There are large marsupial ungulates, marsupial wolves and cats, marsupial mice in the Australian plains. Ecologically, they match the zebras and wildebeests, predatory cats, and mice of the African plains. What is the evolutionary explanation? Both communities evolved to best adapt to similar climates, obviously, even though from different starting materials (the initial floras and faunas).

There are two ways of interpreting convergent evolution of this sort. One is that the individual species of African and Australian grasslands independently, species by species, evolved in response to similar environmental factors to show similar characteristics. The other is that, given similar conditions, only a specific ecological structure to the animal and plant communities could persist. Only the fittest superorganism, perhaps, is found in those grasslands. Assemblages of species might have been replaced through evolution by other, better ecologically adapted assemblages.

The explanations are only subtly different, really just matters of emphasis—either on the individual or the community. Both explanations are consistent with ecological evidence. Tansley, the originator of the ecosystem term in opposition to Clements' ideas, saw it as a fact that there was "a kind of natural selection" for more stable ecosystems. Even into the 1960s, natural selection on groups of species was perfectly acceptable to ecologists. LaMont Cole wrote, "communities composed of interacting populations of several species may also behave as units of natural selection" in 1957.

As late as 1960, a biologist could write "selection may apply at the level of the ecosystem as well as at the levels of the individual and the specific population. Ecosystems can compete, and evolution of the stable ecosystem can be looked upon as a process of learning, analogous to the learning of regulated behavior in the nervous system of animals."

He went on to write, "As to the mechanisms by which selection might take effect at this level, they are of the ordinary Darwinian sort except that the criterion for selection is survival of the system rather than of the individual or even species."

It is today an obscure paper by an author whose name is no longer known, but that obscurity speaks volumes. It was published in *The American Naturalist*, then as now respectable and well read. At the time of its publication there was every reason to believe that species sharing ecosystems might evolve to benefit each other. Examples of symbiosis—mutualism, in particular—are many.

(There is a bit of a semantic maze here, for which ecologists are totally at fault. The term, symbiosis, can technically mean interactions other than those, such as with the algae and fungi that together make up what we call lichens, in which both species benefit. Thus the need for the term mutualism.)

It was when attached to the superorganism that ecologists stopped to reexamine the issue of selection at levels above the individual. Even Eugene Odum's followers found they had to distance themselves from the superorganism concept, although not that of the ecosystem.

"That was really an exaggerated concept that was exclusive to Tom Odum and a few of his students. None of the rest of us (including his brother) ever bought into the exaggeration," Robert O'Neill, then of the University of Tennessee and Oak Ridge, who should know, reminded me.

Although Gene Odum took care not to use the term, his ecosystem concept did have most of the attributes of the superorganism, and it was assimilated by the public as such. As far as the science of ecology is concerned, ecologists may have been out to slay Odum and his followers as much as his dragon.

Save the Cat!

In their study of a coral atoll, the Odums were impressed when they discovered that the algae and coral making up the reefs were single symbiotic units. How did they evolve?

Coral benefits from the algae's "parasitic" presence on it because the coral harvests some of the photosynthetic energy fixed and stored by the algae. The algae benefits through absorption of mineral nutrients harvested from the sea by the coral. It is a situation that might have come about entirely by chance, that mutual benefit. Any adaptation in one species that benefits the other incidentally will also benefit the first. Put together a series of such evolutionary accidents and you have a coral.

The two species could have evolved in concert by having the more mutualistic pairs (coral plus algae) outcompete less mutualistic ones. Or they could have evolved totally independently, each evolutionary change benefiting only one organism at a time. The eventual outcome would have been the same. So could less interdependent species in an ecosystem also evolve together or independently to become the homeostatic ecosystems of Eugene Odum. The well-adapted superorganism could arise through either set of mechanisms. So why was there an issue?

At about the same time that Odum was championing his ecosystem concept, two controversies tied only marginally to ecosystems combined to make it heresy to admit that ecosystems could have evolved along the lines just discussed. They had to do with group selection and genetic load.

Group selection was an outgrowth of a previous controversy, that of density-dependent population regulation. Again, as in almost all things ecological, it all goes back to Darwin. Darwin had made certain predictions against which his theory could be tested. During one of the times when evolutionary biology was being questioned on its logical structure—surprisingly, this was in the late 1970s—defenders of natural selection took to quoting Darwin as having written, "If it could be proved that any part of the structure of any one species had been formed for the exclusive good of another species, it would annihilate my theory, for such could not have been produced through natural selection." This is as testable a hypothesis as any in science. And ecologists had blundered into a test situation.

Darwin had natural selection operating within a Malthusian background of unlimited population growth. Darwin himself calculated that the slow-breeding elephant could produce fifteen million progeny from a single pair in five centuries, assuming no deaths. Although Darwin admitted that what checked this natural tendency to increase was unknown, competition was, nonetheless, the natural result of population growth. Competition between individuals of the same species, Darwin noted, was the most severe and that was where natural selection acted.

A certain logic began to develop concerning the competition of types under natural selection. It was in LaMont Cole's review of *Silent Spring*. A population undergoing growth under conditions of unlimited resources will have no need to compete for those resources. There is enough for all. Therefore it would not be subject to natural selection as far as its ability to obtain those resources. It is only when resources become limited, and population growth becomes regulated, that some types will have a greater chance for survival by grabbing more of the resource. (Actually, current research on finches in the Galapagos Islands suggests that evolution can change direction, in a sense, in going from good times to bad. Natural selection works under both conditions.) The situation in which resources are limited is determined by the relative amounts of organism to resource, what is usually measured as population density. Therefore, natural selection should operate more at conditions of high density than low density. So went the logic. A corollary was that the regulation of population growth should be density dependent. Growth should be slower at high populations than at low populations. Or a population could decline when it exceeded some optimum.

This is not science in its true sense; this is simple Greek syllogism. It may follow logically, but there is no reason for it to be true. Nonetheless a number of ecologists saw all population regulation to be density dependent. Examples of populations being regulated in a non-density-dependent manner became heretical to the accepted evolutionary scheme, even though there is no reason that natural selection might not act, in a much slower, perhaps less effective manner—or just as rapidly and effectively—at low densities.

As is true in much of the history of ecology, it did not take long for ecologists looking into how the sizes of natural populations are regulated to fall into two camps: the density-dependent and density-independent camps. Each camp had data in support of its point of view. Density-independent population regulation became associated with Australian scientists HG Andrewartha and L Charles Birch, while the density-dependent camp was led by Englishmen Alexander J Nicholson and David L Lack.

To Eugene Odum, the controversy was one in which the theories were "correlated with the environment of the theorist." Andrewartha and Birch studied insects in severe environments, while Nicholson and Lack were presumably studying animals and birds in English gardens. The latter were "benign environments," the former stressful. There was a famous meeting, the 1957 Cold Spring Harbor Symposium on Quantitative Biology, held to resolve the issue, at which one practical-minded ecologist suggested that "the controversy would go away if the investigators would simply trade organisms."

Tangled with this thread of competition and density-dependent population regulation is the infamous equation for logistic population growth. Logistic growth, by which the well known "S-shaped" population growth curve is known is shown in the graph on the next page. I have (appropriately) appropriated the graph from *Great AEPPS*. It came out of what was going on in Raymond Pearl's laboratory in the years that Rachel Carson had been working with its rats and fruit flies (a time in her life that is not at all very well documented or investigated by historians or biographers).

It is an overly simplistic equation (we will get to its math later) that was over-optimistically applied below by Pearl and Gould to data that was insufficient for their ludicrous prediction of less than

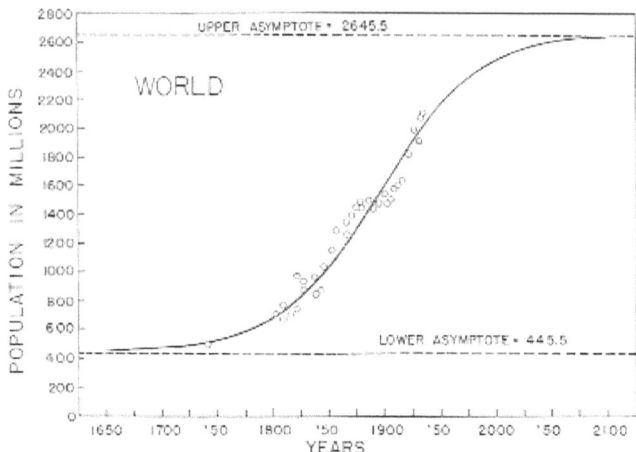

3 billion for the stable population that our species would reach … in … well, not too much time from now. (We have already reached more that two-and-a-half times that population. (Hmm.))

Two very important ideas pop out of its analysis. One is that of the intrinsic rate of increase of a population; the other, the carrying capacity of the environment for that population. It is the carrying capacity, represented by the upper dotted line on the graph, that is relevant here. (Pay no attention to the "lower asymptote," it will only confuse you.) An environment's carrying capacity is how many organisms it can support. Therefore, the logistic equation proposes (or has as one of its parameters) a population density that an environment can support. Population changes can vary about this density, presumably due to changes in the environment. Regulation at a given carrying capacity, however, based on this outgrowth of the logistic curve, should be density dependent. If the population goes above the carrying capacity, it declines. If it drops below it, it increases. High density up; low down.

Except that there is nothing in the theory that disallows density-independent population growth. A population might never reach its carrying capacity due to events extrinsic to it, such as severe weather, for example. It then fluctuates well below the theoretical carrying capacity of the environment for it, or it takes off the way human population growth has. (To then crash to well below that number of carrying capacity?)

Even with the acknowledgment that Darwinian evolution could survive either view—although density dependence would make it more likely than the opposite—the argument over density dependence had practical importance. Arguments that on their surface seem to be entirely semantic often have more substantive issues hidden behind them. In this one, the issue might be stated as how, for example, does a game manager or pest control worker predict population growth and density? Based on weather? Based on last year's density? Based on some estimate of its innate ability to increase and the total resources available for the increase? Different mechanisms of population control require that different data be collected. (The concept of an environmental carrying capacity, even independent of the logistic growth equation has been useful to game managers since the 1930s. It is as evident in Aldo Leopold's *Sand County Almanac* as is a balance of nature. His view of ecology was always that of the enlightened gamekeeper.)

Darwin's theory of natural selection, keep in mind, does not require that population regulation be density dependent. It would have been nice, tough. Density-dependent population regulation would have natural selection acting at its maximum strength at all times. But this is not necessary. All that natural selection needs to operate from Darwin's first explanation of it to now is the differential survival of types having heritable differences related to the environment in which they are found. Period. Neither, take note, does natural selection require greater fecundity (more offspring) in the more fit type. Greater fecundity can be a property of "luxury," not fitness. The stronger type is not necessarily the fittest. Ponder the superior "luxuriant" characteristics of the horse-donkey hybrid, the mule. Then ponder the zero fitness of this sterile creature. Similarly, a wolf that produces a large litter of cubs is not necessarily more fit than one with a smaller litter. Improved parental care might cause more pups from the small litter to survive to adulthood. Fitness is best measured by the number of your grandchildren, not the number of your children. Producing offspring that then fail to reproduce results in the same net fitness as sterility. However, if a population *is* to control its own density, to self-regulate, it needs the capability to respond in appropriate ways to changes in density.

What does this have to do with the superorganism concept of the ecosystem? Odum's ecosystem developed toward a state of homeostasis, or self-regulation. Self-regulation means that its populations need to self-regulate. In self-regulation, instead of dying of starvation or disease at high densities (density-dependent regulation), a population somehow limits its reproduction to keep its numbers below starvation levels. This is, in essence, what Paul Ehrlich asked the human population to do at the time of the first Earth Day. Can it? Certainly the technology is available to do so. Will it? Maybe. Or maybe religious or race wars will reduce the population density below our carrying capacity in density-independent ways.

Can natural populations, lacking the power of the press and prophylaxis of our own species, control their growth? The evidence appears to be yes. And the evidence may be all around. However, theory requires an answer to a second question, before going on to look at the evidence. My hasty yes given above, had it been given in the 1970s, would have immediately raised a flag of heresy above me. It still will in some circles today.

The question that the inquisitors posed to heretics of my persuasion was—and is—can natural selection result in the evolution of self-regulation in populations? How can populations evolve to keep down their own numbers? The answer to that question, yes or no, will, believe it or not, accept or reject the concept of ecosystem evolution.

This is how it works: the heresy to which I have just confessed is that of group selection. Instead of natural selection choosing the most fit individuals, what survive are individuals belonging to the most fit "groups," however they might be constituted. A group could be a family unit such as a clan, a species, or an entire collection of species in an ecosystem.

Group selection is necessary for a population to self-regulate. Self-regulation is a necessity for ecosystem homeostasis. No group selection, no ecosystem evolution. No ecosystem evolution, no homeostasis. No homeostasis, no point in studying ecosystems.

The Modern Synthesis that is our view of evolution today requires a genetic mechanism in order for natural selection to create evolutionary change. Group selection could be that mechanism. Keep in mind that I have been trying to develop an explanation for how ecology split into two camps: ecosystem ecologists and evolutionary ecologists. I am still laboring toward that explanation. The arguments over density dependence, self-regulation, and group selection represent the technical part of it. Messy, but no less an explanation.

The evidence for self-regulation in populations is as follows. The "clutch" size of birds (the number of eggs a bird lays) appears to be adjusted to environmental conditions. The average number of eggs in a clutch is higher in good years than in severe years. This is data. Similarly, rodents react to overcrowding with a cluster of hormonal changes, many of which lead to reduced reproduction. Voles, mouse-like creatures whose populations, like lemmings, can exhibit severe and regular fluctuations in density, were found in nature to suffer from some sort of unidentifiable syndrome at high densities that made them compare badly to similar voles at low densities. This reduction in "intrinsic viability," as one researcher labeled it, was found even after disease, starvation, and stress were eliminated as possible causes. The syndrome was thought to have a genetic basis.

Both examples point to some sort of intrinsic population self-regulation, although one could call the evidence equivocal, and some did. However, taking the evidence at face value raises a serious question: What is the mechanism of inheritance of the observed behavior?

Natural selection would favor the bird that managed to squeeze out one more chick under bad conditions or good. The vole that would not die when the Longworth trap door snapped shut on it during high density conditions should have a considerable fitness advantage over the poor creatures I observed exhibiting this unfortunate behavior. I at one time had pretensions of being a mammalian population ecologist. I did become an ecologist and I am a mammal, but my pretension was to become an expert on the ecology of mammal populations. Lions eating wildebeest was what I had in mind. Neither being in evidence around Ithaca, New York, I settled for meadow voles. I was soon disaffected from my pretensions by the difficulties involved in collecting population data on mammals, even wee, abundant ones. You have to find the little critters to count them, and that requires traps. Longworth traps are rectangular metal boxes, one end of which has a spring-loaded trap door, the other, closed end, a peanut-butter-covered mechanism for springing the door. It is set into the tunnel-like paths, called runways, that voles make in grass. The device is a humane version of the "museum" mousetrap once common in all households. Instead of having its neck snapped on touching the cheese, the mouse survives to munch on the peanut butter, presumably in contemplation of why it was now shut in.

Actually, though, many did not survive. It alarmed me how many of the little creatures would be dead in the traps by the time I came to collect them the next day. I needed them alive, because my intent had been to breed them in the laboratory in hope of developing a method to determine the ages of voles, a useful statistic in population studies.

With determination, I resolved to check the traps more frequently. Doubling my effort, however, still left me with dead voles in traps.

Finally, there came discovery. On setting one trap into a runway, I took no more than two steps away from it when I heard the telltale snap that signified a trapped vole. Immediately turning back to collect the specimen and reset the trap, I was shocked to find a dead vole, showing no sign of injury. Once the trap door shut, apparently, they could be dead in a manner of seconds, succumbing to nothing more than surprise.

Was this the result of evolution? It had to be. Was it adaptive? Not to the dead vole, certainly.

It was all part of a syndrome the population was undergoing. How then can such apparently nonadaptive characters be maintained in a population undergoing natural selection for fecundity and survival?

To British scientist Wynne-Edwards, the answer was that natural selection did not necessarily act only on the individual. Natural selection could select for groups possessing traits that, although making their individual possessor less fit, make the group more fit. This is the heresy of group selection in a nutshell.

In theory, as in the argument over density dependence, there should have been no controversy, for the theories were not really conflicting, but only questions of emphasis and interpretation. But isn't that the rule for all heresies? The smaller the differences, the more severe the schism?

The observations on which group selection was based can be easily explained through individual selection. Why shouldn't populations evolve to regulate themselves in a Darwinian way? If an individual bird should increase its own survival by rearing a smaller clutch, it has the advantage of being able to reproduce some other day over the bird that adds an extra egg to its clutch, only to have that be the last breeding act of its life. Plausible? Of course. This was essentially the explanation for variation in clutch sizes given by UCLA bird ecologist Martin F Cody, student of Robert MacArthur, student of GE Hutchinson. Is Cody's explanation true? Who knows? Who cares? Both group and individual selection seem plausible, giving the same results. Isn't that what matters?

The principle of parsimony, however, says that we should accept the Darwinian explanation (even though Darwin never required it) and get on with our work. The same semi-scientific principle, sometimes called Ocham's Razor, requires that the simpler, Darwinian explanation should be preferred over the more complex ecosystem selection hypothesis. This logical rule attributed to William of Ockham requires that the simplest of competing theories be preferred to the more complex, that explanations be first sought from the realm of known quantities, and that assumptions should not be unduly multiplied. (Note the varyation in spelling. In some ways, life was simpler before spelling was standardized. Denial of written statements might have been possible on the basis of misinterpretation of an aberrant spelling.)

But Occam's Razor will not guarantee reaching the truth. EO Wilson called it the "Fallacy of Simplifying the Cause." The simplest explanation is not always the correct one according to him. Yet others, population geneticist Richard C Lewontin and, prior to him, biologist George C Williams, argued influentially that this very fallacy be used on how natural selection operates.

"The ground rule—or perhaps *doctrine* would be a better term—is that adaptation is a special and onerous concept that should be used only where it is really necessary," Williams wrote in his 1966 book, *Adaptation and Natural Selection*. "When it must be recognized, it should be attributed to no higher level of organization than is demanded by the evidence. In explaining adaptation one should assume the adequacy of the simplest form of natural selection, that of alternative alleles in Mendelian populations, unless evidence clearly shows that this does not suffice."

Unless a specific mechanism of inheritance can be proposed that works only through ecosystems, the evolution of ecosystems need not concern ecologists, the statement could be interpreted as saying. And we know that traits are passed on through individuals. Interestingly, another evolutionary phenomenon that on its surface required some sort of closely interactive evolution between different species, Ehrlich and Raven's theory of coevolution, the process that might have resulted in the entity we call lichen, was not called to submit to Occam's test. Research on coevolution was becoming rampant at about the same time as Odum was perfecting his superorganism ideas. Both ecosystem evolution and coevolution were presented with no specified evolutionary mechanism. Only one was found theoretically wanting, however.

Although Williams's book was the least tolerant statement on selection operating on the level of the individual, Richard Lewontin is often credited with having the last word on it in an influential article in 1970 and later in his book, *The Genetic Basis of Evolutionary Change*. Professor Lewontin actually identified the totality of interacting genes possessed by an individual, the genome rather than the individual organism, as the target of selection, a highly technical distinction that need not be pursued here.

He also had this to say: "At yet higher levels, the species and the community, natural selection obviously must occur."

And also: "The same is true of communities whose stability of composition depends upon the interaction among their constituent species."

It is not surprising that Lewontin could see the issue both ways. He is one of the more conflicted personalities in the science of the 20th century. His being at the absolute forefront of his science, proper evaluation of his career must probably wait until the 22nd century. Let it suffice for now to note that the impression he gave was that of the asthmatic kid with glasses who was smarter than everyone else and knew it—as did everyone else. His personal ways could be rather off-putting. At one time he had a habit of correcting whatever pronunciation of his name was used. LEW-won-tin or Le-VON-tin, either would be disparaged as incorrect. I can only guess at his motivation. Was it to put people at ease? He was the only person that Robert MacArthur admitted could make him sweat. Unlike Hutchinson, who failed at his first post, then settled into a lifetime at Yale, Lewontin went from North Carolina State to the University of Rochester and then to the University of Chicago, before settling into Harvard as its Aggasiz Professor of Zoology. It is an upward progression without doubt, but it hinted at a difficult personality. A geneticist by expertise, he will reappear later in ecology's story.

I suspect I know why Lewontin came down on the genome (contained in an individual) as the target of selection. even though he showed obvious sympathy for species and ecosystem-level selection. Keep in mind how radioactive tracers and digital computers had given scientific legitimacy to ecology—and financing. Close on the heels of the ecologists in chasing after government funding, population geneticists also unveiled a tool to separate them from the butterfly collectors. This was the technique of electrophoresis, in which individual proteins, most usually enzymes, could be separated out from a mixture by their charge-to-mass ratio. A useful gadget for checking the purity of enzymes or purifying very small quantities for research purposes, it paid an extra dividend: it was discovered through electrophoresis that seemingly identical proteins could have slight variations in charge-to-mass ratios. The notion of how genetics acted at the molecular level at that time was that the DNA representing each gene coded for one protein, usually an enzyme. So each different protein represented a variant of a gene, called an allele. (The blue-eye allele my son inherited from my wife combined with the brown-eye allele he inherited from me to give him his brown eyes.) Lewontin, who built an outstanding career as a population geneticist, made a huge impact in the 1960s by uncovering the amount of genetic variability there was in natural populations. In particular, using the then-new technique of electrophoresis, he studied variations in enzymes found in wild fruit flies. Soon after, Robert Selander and his colleagues at the University of Texas found similar amounts of electrophoretic variation in wild species of mice.

  It should not go unsaid here that the electrophoresis apparatus was considerably more sophisticated—and expensive!—than a butterfly net and pinning board. The government paid the costs. The amount of variation found by Lewontin and Selander and colleagues was staggering, so much so that, at first, the electrophoretic results could only be accommodated in then-current evolutionary theory by having the variants be selectively neutral compared to each other or, somewhat later, almost neutral. The enzymes were different, the argument went, but not in any important ways.

  The need for these enzyme variants to be neutral, to have no effect on fitness, was an outcome of another, previous conundrum in human evolution and genetics, that of "genetic load." This concept stems from the idea that polymorphisms (situations, like eye color in humans in which there is more than one allele controlling a trait) are maintained by what geneticists call "overdominance" or "heterozygote superiority." (Think about hybrid corn here.) Gene combinations with unlike genes (unlike alleles) for the same trait were thought to be more fit than those with identical genes. If not, then the laws of chance, in the simple bean bag formulation that has been disparaged only a few pages before, demand that one of the alleles be lost from the population. Less fit, they are less likely to persist. As part of a more fit heterozygote, they can persist. This is the explanation of the persistence of sickle-cell anemia.

A consequence of heterozygote superiority is that half of the human population must, by the laws of Mendelian inheritance, be walking around with suboptimal genes. Think of death from anemia or malaria as a price to be paid by half the offspring born to a population in order to maintain the optimal heterozygous types. A population can not be made up of only heterozygotes, for the laws of genetics and chance predict that only half of their offspring can be heterozygous. The other half must be the less fit homozygotes. A maximally fit (heterozygous) population has to be succeeded by both homozygotes and heterozygotes. There is no escape from genetic load. And this would be only at one "locus," or for one trait. Add up the statistical probabilities for thousands of traits and one wonders how anyone can get out of bed in the morning under the ponderous genetic load that can be calculated. Lewontin, using what were reasonable numbers for the variability and fitness of genes in the fruit flies he worked with, estimated that the average fruit fly was less fit than a completely heterozygous fly by a factor of 10^{-43}. In simple terms, an "average" female fruit fly would have to lay 10^{43} eggs—an astounding number! One followed by 43 zeros—to match the one egg of the "heterozygous" female in fitness. Clearly the whole issue is utter nonsense, but it made up the careers of quite brilliant scientists by snaring them by the mathematical precision with which the concept led to a seemingly impossible paradox.

Besides the intellectual appeal of paradoxes, I have been dropping hints that there was also money involved in genetic load. As with any investigation, "follow the money" is good advice. The Atomic Energy Commission, concerned by the impact of radiation-induced mutations, funded a large number of studies on genetics, especially that of populations. The great majority, by far, of mutations are sub-optimal in fitness. What might be the impact on populations of increasing their mutational load? Those studies generated the need to fund other studies, Lewontin's among them. Without an atomic bomb being set off at Alamogordo, the ecology of today might never have been.

Lewontin, beneficiary of such funding, could only conclude that the object of natural selection was the gene, contained within an individual. The gene, after all, is what a geneticist, even an influential population geneticist like Lewontin, studies. Genetics, however, only provides the material for natural selection. Geneticist Lewontin was not allowed the last word, however. Evolutionary biologists turned to Ernst Mayr, a bird systematist (someone who really, really knows his birds) and one of the grandfathers of the Modern Synthesis, for an opinion. He also concluded in favor of selection targeting individuals. He had been leaning in favor of selection at the species level, but there was more trouble brewing in evolutionary heresies of a different type that kept him quite guarded toward group selection. He had, not his funding, but his theory to protect.

In a word (or a few words), the bird with the smaller clutch, therefore, *was* deemed to be more fit. It could take care of its brood better and see them to maturity.

Reduction in clutch size might be explained away in terms of individual fitness, but there are other seeming aberrations having no obvious benefit to the individual, that in fact reduce an individual's reproductive potential or increase its risk of mortality. Most are argued away as examples of something called kin selection. One example of that is of the sterile castes of insects. They are clearly less fit than the hive's queen and drones, but by giving up their chance at reproduction, they promote the fitness of the genes that they share with their siblings. Similarly, someone might put himself at risk to save the life of his brother, not out of love, but due to kinship. That brother has some of his genes. This is Richard Dawkins's *Selfish Gene* in action. (That the brother who puts his life at risk seemingly has to do a mental calculus to evaluate the chances of the two of them surviving their ordeal, as opposed to doing nothing and saving the genes in his body directly, is no more far-fetched than the idea of some poor meadow vole with unexplained genetic inviability.)

If we include human behavior, by considering the possible mechanisms that could allow populations to regulate their own rate of growth (but that still had to have come about through Darwinian natural selection), we come up against another conundrum. A real difficulty arises when considering humans. Humans are the only species believed to exhibit altruism, that is, behavior that benefits another, even a stranger, at the risk of harm to one's self. How could evolution through Darwinian natural selection have let such a thing happen? We know that firemen risk their lives not for the munificent pay and widow's pension society provides them, although that might be in keeping with the Darwinian concept of fitness, for by their death, they provide for their children. No, the fireman goes into that burning building out of true altruism. It may be confounded by the satisfying high of surviving danger, more intense when you remove a fellow creature from danger, particularly a child, but it is altruism nevertheless. The hormonal responses resulting in that high may have evolved for just that purpose. How that is is explained by a theory called reciprocal altruism. By risking your life to save a stranger's child, as one example, you are perpetuating a society in which a stranger may some day risk his life to save your own child. The exact mathematics of the effect on fitness of the altruistic act is tricky, but it can be shown that altruism can work within the framework of Darwinian fitness. (Darwinian fitness, recall, is individual fitness.) Presumably, if the child is highly in peril, but the risk to the stranger is slight, altruistic action will be returned through reciprocation to raise, on average, the altruist's Darwinian fitness. In cases of altruism, Darwinian fitness encompasses what is called an individual's *inclusive fitness*. If you assist a stranger with whom you share genes, you are helping to perpetuate those genes, thus increasing the fitness of those genes and your own inclusive fitness. Do not try to work out the math. Personally, I do not believe that evolution is that efficient. Like the human appendix, human altruism may be a relict from some previous form of society in the history of human kind. Or, considering the fact that it is a form of social behavior, it may have a totally non-Darwinian explanation. We do not, for example, stop to consider how a serial murderer is increasing his Darwinian fitness through his actions.

Has there been a fireman who died in an attempt to rescue a cat or a dog? I know of one who perished in the rescue of horses from a burning barn. Horses, of course, are property. They have an economic value aside from being pets. Yet we know that firemen will not risk their lives to enter a burning building only to save the building. In crowded locations, they may even sacrifice one property to the flames in order to save the properties around it. Property and economic value are secondary for firemen when considering the risk of their own lives. The risk-benefit calculus in the case of a strange family's tabby, then, should be amenable to solution with precision: zero risk matches the benefit perfectly. Yet the tabby will be saved.

The Darwinian evolution of reciprocal altruism cannot be as simple as its proponents suggest. Human learning mediated through culture can accomplish changes in behavior that explain all altruism in humans without need for resorting to genetics. Some have, however, seen culture as representing the cumulative effect of behavior that maximizes inclusive fitness. This is sociobiology. It brings us back full circle.

In the Darwinian view, therefore, a person places his own life at risk to save another—often a child—because by doing so, he can increase his own inclusive Darwinian fitness, given that his act is reciprocated by some other unrelated person to his benefit—or to that of his descendants. This is the stuff of legend, or perhaps what makes certain legends necessary. Horatio Alger heroes, I'm certain have many historical precedents in literature and folklore.

Buried with Full Honors

A recap is now in order. Ecology is the study of evolutionary adaptations. It has been so even before the term was coined. An early textbook of ecology that helped to define the field states that "...the growth of ecology and evolution were so inextricably woven together that it seems artificial to separate the two." The working out of adaptations that are under the process of natural selection was once ecology's only raison d'etre. Natural selection requires that individuals compete with each other for resources, resulting in the differential survival of the individuals. Competition is greatest when resources are in shortest supply, hence population regulation must be density dependent. There is even a mathematical expression for population growth to support density-dependence. But even if population regulation is not density-dependent, selection pressure is born only by the individual due to the logical application of Occam's razor. Therefore, only the individuals in a population face selection and only the population can evolve in response to selection. No selection of populations, per se, no evolution of ecosystems. It is the species in a community that evolve, not the community itself. Period. No superorganism.

Need we follow this logic? No. We do not. We can see what the evidence looks like.

Evolution in populations due to the natural selection of individuals has not often been confirmed in nature, but when it has, the confirmation has been spectacular, at least in the most famous example. This is the case of industrial melanism in *Biston betularia*, the Peppered Moth. In England, the moths have changed in frequency from light to dark—then back to light again in the latter part of the 20th century The change has been in response to changing background colors of tree trunks due to increased, then decreased levels of pollution. It is striking to see museum specimens representing collections a century apart. The moths were evolving in direct response to a specific change in their environment. The light colored morphs were better adapted to resting on the lighter bark of trees that were unbesmirched by soot. Morph is a term for a genetically determined, visually identifiable type, such as a blue-eyed individual, but is not, unlike that case, necessarily controlled by a single allele. Light morphs were collected almost exclusively by early- and pre-Victorian naturalists. The dark morphs were better adapted to avoiding predation while resting on the polluted trees of the English Midlands, and they began to dominate the collections of the 20th century. A series of famous experiments conducted by amateur Lepidopterist HBD Kettlewell, under the guidance of British ecological geneticist EB Ford, demonstrated that birds fed preferentially on dark morphs on light backgrounds and light morphs on dark. The prediction that gene frequencies would return to pre-industrial frequencies for the white morph once air pollution was abated is the only case on record of an evolutionary change being predicted in advance. (We must except the prediction of species extinctions.) There are other good examples of natural selection in action, but they are a bit more equivocal, with post hoc arguments and without the field testing done for the industrial melanism example.

Given the beautiful way the data on industrial melanism demonstrate individual selection, can we therefore eliminate the group selection hypothesis? Of course not. Individual selection and group selection are not mutually exclusive. Only if there are a great number and variety of cases that have been worked out, each of which points unequivocally to individual selection, might the concept of group selection need to be abandoned.

Actually, there are models and data to also support the idea that levels of selection above the individual might well exist. Two notable evolutionary biologists (one of whom was Stephen Jay Gould) have proposed that species, in competition with other species, can be the objects of natural selection over evolutionary time periods. Of course, the proposal was not made without starting an entirely new controversy. It was the one that caused Ernst Mayr to back away from group selection. Gould and his collaborator, Niles Eldridge, needed the concept of species being replaced through competition in order to explain certain features of the fossil record. The controversy they engendered was not so much on the idea of there being some form of "species selection," but on points that, interesting as they are, are far from the current focus of attention. Nonetheless, their work had that effect on Mayr, then the ultimate living authority on modern evolutionary theory.

Paradoxically, Richard Lewontin, already noted as being in favor of the individual as the unit of selection, had earlier described a well-founded case of group selection in house mice. Although having to do with a bizarre group of lethal genes that could not possibly contribute to their bearer's fitness under normal conditions, it is still as solid an example for group selection as industrial melanism is for individual selection. The traits involved were named *t-alleles*, because the mice lacked normal tails in the heterozygous condition. In the homozygous condition, many of the alleles were lethal. Strangely enough, though, sperm carrying the tailless gene were found to have a huge advantage over the wild type. (Geneticists substitute "wild type" for normal. In nature, it is not clear what is normal.) As might be expected, lethal t-alleles eventually wiped out whatever local populations in which they were found due to the combination of lethality and sperm advantage. Tailless sperm were much more successful at fertilizing eggs. However, because the mice existed on the whole in numbers of isolated populations, called *demes*, a new deme would form by colonization to replace the one that went extinct. The new deme would then become infected by colonists from demes that had not yet been driven extinct by the t-allele. A balance was thus found to result between the rate at which the t-allele could drive a deme to extinction and the rate at which new demes lacking the allele could be formed. By having its population fragmented, the mice could not be wiped out in toto by the tailless allele. By having interchange between the demes, the t-allele could not go extinct, as there was always another deme it could "infect." The "patchiness" of its habitat resulted in the same sort of equilibrium seen in predator-prey or competition interactions that would otherwise be expected to lead to the extinction of both. (Or is there a species or group selection at work in this latter case that results in a "prudent predator" that does not consume all of its prey, or in competitors that find ways to coexist?) Bizarre yes, maladaptive also, but the t-allele is just as much the work of natural selection as are cryptically colored moths in the British Midlands.

Mathematical models and experiments that support group selection were carefully examined during this time, found to be legitimate, and then were quietly ignored.

"Prior to 1966 it was a widely accepted concept, supported by some of the most prominent evolutionists of the day," wrote David Sloan Wilson shortly after the controversy had resolved itself into the status of non-issue. "It also had many prominent critics but, as with most major controversies, both sides threatened to persist indefinitely."

Change the dates and subject matter and Wilson's words could be used to describe the ecosystem concept.

"Then," Wilson continued, "with the publication of Williams's *Adaptation and Natural Selection*, the concept of group selection was discredited. The fatal blow came, not from a crucial experiment, or even from a new theoretical development, but simply from the elegance and clarity of Williams's thought in interpreting developments of the previous three decades."

The chips from Williams's blow carried far, for they took the superorganism with them.

"For the next decade, group selection rivaled Lamarkianism as the most thoroughly repudiated idea in evolutionary theory. Then it mysteriously rose from the dead. The modern version of group selection is supported by a variety of internally consistent and biologically plausible models," Wilson, who himself contributed to those models, concluded.

No longer illegitimate, by 1983, the ideas were nonetheless not ones pursued by many ecologists, although "decent folk" could again discuss group selection, as Wilson quoted another putting it. In between, according to Wilson, there "has not been a logical and orderly progression of ideas but rather a haphazard and often amusingly convoluted process with many internal contradictions."

Nineteen eighty-three was about the time that Odum was lamenting that ecologists had stopped reading papers about ecosystems.

What had been needed for acceptance of group selection ideas was not just an example of it in nature that was at least as well documented as the example of individual selection represented by the Peppered Moth. The tailless house mouse had been just such an example, but it was seen as an aberration, instead. Nor did lack of a legitimate mathematical model for group selection equivalent to those worked out by population geneticists for the Modern Synthesis hinder its acceptance. Sewall Wright, one of the founders of the Modern Synthesis, had just such a model in place within the mathematics he developed to describe gene frequency changes in populations.

The acceptance of one example, the Peppered Moth, and rejection of the other, the t-allele, as being instructive had little to do with science, but very much to do, perhaps, with scientists. It was not simply that ideas were in conflict, but that they were at the intersection of two different disciplines—ecology and genetics—having to make slightly different sense of the same phenomenon that prevented accommodation between them. The principle of parsimony rallied the geneticists behind it to put a logical conclusion to the group selection controversy that was good for their field. They killed it. To face what was often a world hostile to the idea, the theory of evolution needed to be broad, direct, and unambiguous. Loose ends and heresies such as group selection, sympatric speciation, punctuated equilibrium, Goldschmidt's "hopeful monster," and Lamarckism, were not to be tolerated. There were fundamentalist Christians with their own theory, directed evolution, watching the proceedings. Solidarity and decorum had to be maintained. Ecologists were unwittingly swept along by population geneticists, away from an issue that for them needed a more substantive resolution.

In between, the concept of superorganism was brushed aside. A statement such as "an organism and its niche evolve together," from an ecosystem scientist was dismissed as requiring group selection.

However, similar ideas and models "when stated in a way that did not invoke the concept of group selection," according to Wilson, "were warmly accepted by evolutionary biologists."

When a general mechanism of community evolution was presented, one reviewer found that models of community-level evolutionary interactions were "a bit shallow" and "loosely argued." More credible, one has to conclude, were the models of bean bag genetics in which alleles acted independently of each other, were not subject to regulatory changes in development, and existed, apparently, independently of their environments.

More credible, too, was a coevolution model presented by Paul Ehrlich and Peter Raven having no evolutionary mechanism other than, one presumes, whatever mechanisms were acceptable to evolutionary biologists.

Gene Odum, along with a colleague, finally had to concede about ecosystems: "They are not superorganisms," but their "structure appears to mandate ecosystem level coevolution."

Like congressional orators on C-SPAN giving special-order speeches, however, they were speaking to an empty house. Ecosystem ecologists were by then communicating only with each other. Occam's razor had proved lethal to the superorganism.

Ecology had already split into ecosystem ecology and evolutionary ecology by Earth Day 1980. Group selection had been the wedge. Of the two ecologies, one was the highly theoretical and romantic study of natural history that is the subject of following chapters and whose proponents could be found in the Eastern universities and bastions of civilization elsewhere, such as certain West Coast campuses. The other ecology was the descriptive analysis and computer modeling of the thermodynamics and chemistry of ecosystems, all the things that biologists usually chose their major to avoid. Its practitioners, and there were many of them, could be found in Georgia and Tennessee, or in campuses with agriculture, wildlife management, or forestry schools.

The two ecologies would not stay apart, however, for just as they were splitting around the First Earth Day, society was beginning to call on ecology to be the major player on environmental issues. Evolutionary ecologists would have to show their relevance in order to stay in the forefront onto which Earth Day had thrust them.

Aside from the issues of the superorganism and group selection, the ecosystem concept was vulnerable in one other, more fundamentally important way: that of there being a lack of a true metric for ecological change. Ecological interactions are so complex that analytical models are hopeless in mimicking them in any but an oversimplified way. The systems analysis models, promoted as doing better by dealing with biological data about the whole system, were better only by effectively including energy and nutrient flows. The productivity of various forest stands in Yellowstone, however, typical of data with which systems modelers work, can do little to inform a wildlife manager of the impact that doubling the number of wolves in the park would have on rodent populations. Neither would ecosystem productivity studies predict how the population response of the rodents would feed back to wolf populations.

Like Newtonian physics and Mendelian genetics, ecology needed a new form of mathematics. Newton invented calculus (or co-invented it with Leibniz) in order to describe physical processes. RA Fisher developed statistics in order to deal with the chance events that underlie so much of biology. The statistics of bioinformatics has made the newly elucidated human genome immediately productive. Ecology, however, had trouble finding a productive metric.

Still, the group selection controversy may have been the death knell for the superorganism, but it did not kill the ecosystem concept. Ecosystem practitioners kept it alive by distancing it from the superorganism concept. Systems analysis, they argued, is based on a "machine analogy," rather than an "organism analogy." Machines, we know, do not evolve in any way that requires genetic transmission. However, even given the machine analogy, one of its major practitioners, Robert O'Neill of Oak Ridge, came to wonder if it was "time to bury the ecosystem concept." Not, of course, "without full military honors," he added.

"Limitations in the concept are becoming more apparent and leading to a vigorous backlash toward ecosystem concepts in particular, and ecology in general," according to O'Neill. "Part of the backlash results from the apocalyptic fervor of the environmental movement over past decades. Ecology oversold its ability to predict doom, and is now seen as unnecessarily constraining human freedom and economic growth."

Besides an ideological backlash, O'Neill pointed out a number of difficulties with the concept, such as vagueness about its boundaries and stability properties, its assumption that species can all be lumped without need to consider each individually, that it excluded the work of natural selection, and that the relative size of disturbances did matter, meaning that so did the size of an ecosystem. None of these difficulties are, of course, insurmountable.

"Many would seem to be addressed by sophisticated developments in ecosystem theory involving, for example, non-linear dynamics and fuzzy set theory," O'Neill continued. "Unfortunately, the developments make the ecosystem theory more intriguing for mathematicians, but less useful and intuitive for biologists."

In the end, according to O'Neill, "We are putting splints and patches on an old horse."

Maybe, like an old soldier, the ecosystem concept should be allowed to fade away. Should it be buried?

"Probably not," O'Neill concluded, "But there is certainly need for improvement before ecology loses any more credibility."

O'Neill's pessimism about the usefulness of ecosystem science may not have been warranted. If nothing else, an ecosystem point of view had become indispensable in the vocabulary of how we understand our environment. Its practical significance makes up some of the chapters that follow.

Nor were we truly at the end of the life cycle of the superorganism, neither scientifically nor philosophically. It came back to life scientifically as the Gaia Hypothesis and philosophically as Deep Ecology.

JE Lovelock's Gaia Hypothesis is the organism analogy pushed to the level of the entire biosphere. It fits some observations perfectly. The earth-space boundary is an absolute one for life as we know it. Sunlight and heat can cross the boundary; there are the occasional additions of water through comets and amino acids in meteors, but granting these possible exceptions, life begins and ends on earth. Lovelock made a convincing argument that the earth's environment is perfectly suited for the life that is on it. Meanwhile, Boston University (for most of her career) microbiologist Lynn Margulis was also convincing in arguing that symbiosis is so common in nature that it had to be a driving force for evolution. Little troubled by the group selection argument (possibly even unaware of it, not being ecologists or population geneticists), Lovelock and Margulis extended the superorganism concept to its ultimate.

Margulis, who started publishing her ideas as Lynn Sagan, wife of Tonight Show celebrity and astronomer Carl, studied specialized structures (organelles) such as mitochondria, chloroplasts, etc, that were responsible for a number of important cell functions. In the view of a cell biologist, they are somewhat isolated from the rest of the material of the cell. They also have peculiarities that appear "foreign" from the rest of the cell. These organelles are found in animal and plant cells, but not in certain algae and bacteria. Margulis resurrected an old idea that they were endosymbionts, as is algae a symbiont in a lichen, even though being technically outside of its fungal host. The symbiosis between cell and organelle is such that there is no longer a separate identity to cell and symbiont.

"Mitochondria live inside our cells but reproduce at different times with different methods from the rest of our bodies' cells," in the anonymous words of a web author. "They are descendants of ancient bacteria. Either engulfed as prey or invading as predators, these bacteria took up residence inside foreign cells, forming an uneasy alliance that provided waste disposal and oxygen-derived energy in return for food and shelter. Without mitochondria, the nucleated plant or animal cell cannot breathe and therefore dies."

Margulis's contribution was in the clarity and detail with which she presented her argument and the abundance of evidence that she marshaled in its support. Nonetheless, her first book, now a classic, was criticized for having occasionally strained logic. For example, one reviewer complained that an idea "does not appear to be the simplest explanation for this datum, and no grounds are given for choosing it over the simpler alternative."

This is the same principle of parsimony we just saw trivializing group selection. A decade later, another reviewer, although praising her first book, took issue with her discussion of the Gaia Hypothesis in the later version he reviewed.

"This is group selection writ very large," he complained, apparently assuming that that was enough to dismiss the idea. Obviously, when you see organisms as the evolutionary result of collections of symbiotic cells, as Margulis did, group selection is not so heretical.

"It is the scientific establishment that now forbids heresy," Lovelock wrote in preface to the second edition of his little book.

Of heresies, Margulis has said, "I'm not worried about my reputation; I'm worried about the science being correct."

Lovelock, a physical scientist and inventor originally trained in medicine (PhD), became interested in life on Earth by way of searching for life on Mars. He was struck by the thermodynamic improbability of the concentrations of certain gases in the Earth's atmosphere. Lovelock saw a disequilibrium on a scale that suggested "that the atmosphere is not merely a biological product, but more probably a biological construction." The Earth was a complex entity with all of its systems comprising a single cybernetic system capable of homeostasis. A sort of superorganism.

Life evolved under conditions hostile to it, but modified its environment so that it was optimal for life. Slightly lower or higher carbon dioxide concentrations in the early history of life on earth, for example, would have led either to a "white frozen sphere" or temperatures "well above the toleration of life." Oxygen, fatal to early life when it first appeared in the atmosphere, now seems to be regulated at a fortuitous 21%. Below 12% oxygen in the atmosphere, fires will not ignite; above 25% they will not go out. More importantly to Lovelock, birds can fly and humans think in an atmosphere of 21% oxygen.

Lovelock's thinking, if heretical, is so by way of that damned "razor" again. There is another version of that principle, this one from physics, that is more appropriate to Lovelock's planetary science. Physicists who become overly enamored with the coincidences of life in our cosmos are usually brought back down to Earth by being reminded of the "Weak Anthropic Principle." This pretty much says that if conditions were not what they were (suitable for life), we would not be here to worry about the situation. All that needs to be recognized by science is that the range of conditions in which we find ourselves, as statistically improbable as these conditions might be, is physically possible. Anything more admits to some sort of directed change in the universe. This is the heresy of teleology or—worse yet, as we have seen—religion in science.

Deep Ecologists love the goal-directed original Gaia. Gaia allows for a religion of science, ecoscience, perhaps. Lovelock has modified, not with complete success, his original concept to remove its teleological aspects.

Of Gaia, Margulis says, "It has been perverted by a lot of people, but does that mean that it's wrong?"

Apparently not, for according to her, "Scientists are using Gaia in their everyday work, but they are not calling it that because of this goddess business. They call it Earth systems science or biogeochemistry or geomicrobiology. They use more words that sound much more scientific, but it sounds like basically the same thing."

While it is safer for a scientist to accept the Weak Anthropic Principle and get on with research, sometimes there is a very important practical reason not to. Lovelock had argued for such a reason: Gaia's response to man's changes. He found much about which he could be optimistic. For example, in his view, the resilience and homeostasis of Gaia is too great for man. He concluded that "The evidence for accepting that industrial activities either at their present level or in the immediate future may endanger the life of Gaia as a whole, is very weak, indeed." A few pages later, on contemplating the ice ages, Lovelock concluded that Gaia can tolerate the loss of "30% of the Earth's surface." This was in 1979.

A few years later, ecologists estimated that human activity had "appropriated," if not exactly destroyed, 44%—not of its surface —but of the Earth's productivity.

A Mathematician Who Knew His Warblers

In 1958, a young ecologist published a paper that, like much of his graduate-student investigations, addressed a question that long intrigued biologists. That same year, his mentor and graduate advisor asked the same question in addressing the American Society of Naturalists, of which he had just been elected President. Published the following year under the title "Homage to Santa Rosalia or Why Are There so Many Kinds of Animals?" it and the paper by the graduate student became immediate classics, making their authors leaders of their field and causing a change in the direction of their science. Theirs is a question that makes up a large proportion of ecological research today. It fits under the modern term, "biodiversity."

Like Lindeman's ecosystem paper, these two papers set ecologists off on a new, modern direction. It led to a tortuous path that eventually resulted in the science of ecology of today.

The young scientist, Robert Helmer MacArthur, and his advisor, GE Hutchinson, the kindly Englishman of previous chapters, were not the first to address the question—like most ecological ideas, a version of it is found in the writings of Charles Darwin—nor to assay an answer. One answer had looked toward the divine. Hutchinson recounted it in his paper.

"There is a story," he wrote, "possibly apocryphal, of the distinguished British biologist, JBS Haldane, who found himself in the company of a group of theologians. On being asked what one could conclude as to the nature of the Creator from a study of his creation, Haldane is said to have answered, 'An inordinate fondness for beetles.'"

This story is no doubt apocryphal. The cleric is sometimes the abbot of a monastery or not clerical at all, and "a special preference" replaces "fondness." There are, however, an inordinate number of species of beetles. Without quibbling over exact numbers, which are unknowable, beetles comprise from one-quarter to one-half of the million and a half species known to exist, according to EO Wilson, who has been devoting his energies to making sure that the known species and the ones that must exist, even though we have yet to identify them, continue to exist.

The paper that Robert MacArthur published was based on his dissertation work at Yale University. Although I hadn't read it when I set off on my own ecology research, it was, in fact, an exact model for what I had attempted to do in becoming an ecologist, but MacArthur seemed to have pulled it off. What he did was study an attractive group of organisms in beautiful locales while relating his observations to theoretical predictions. The organisms were the warblers, tiny insect-eating birds with colorful plumage, that his brother John W MacArthur and he had enjoyed watching in the woods of Vermont, a place that held a special attraction for Robert. For his research, the two brothers extended their observations to Mt Desert Island in Maine, a spectacular bit of East Coast real estate.

On rereading the work recently, I was struck by a number of things. It truly is excellent work even six decades later. It holds up. MacArthur really did know his warblers. Partly through the paper, though, I suddenly pounced on what seemed to be an incredible flaw.

"Aha!" I had shouted to my dog, which was lying at my feet where she had me trapped in my chair. "He's trying to show how the warblers subdivide their New England environments in order to limit competition, while all along these are migratory birds!"

Or something like that, which made my dog wag her tail at me. She knew that an intellectual frenzy would often send me outdoors with her for a long walk. She was as unimpressed with the logical fallacy I thought I had discovered as she was oblivious to the threat of having my desk chair roll over her tail.

But no, the dog would have to wait for the walk and her tail was safe. Scanning forward a few pages, I was reminded that MacArthur had spent the Christmas holiday prior to completing his dissertation watching the same little birds in Costa Rica. Not exactly hardship duty, but the data he gathered there was vital to his point. The birds also managed to avoid competition in their tropical wintering grounds through differences in habitat preferences and behavior.

The other thing that impressed me was how much of other people's data he used to set up and support his hypotheses. Although the observations that Robert and his brother made on warblers were crucial to his study, filling in important gaps in knowledge of the birds' ecology, such as their winter habitat use, many of the details on which his study rested were those of others.

This became a pattern in the ecology of Robert MacArthur. He became a theoretician, using the data of others—or no data at all. There was little fieldwork or experimentation in his later research. What MacArthur accomplished was to take his birding hobby and his training in math and make a career out of the two. Coming from a family in which both his father and mother were biologists (his father was a professor at the University of Toronto for almost all of Robert's childhood) and his brother was a physicist, it seemed to be a combination that was inevitable. It was his brother John, however, eight years the elder, who probably instilled the interest in bird watching in him. John recalled for me how he started to watch birds with a group of boys his age in Toronto (one of whom, he had been pleased to tell me, he still went birding with) before Robert was born. The younger Robert began to tag along with the older boys as soon as he could. Years later, it would be John who would tag along after Robert during the bird studies that resulted in the paper mentioned, among others.

Robert's math ability seemed to be innate. There is a family story about it.

"Robert must have been in about first grade at the time," John MacArthur recalled for me. "One night my father was trying to read a bedtime story or something to him and ran into resistance from Robert."

"I don't want a bedtime story," Robert was remembered to have said. "Send John up. Tell him to teach me algebra."

It must not have gone over very well, I had remarked. Experts in schools of education had me convinced that abstractions such as algebra can't be learned until thirteen or fourteen years of age.

"Oh no," John had said. "He seemed to have caught on fairly well to what I was trying to explain. He always had a talent for math."

Robert followed that talent by majoring in mathematics at Marlboro College, a small, experimental college in Vermont that let him plan his own curriculum unhindered by the requirements of disciplinary structure. The math major let him pursue his science interests in a very unstructured way, according to John. The overriding educational philosophy of one of the founders has been said to have been "a teacher on one end of a log and a student on the other." No doubt Robert discovered that a new college was being launched on the southern spine of the Green Mountains while vacationing with his family at Newfane, a short distance from Marlboro. John remembers a family friend from Toronto introducing the parents to Newfane, a tiny hill town of lovely houses with white clapboards set into the same irregular topography as Marlboro. It must have looked very much the same then as now; those clapboards are all on buildings that date from the 19th century or before.

I will insist to my dying day that, like me, Robert took one look at the Vermont landscape and decided that he would have to spend his life there. Right there close to where his family summered in Newfane, in fact. That is how I see his decision to attend Marlboro in its first year. Neither his brother, John, nor his widow, Betsy, could corroborate that for me, however.

"He chose Marlboro because it accepted him after only four years of high school," she told me. Canadian high schools in that time (and possibly now, for all I know) required five years for a degree. Other than summers, Robert had been living and attending school in Toronto.

A photograph of Marlboro's graduation ceremony in 1948 found its way into *Life* magazine. The smallest such ceremony in the country, it had had exactly one graduate, a GI returning from World War Two, already having credits. The school had been created to take advantage of the GI Bill. Although he enrolled in the college's first full class, his brother John recalls Robert's contribution to the new institution as providing the school a warm body, rather than his being involved in building it up. His enthusiasm for the school was sufficient, however, to entice both his father and brother to join its faculty soon after his enrollment. His brother John spent his career there.

It was in all respects a typical Vermont experimental college. (I write that as though there are dozens of such places scattered throughout Vermont, while there really were only a handful, most of which have closed by now.) The campus had been the farm and summer home of its first President. With the addition of the adjoining farm, the campus consisted of farmhouses, barns, and outbuildings on a scenic hilltop. Some not only still stand, but, marvelously maintained, were still in use almost yesterday. When the college opened on September 25, 1947, with Robert MacArthur in its first class, there were no classrooms, just a large barn, "Dalrymple Hall," which later served as the dining room and student center. The students helped with maintenance and construction. Governance was fully democratic, with student representation at Town Meetings that survived to its very end.

That first class consisted of 56 students and 7 instructors. Sometimes the instructors, like Robert's brother John, who had to work as a consultant for a machine tool company, did not get paid. Among the trustees, however, were Ellsworth Bunker, Dorothy Canfield Fisher, and Robert Frost. There were rocky times, but the college survived and prospered. John MacArthur had stayed with the college, retiring in 1988 after 40 years, but was still active when I spoke to him in the same school at which both his father and mother taught (Olive also became a Dean of Women) and his brother Robert launched an academic career.

After Marlboro, Robert went off to Brown for a master's degree, then in 1953 to Yale in pursuit of a PhD in math. In parallel to the path to ecology of Howard T Odum, MacArthur, too, found himself in the laboratory of GE Hutchinson, who was then deeply involved in the question of why there were, for example, so many species of beetle.

Another coincidence, of course, is that it was in Hutchinson's laboratory at Yale a decade before that Raymond Lindeman had devised his trophic-dynamic concept of the ecosystem, later championed by the Odum brothers. MacArthur's paper on warblers, which won the Ecological Society's Mercer Award as the best ecology paper published that year, was as influential as Lindeman's paper had become, impressive enough to give his more mathematical papers prestige by association, if nothing else. Young ecologists became more than willing to follow him away from the ecosystem ecology that had once seemed to be the only form of ecology with enough rigor to it to impress engineers and physicists.

The rigor of the mathematics cannot be denied. Before the warbler paper, MacArthur had already published two papers applying mathematics to ecology that became rallying points, of sorts, in the decade ahead. In the first, "Fluctuations of Animal Populations, and a Measure of Stability," published in *Ecology* in 1955, he tried to prove mathematically that stability in nature is related to diversity. It was a result that Hutchinson used in his "Homage." Student and teacher were definitely in a situation of synergy. In the other, "On the Relative Abundance of Bird Species," published in *Proceedings of the National Academy of Sciences* in 1957, MacArthur made a mathematical attempt at describing the process through which related species, such as his warblers, subdivide their habitats. It was both an outgrowth of and a contribution to Hutchinson's new way of thinking about how an organism's niche might be related to that question of why there are so many species. The papers were influential enough and important enough to let us call them MacArthur's Three Influential Papers for convenience. If that sounds an awful lot like the "Three Easy Pieces" of the concert pianist, so be it.

There was also an unfortunate side to the MacArthur example. This was a flaunting of scientific tradition, no doubt not deliberate. It might have understandably resulted from his background of mathematician-naturalist, rather than scientist-experimentalist. He had still been a math major when he found himself at Hutchinson's lab. Math and math research were then on separate shores from ecology and its research, especially in the way results were presented. MacArthur would often fail to acknowledge the work of his predecessors, for example. Where a paper by him in *The American Naturalist* might have as few as four or five citations of previous work, similar papers by others might have several dozen. There being essentially no data in mathematical research, mathematical theorems and proofs are totally self-contained works of logic not requiring the marshaling of similar views for support. They stand or fall on that logic. They do have a starting point and they do often use tricks invented by others. These are acknowledged, but there is no reason in mathematics to survey an entire subject or go over the failed attempts of others. Ecology, meanwhile, was then awash in a sea of descriptive data at a time that ecologists were trying to make sense of using what is know as an inductive method. In this mode of research, scientists gather data until they can begin to discern a pattern. One not only had to cite the work of others in a theory paper in order to marshal their evidence in one's support, but also show that one's own theory was superior to those of others purporting to explain the same evidence. It was dueling with evidence. A long list of citations was unavoidable. MacArthur, however, was using a hypothetico-deductive method that had rarely been employed in ecology before. He posed his ideas in mathematical terms, then he found data in their support. It was backwards in a sense from the inductive approach. And being essentially mathematical logic, there were few prior studies to list in the bibliographies of his papers (although there were more than he chose to acknowledge).

 I had an opportunity to examine what most of us would recognize as MacArthur's senior thesis, but what in the jargon of the school's curriculum was called his Plan of Concentration, "Mathematical Foundations of Boundary Value Problems." It is a slim volume, 14 pages in all, typed on the onionskin paper of the time. There are ten references, several in German. On the front page, "MacArthur" has an extra space, "Mac Arthur." There is no date. That is found on the binding. Even though the corner is marked,

A Mathematician Who Knew His Warblers

"Excellent Piece of Work," over the initials, AHP, clearly the advisor on it, I begin to grow suspicious.

"I bet it's sloppy," I say under my breath in the Marlboro College library, where I was alone, but peacocks would have drowned me out anyway. "I bet the math is pretty basic."

The reason for my suspicions is that the mathematics in MacArthur's works seemed too simple compared to the relatively sophisticated math I was used to in classes I had taken in population genetics and engineering. It turned out that that was a fault (if it can be called that) of ecology, not MacArthur. MacArthur's undergraduate work was impressive and not just for an undergraduate at an experimental college in Vermont. It is dense with equations. His proofs are clearly the work of a mathematician. There are Lebesque integrals, Cauchy's inequality, Hermitian and Sturm-Liouville type equations, Rodrigue's formula, Legendre and Laguerre polynomials, and Bessel's inequality. Symbols, such as for partial derivatives, available to us through software, are neatly penciled in by hand. And symbols dominated the pages; what little text there was receded to the background.

Although the equations had to do with "boundary value problems in mathematical physics," in his own words, what the physics was, MacArthur never explained. The text part of the document consisted of two short paragraphs, each less than 100 words long, bracketing the dense math. Finally, I thought, a mark of the sort of laziness I expected in someone who matriculated at an experimental college in Vermont in its first year. Perhaps. More unmistakable, however, was that it was the work of someone who loved math. Real world details were unimportant.

This attitude may have surfaced later. MacArthur has been described as being totally undistressed by the possibility of error in his use of deductive method.

"His papers began in speculation and ended in data, instead of the other way around," one of his admirers, Stephen D Fretwell, has written of him. "The math was often fuzzy or incomplete, and the data oversimplified and limited."

There is a story of one of his influential papers having been submitted to *The American Naturalist* with a mathematical error. Before the referee could review the paper, MacArthur had found and corrected the error, then revised and resubmitted the paper. The referee had turned down the paper as having conclusions that were too intuitively obvious, except that the conclusions had been based on the erroneous math, and had now been reversed, leaving the editor no choice but to publish it. According to Erik Pianca, MacArthur's preference for the *Naturalist* stemmed from its editors' frequent willingness to publish his papers with a minimum of changes. For many of his papers, MacArthur eschewed the refereeing process entirely, having Hutchinson (a member) transmit his papers to the unrefereed journal then put out by the National Academy of Sciences.

"Having bypassed the normal review process," Fretwell concluded, "he was well on his way to being famous."

A misfortune for his legacy is that his style was more often emulated than his substance.

"MacArthur's primary contribution to ecology," an ecologist concluded not long after his death, "if one must be singled out, was one of philosophy and attitude. He brought a fresh breeze of creativity, theory and hypothesis formulation blended with field study to a somewhat enervated science, thereby fanning the imagination of many a young worker."

The ecologist was writing a review of a symposium volume dedicated to MacArthur soon after his death. MacArthur's colleagues, disciples, and admirers—some might facetiously be called "groupies"—contributed to it.

"It is also appropriate that this should be a most frustrating volume in its lack of data and supporting evidence for hypotheses presented," he continued his review. "While MacArthur was a careful worker, the school of ecology he founded has become notorious for theory supported by only a minimum of data, transformed, manipulated, and massaged to fit expectation."

A Mathematician Who Knew His Warblers

The volume under review, *Ecology and Evolution of Communities*, was a Fort Sumpter in a war.

Another misfortune was that MacArthur had not followed his teacher more closely. In his "Homage," Hutchinson had pointed to food relationships of various sorts for the answer to his question. MacArthur's 1955 paper, used by Hutchinson to support some of his own ideas, built on Lindeman's and Odum's work concerning energy flow through a community. It appeared as if an edifice was about to be constructed at that time that would marry the population dynamics of species to the energetics of ecosystems. It never happened. Instead of marriage, it was divorce. Energy disappeared from MacArthur's thinking in all but a very perfunctory way. Ecology split into two schools of thought, one led by Odum, the other by MacArthur. MacArthur's turning away from energetic considerations was the precipitating event.

"The reason for this appears to have much to do with personalities," one ecologist was later to write.

What the specific personality aspects were was left unsaid. Species identities, however, did not fit well into the energy flows studied by the Odum school. Neither did energy fit well into the elegant mathematical framework the MacArthur school was building. Like Howard Odum, seduced by his analog ecosystems until he could no longer see the forest for the circuitry, perhaps Robert MacArthur let himself be seduced by his love of the mathematical forms he had studied in Vermont and began not to see the birds for the equations.

Besides the aura of intellectual excitement that seemed to surround MacArthur's work, there must also have been some sort of personal magnetism, some dynamism in his personality that attracted acolytes. One ecologist affected and influenced by him pointed to MacArthur's "lifestyle and attitude" as responsible for the indirect, but personal effect that he had on people.

Unhappy with life in the city, according to his brother John, MacArthur was gone from Penn, where he taught briefly after obtaining his degree from Yale, by the time I got there, He was already at Princeton, where his brief career was prematurely brought to an end by renal cancer. He left an impalpable presence at Penn, though, even before death could make him a legend. The department was then full of young MacArthurs, Martin Cody, Robert Ricklefs, and Henry Hespenheide, foremost among them, who would go on to make an impact of their own upon ecology by continuing in the tradition of their mentor. From my perspective as an engineer, I found their view of the world of biology infectiously appealing. I would have followed them anywhere and did, in fact, right out of engineering and into ecology. Yet it was not anything in the personalities of any of those individuals that attracted me, but something in the air around them, something in the form of ideas. With MacArthur, however, it might have been the man himself, as much as his ideas, that drew people to him.

In the formal photograph of MacArthur that is available, I see a handsome man, although not extraordinarily so. I fail to see in the physiognomy preserved in the photograph anything that would make me think that this is a man I must follow. Nowhere evident is the steel determination of a leader or the playful countenance of great brilliance. But there is another photograph that I chanced upon. This one is not formal. It is a snapshot rather than a portrait. The collar of his casual shirt is open. The camera is angled slightly upward and is close, a Brownie, perhaps. MacArthur's closed smile glibly curls the corners of his mouth. His hair is disheveled; a wisp falls across his forehead. What is he coming from? Is it a hike? A canoe trip? The face is of a companion I would have valued for either.

Unfortunately, I was unable to learn the circumstances of that photograph. I had found it on the cover of an issue of Michael Rosenzweig's *Evolutionary Ecology Research*, now a journal that is totally on the Internet (and apparently free, although with a tab that is ominously labeled "Final Editions"). Rosenzweig had been one of MacArthur's graduate students at Penn. I thought he might have taken that candid photo at some time during those days. He was not helpful to me about its circumstances, however.

A Mathematician Who Knew His Warblers

"Mike got it from a book dedicated to MacArthur," his wife replied to my question. "I don't think he knows much more about it. We corresponded with his widow Betsy (an old friend of ours) and she had no other pictures of him."

The book is the volume I have just compared to Fort Sumpter. I have that photo. It is a formal portrait of MacArthur, rather than the candid. When I pointed this out and suggested that perhaps Rosenzweig had taken the picture himself, she emphatically denied it.

Neither photo of the elusive Professor MacArthur is at all like Lenin or Einstein looking out at me from the page. It is a soft face. Ronald Reagan, perhaps. Nor did I ever have an opportunity to hear him speak, but I don't imagine that the magnetism was there, either.

EO Wilson, with whom MacArthur collaborated famously, described him in his autobiography, *Naturalist*, as being "a thin, diffident young man" when they first met, "who spoke with an American accent but in the British style of cautious understatement, perhaps acquired at Oxford."

(Although he had spent a year in England working with David Lack, a biologist most noted for his studies of finches on the Galapágos, MacArthur had more probably acquired the accent through a childhood spent in Toronto. Had Wilson found himself almost anywhere in Southern Ontario, he would have been surrounded by just such accents.)

Stephen Fretwell, who wrote the rather peculiar elegy on MacArthur from which I have been quoting, found that this patience and tact, as he identified it, made "conversation with him seem especially clear and significant."

"I rarely thought so clearly as when I talked with MacArthur," Fretwell concluded.

"Of medium height, with a handsome rectangular face," Wilson continued in describing MacArthur, "he met you with a disarming smile and widening of the eyes. He spoke with a thin baritone voice in complete sentences and paragraphs, signaling his more important utterances by tilting his head slightly upward and swallowing. He had a calm, understated manner, which in intellectuals suggested tightly reined power."

Perhaps. Perhaps in the Harvard milieu in which Wilson found himself at that time, he might attribute understatement to intellectual power. In some other intellectual environment, calm understatement can be attributed to all sorts of other things. Perhaps Wilson, from a neglected though not impoverished Southern childhood, was attracted to the more privileged background he saw evidenced in MacArthur's lakeside retreat in Marlboro. The two days Wilson spent there in the company of other biologists impressed him enough to label the group the "Marlboro Circle," even though it was a group that never met again, at least not with Wilson as part of it.

The lakeside cottage so important to MacArthur's "lifestyle" belonged in actuality to the family of his wife, Betsy Whittemore. Her father had purchased half the lake during the Great Depression.

"He bought up a lot of land then," Betsy told me. "It was cheap."

Betsy Whittemore MacArthur is the daughter of one of the three founders of Marlboro College and remained in the area after Robert's death. She was also the brother of Robert's roommate when she and Robert met at a local square dance, although according to Betsy, Robert had not been aware of that fact at the time.

The lake, South Pond, a small, glacially formed body of water, is distinguished from most other Vermont lakes by having no inlets and almost no shoreline development. Similar lakes can be found close to the same altitude (according to John MacArthur, it is at 1700 feet above sea level) in the wilderness areas of the Green Mountain National Forest, but South Pond had been entirely held in private. According to John, Robert was instrumental in placing the shoreline and surrounding land into a conservation trust that would forever keep it undeveloped. It also keeps it relatively unvisited. There is a boat launch maintained by the state that allows access to all, but no signs to direct someone who does not know the way. Except from the end of the access road and a second, gated road leading to land held in common by the other group holding it in trust, the pond is not visible.

Robert would make his way to one of the two rustic cabins on the lake by canoe or by a mile-and-one-half hike through the woods from his mother's house. He and Betsy had built their cabin shortly after they married in 1952. They spent every summer at the lake with their four offspring, according to Betsy. Supplies, such as the large propane tanks needed for cooking and, I would imagine, washing diapers—there is no electricity—were all floated in by canoe.

According to John, who had attended only a few of the "'Circle' meetings," as he called them, "they happened to be not at the pond, but at our Mother's house," a small wood-shingled cottage that would be more at home on the Maine coast than Vermont.

"They probably met at my parents cabin above the lake," Betsy disagreed.

Neither could it be established that the Vermont data for Robert's famous warbler paper had been collected in the woods around the lake. John had no recollection of where data had been collected for particular papers, even those he had assisted with.

"I think that that work was done at Mount Desert Island," was Betsy's only recollection.

Still, as my canoe touched the shore close to the cabin, I had the feeling of having come upon an ecologically sacred place. Aldo Leopold's Sauk County cabin sprang instantly into my companion's mind, who immediately insisted that we arrange for a stay at the cabin during which we could live the life of Robert MacArthur, I assume. Without the four children, of course.

Wilson characterized MacArthur in a way that is at odds with Fretwell. To Wilson, MacArthur was "basically shy and loathed being caught in a careless error." His observation that MacArthur "was generous by instinct and capable of lavish, almost Hutchinsonian praise during private conversations for work he thought important" may also provide a hint to unraveling the intellectual attraction of GE Hutchinson. They were men of genteel breeding, Hutchinson and MacArthur. It may not have been common then in the sciences. Currently, as in baseball, "Nice guys finish last," to quote Leo Durocher, someone who was definitely not a nice guy, but who did manage to finish last on occasion, undeservedly so, no doubt.

David Quammen, in his book about island biogeography, *The Song of the Dodo*, describes MacArthur as having become the James Dean of ecology.

"A direct or secondhand association with MacArthur," according to Quammen, "is now a form of credential. It's like being able to say you once had a cup of coffee and a political argument in Zurich with an irascible Russian expatriate named VI Lenin."

Quammen's revolutionary could have bolstered his credentials better by claiming to have had coffee with VI Ulyanov, or even N Lenin, in Zurich. These were names by which the irascible Russian was known before he became famous, but the analogy is still solid. MacArthur and his followers sincerely believed that they were revolutionizing ecology for the better.

Emotionally overt James Dean, though, is more apt of a comparison to MacArthur than cold Vladimir Lenin.

"MacArthur, in most of my personal encounters with him, was a very loving person. He was patient, tactful, joyous, responsive, openly human," is a description from Fretwell, again.

"Robert clearly loved most of the things he encountered in life," Fretwell continued. "He loved his family, nature, and any exchange of ideas that sought to resolve some honest and tangible confusion. He loved elegant mathematical structures, and patterns in nature."

Neither Lenin nor James Dean is described in the words above. I can not rid my mind of a different analogy to MacArthur, this one literary. Although literary devices are poor substitutes for the actual ideas of a science, science is something made by humans, with all of the frailties that implies. Fiction does convey things about human nature that are timeless and universal. Great literature can provide more understanding of the human predicament than a host of scientific studies. And it is too often very human motives that cause a seemingly dispassionate science to pursue a certain idea or to follow a particular leader.

The published impressions of MacArthur by his contemporaries and the characteristics of those who tried to emulate him bring to my mind a personality that pervaded Russian society during the first part of the 19th century. It was found in the Byronic individuals whose intellectual descendants became the Marxist revolutionaries of the late 19th and early 20th centuries, Belinsky. Herzen. Bakunin. In Russian literature, they inspired the tradition of the "superfluous man." His appeal was similarly inexplicable to that of MacArthur. There are other parallels that might be drawn. For example, in Mikhail Lermontov's *A Hero of Our Time*, superfluous man Pechorin, the first character of the literary tradition, seems to charm almost solely by a dashing indifference. In Pushkin's *Eugene Onegin*, substitute MacArthur for Onegin and Tatyana for ecology. Superfluous man Onegin trifles with Tatyana from not much more than the need to escape boredom, almost destroying her. Onegin, like Pechorin, financially well off thanks to the serfdom that supported his noble lifestyle, also like Pechorin was infused with the Byronic ideals that a Western education then provided. Those were ideals that fit poorly into the reality that was Russia, however. Onegin's flaunting of convention, for no other reason than to create some sort of stylish exhilaration, causes him to kill a friend in a duel that had resulted not so much from rivalry over a woman, as over style. Tatyana, being stronger and more sensible that Onegin, salvages her life and survives without him. Was the friend that Onegin sacrificed Odum-style ecology?

Not just a literary device, the superfluous man was an actual part of the society. Both Lermontov and Pushkin can be said to have lived the lives about which they wrote.

Might the hypothetico-deductive process championed by MacArthur have been equally "superfluous?" It proceeded somewhat along the following lines. Formulate a theory (more often called a hypothesis, whether testable or not) that is described by a mathematical equation. Try to deduce what you can of the real world from the theory. Check it against observations. Does it make sense? Whether it does or not, go on to another theory or another set of observations. It is the deductions that are the fun part of the process, after all.

"I had the distinct impression that MacArthur worked on ecology mostly when his family got tired of his hanging around," Fretwell noted. "I suspect that he had the freedom to do what he did for ecology, going against all convention because it was all secondary to him anyway."

Wilson is in agreement.

"He placed family," according to Wilson, "above all else. After that came the natural world, birds, and science, in that order."

All those summers at the lake. It almost sounds as if it is a dilettante that is being described.

My imagination may be getting out of control, but only because there is quite fertile ground for it here, especially if I let MacArthur transform himself into a composite, a "MacArthurite." This is a term once in use for those who showed all of the less desirable traits of MacArthur's legacy and did so to excess. If it sounds all too similar for comfort to the pejorative then in use in liberal circles, McCarthyism, that may have been the intent. Giving my imagination free reign, then, in Ivan Turgenev's *Fathers and Children*, a continuation of the same literary tradition, the MacArthurite would not be Bazarov, the prototype nihilist. Bazarov found more meaning in looking at the internal anatomy of frogs than in what he saw as the unsupported theories that were used to justify then-current social, historical, and scientific structures. Bazarov would have to be the anti-MacArthur. A MacArthurite in Turgenev's great work would have been the foppish Pavel, for whom the style of the idea may have been more its essence than its truth. Who Bazarov would have been in my flight of fancy is not clear. It would not have been Eugene Odum. Perhaps, Daniel Simberloff, who we will meet soon.

(By the way, that whole literary tradition ended badly in Ivan Goncharov's Oblomov, who was too psychologically moribund to get up from his morning bed for the first 200 pages of the novel by the same name. There is probably little insight into ecology to be had by continuing my fanciful analogy further, but the point that I tried to make is that a sociological study of ecology is overdue, particularly for this era. The list of authors publishing in *The American Naturalist* or *Ecology* around Earth Day does not include many women's names nor those of immigrant or other minorities who were then bootstrapping themselves through other professions. It is what Auerbach noticed on joining the science. The innocent explanation is that, in terms of financial gain, a career in ecology was about equivalent to one in medieval history until the advent of Big Science after World War II and an environmental consciousness after Earth Day. Duke University's William Schlesinger, as one late-20th century example, oversaw a research budget in the millions of dollars. Robert MacArthur had peanuts in comparison. There just was no appeal then in ecology to those bootstrapping their way out of a lower social class. I can say from my own experience, however, that personality, social status, and the right connections, although important in all sciences, seemed to have greater sway in ecology than in the engineering or medical sciences.)

    Robert MacArthur's promise of rigor in ecology began its ascent in the field just at the time that Eugene Odum was proclaiming a new ecology based on the ecosystem. Ecologists were now given a choice: ecosystem energetics or evolutionary theorizing. Putting aside issues of personality, MacArthur's way toward rigor was more attractive, perhaps because more direct. The rigor of ecosystem science required biological, thermodynamic, and chemical data, lots of it, obtainable only through difficult field studies and experiments, before the math used could be validated, often through tedious computer models. MacArthur's science seemed to require only that the simple math results made some sense based on whatever data was available or could easily be obtained. Often, it required nothing more than a trip to some exotic locale and a census of species—or, lacking time or money for such, the use of someone else's trip results. Sometimes it required no math, at all. Just clever ideas.

One way appeared to promise drudgery, the other glamour. The choice was an easy one for many.

After the rise of the MacArthur-style hypothetico-deductive method in ecology, whatever faults there were in the ecosystem concept were pounced on and used to discredit it. In particular, it was proclaimed that there were no emergent properties to ecosystems, nor could they evolve adaptations that would make them worthy of ecological study. Ecology is, after all, the study of adaptation. All of MacArthur's ecology rested firmly on the study and explanation of evolutionary adaptations. Absent the superorganism, Odum's ecology only had a basis in the physical sciences. And MacArthur's math had a rich tradition behind it. Those differential equations had been developed in the physical sciences and polished by successful applications in chemistry, engineering, genetics, and economics. At around Earth Day, there had seemed to be a race on in applying the methods of stability analysis, for example, to as many new situations as possible. I recall crossing the Penn campus from a graduate course in population genetics to the office of the engineering professor with whom I had taken a graduate course in reactor kinetics. Liapunov functions were of importance in each. I had hoped that the engineering professor had finished writing the text he had been preparing so that I could get a little more light on the subject than I was getting in the population genetics class.

No, he had not completed the book. But he surprised me with his interest in the use of stability analysis in genetics, questioning me on it and asking for references to the work.

"I am always on the look-out for new applications," he explained.

But he was an engineer! I thought. Not a geneticist. What business was it of his?

Actually, he had a lot to gain by poking those instruments into unfamiliar territory, as will be learned by the story of how Robert May became The Right Honorable Lord May of Oxford.

Worse yet for the ecosystem, its study also appeared boring to ecologists in comparison with the more exciting ideas generated by

MacArthur-style hypotheses. Ecologists, bred to appreciate natural history and field observations, could dismiss ecosystem studies as being simply uninteresting to them. They could admire MacArthur because he not only knew his birds, but he also developed theories for them. On the other hand, Eugene Odum, who it turns out also knew his birds, promoted studies in which the actual species identity of a bird was inconsequential. (It was around this time that one noted mathematician-turned ecologist proudly displayed the scars he had obtained in a fall during an episode of trying to know his barnacles.)

The avoidance of boredom represented in MacArthur-style theoretical excitement became justification enough for a research program. Time and time again in the 1970s I heard ecologists praised for asking "interesting questions." It was praise that in my memory was particularly directed toward GE Hutchinson and Daniel Janzen. Janzen, for one, managed to overcome the handicap of earning respect only through asking interesting questions and redeemed himself with quite useful conservation work. The questions he later came to ask may no longer have been intellectually interesting, but they are important in a practical way. He asked such things as how to obtain funding to secure a piece of tropical habitat and how best to manage it on a day-to-day basis. MacArthur, of course, also reaped praise for the questions he posed, much of it posthumous. Meanwhile, there was lots of important work being left undone. In addition, some of the brightest minds of the Earth Day generation of ecologists were being turned away from ecosystem studies. In 1973, the Graduate Field of Ecology and Evolutionary Biology at Cornell admitted some 20 new students; the year before it had been 5. They were the fruits of seeds planted by Earth Day. The best and the brightest.

The Climax of the Ecosystem

Ecologists may have killed the superorganism, but the ecosystem approach lived on to the present day. In some ways, it prospered. The ecosystem has today returned to a position of primacy in environmental policy. Before it could, however, ecology had to play out a drama unlike that in any contemporary science. Except for much of it being played out with mathematics and data, it was a conflict worthy of literature.

While that drama was being played out, ecosystem studies continued unabated. There was just too much about the ecosystem approach that was so basic to the science of ecology—and to science in general—that it could never be abandoned. It may have lost its ascendancy. It may have become no longer glamorous. But it never left the thinking of ecologists, any more than could natural selection.

None of the properties of ecosystems require that natural selection act at any particular level, nor that evolution of ecosystems occur in anything but a Darwinian manner. That argument was a smoke screen. The systems scientists' black box is just that: a black box. During the era following the superorganism's demise, that black box was full of mathematical equations as much as anything else. As many of those equations found their way into the black box by way of the evolutionary theorists of the MacArthur school as they did by the systems analysts inspired by the Odums. Some of MacArthur's equations were, in fact, systems science equations made somehow more palatable.

There are 24 ecosystem attributes identified by Odum. They can be organized into five categories. The first is energetic. Ecologists can measure an ecosystem's production and respiration. They can measure the flow of energy through ecosystems and its storage in standing biomass. They continue to do so.

A second category is the cycling of nutrients. In ways quite similar to those for carbon and energy flows, ecologists can measure the flow through ecosystems of non-carbon nutrients, the nitrogen, phosphorus, and potassium compounds of N-P-K fertilizers, for example, and the amounts of nutrients locked up in standing biomass.

The Climax of the Ecosystem

Another has to do with an ecosystem's structure. Analysis of how ecosystem parts combined with each other most closely brought together the theoretical evolutionary ecologist and the ecosystem scientist. Ecologists had long worked at determining the number of species present and their abundances in a given area and what caused the patterns found. Some sense could be made out of the patterns when they were looked at in terms of functional groups ranging from trophic levels to guilds (groups of organisms doing the same things in a finer sense than those of the same trophic level) to niches (a single species' place). Whether analysis came from a systems perspective or an evolutionary one, its math tended to be shared. Both groups worked with similar equations describing unnamed species interacting with each other.

The successional changes of ecosystems, although now stripped of the idea of being some sort of physiological development, are undeniable observations that form an important foundation to that part of ecology labeled plant ecology or community ecology. Superorganism or no, there are changes observable over time in almost all habitats.

Finally, there are the cybernetic attributes of ecosystems. These bring together all of the other attributes in an analytical way. Even granting that they are not self-regulating, ecosystems can still be characterized according to their diversity, connectedness, or complexity, all cybernetic modes of analysis. As with structural attributes, the ecosystem scientist and the theoretical evolutionary ecologist can work hand-in-hand on how these qualities are related.

To understand how the ecosystem concept can be both dead as a Dodo and very much alive and in use requires a trip into a forest. Imagine your favorite. Maple-beech-hemlock does it for me where I live, but what is a maple-beech-hemlock ecosystem really? It could be looked at as the state toward which that ecosystem was developing or had reached in its maturity or "climax." Odum's 24 ecosystem attributes all changed in specific ways in ecosystems that approached and reached their "climax." Except that there is no "climax." That was a part of the life cycle of the now-dead superorganism. It never existed except as a mental construct. Wander through that maple-beech-hemlock forest all you want, you will not find its climax. You may find some very mature trees (if you are very lucky), but you will find no ecosystem that has reached its climax.

As with almost every idea that can be found in ecology, as soon as it was proposed, there was someone to negate it. The superorganism was no different. When Clements first proposed it early in the 20th century, Henry A Gleason was there to negate it. There was no climax and there was no superorganism, only change and flux. To Gleason, and to ecologists today, succession was the simple result of the behavior of individual plants, often dependent on chance events such as seed dispersal. What seeds happen to fall on an area have as large a role to play in the composition of the vegetation as do any interactions between the plants found there. Individual species respond individually to environmental differences according to Gleason, even when those differences are the result of the presence of other plant species.

Although ecologists of the time polarized into ecosystem-as-organism (Clements) and random-association (Gleason) camps, the two views were not completely incompatible. Proponents of each sharpened the differences, as so often happens in any disagreement, as happened in the split between Odum- and MacArthur-style ecology. In his writings, Gleason made it clear that the areas of relatively similar vegetation which ecologists such as Clements were naming, categorizing, and trying to understand did, in fact, indisputably exist. There were maple-beech-hemlock forests. In Gleason's view, however, environmental differences and chance events were more important in forming those vegetation groupings than the facilitation between species necessary for Clements to look upon them as superorganisms. That those tree species grew where they did was because each was adapted to grow there. That they seemed to be found together in what might be imagined as a single unit was purely coincidental.

Redwood groves, thousand-year-old trees capable of holding the Pacific fog in order to perpetuate the misty environment they need to thrive in what would otherwise be too arid a climate, are perhaps the most common misconception of a climax forest. The misconception is that once that thousand-year-old stand of redwoods is reached along the flats and hillsides of the rivers of California's North Coast, ecological succession stops, and the sites continue perpetually as old-growth groves of redwoods, having modified their environments to suit that growth. Lightening, disease (not too common among redwoods), or the storm of a century might knock down some venerable giant to leave a gap to be filled by younger recruits, but the character of the stand would remain essentially unchanged, it might seem.

Coastal trees such as redwoods and Douglas firs have evolved to wring moisture from coastal fogs, but they are no more responsible for them than is the city of San Francisco for its fogs. Local meteorology is. In addition, if redwood groves were truly climax vegetation in the Clementsian sense, then they would not be as endangered by logging. Other than the matter of the thousand-year time scale involved, they should return to that climax condition after disturbance without intervention. (An argument can, of course, be made that too much logging can kill superorganisms, but a dead superorganism in this case means a dead concept.) A more reasoned view of redwood stands is that they are relict populations from previously suitable climates, surviving in pockets to which they are still well adapted. The long life spans of the trees allow them to out-compete other species in those pockets by simply out-living them, while also cloaking them in an air of timelessness compared to relatively short-lived species such as ourselves.

That self-perpetuating grove of Redwoods, the climax condition toward which the Pacific Coast redwood forest superorganism matured, simply did not exist if we apply the ideas of Gleason. They are as unlikely an endpoint to ecological succession as a society dominated by centenarians in geriatric care facilities might be for human communities. Even when not looked at as the end product of the physiological development of an ecosystem, juvenile to adult, there is no such thing, in Gleason's view, as climax vegetation. Redwoods will not remain forever in the stands along the California coast absent human disturbance. They are there because, at the moment, the environment is by chance suitable for them. Climax communities are not really in a state of equilibrium. That had been one of the observations that caused Arthur Tansley to invent the term "ecosystem."

Eugene Odum straddled the fence on the issue. His idea was that succession culminated in a stabilized and symbiotic ecosystem. Intentionally or not, he did put the superorganism that Tansley excised from it back into the ecosystem. For Odum, the properties of the climax state were what was important about ecosystems; for other ecologists, it was the Sisyphian task of getting there.

Research on landscape-sized portions of the environment, rather than small experimental plots of forest or grassland, has revealed to us that ecosystems are mosaics of patches in different stages of succession. The amount of patchwork and what is found in the patches depends on the degree of disturbance the landscape has faced. There has been almost universal agreement on this since the first Earth Day.

Patches in the landscape have various sizes and histories. Some take longer to develop (the "climax" vegetation), but none represents the endpoint of succession for any landscape. Maple-beech-hemlock may appear to dominate the late stages of succession, but stable they are not. They are merely longer-lived species than the ephemeral species we see in the early stages after disturbance. Senescence and disease will bring them down, even when no physical disturbance does.

Studies of fossil pollen has also shown us that, whatever those associations of trees might be, they are not staying put. Many North American tree species are in the process of moving northward. They have been doing so since the last Ice Age ended some ten thousand years ago, more so, probably now, given the acceleration of climate change by human activities.

This makes things doubly difficult for those who have been charged to manage ecosystems. There is no stable, unchanging state to manage or to manage toward. Change is what is natural. Disturbances play a key role in ecosystem dynamics. And assuming that the higher trophic levels of an ecosystem, its animals, are strongly affected by the dynamics of the vegetation on which they depend, an ecosystem point of view is as likely to pose new questions as to provide answers about the dynamics of species in those upper levels. It may not have been the management tool that Gene Odum had boasted it to be.

The management of Yellowstone National Park is a case in point. Alston Chase has claimed that the Park Service has failed, in all ways imaginable, in its management of wolf, bear, and elk populations. The failure, according to him, was one resulting from bad science, that of ecology, Odum's ecosystem science, in particular. Chase carried this argument even further in his assessment of management practices in government-controlled Pacific Coast forests. He subtitled one of his books, *The Fight over Forests and the Rising Tyranny of Ecology*. He meant it to be provocative.

Incompetent management need not mean bad science. However, the fire suppression policy that led to the disastrous fires of 1988 was an application of basic scientific principles as they were presented by the practitioners of the science. The principles were those of the science of ecology. The policy derived from them was that, unless structures were threatened, natural fires (as if one can identify with assurance what is natural and what is an accident of man) were generally left to burn, rather than be subject to immediate attempts at suppression.

The policy cannot be traced to Odum's ecosystem concept, as Chase concluded. The balance of nature implied in Odum's homeostatic ecosystem is hard to reconcile with the view of landscapes being naturally in flux from disturbances. Letting fires burn seems more in tune with one of Barry Commoner's laws, "Nature knows best."

Now, more than two dozen years after that fire, Yellowstone's forests still bear the scars. They are impossible to ignore. Although no more than 40% of the park suffered in the fires of 1988, probably

75% or more of that acreage was forestland. Even though I know that recovery from fires is an ecologically natural process, on my latest visit there I found myself gravitating toward the park's meadows. They recovered much faster than forest stands. I also sought the shade of the few fully green islands of trees that were spared. I spent little time in areas that had been burned.

Looking down—and up—from highway switchbacks, I wondered how I might model the recovering hillsides on my HO train layout. It would be easy, I decided. Greenish-brown ground cover below a forest of charred pipe cleaners. I could just dip the pipe cleaners in black paint. But why would I want to model something like that?

Should forest fires be suppressed? There is science that points to letting them burn. In Yellowstone, the ultimate outcome was the complete destruction of the aesthetic value of great portions of the park, albeit, temporarily. The burned acreage may indeed be natural, but to my eyes, those ghost trees, charred a dirty black, were, well, ugly. Natural is not always synonymous with good or attractive. Malaria, I suppose, like smallpox, is a natural affliction of man. I could personally do without it, as I probably could do without the entire order of Diptera. (Think houseflies, horseflies, deer flies, black flies, tsetse flies, bot flies, blowflies, and the ever-popular mosquito. What might Gandhi have done to avoid insect bites?) Piles of buffalo dung are no doubt natural, but I would not go out of my way to look at them. (Yes, Virginia, a bear does defecate in a forest. Please watch your step.)

As interesting as the results of ecosystem science are when applied (or misapplied) in Yellowstone, in truth it is difficult to see what lessons are to be learned from them for the science. If, as in my opinion, the fire suppression (or lack thereof) policy led to seriously harming the aesthetics of Yellowstone, there is no ecosystem or other ecological theory that can contradict me. How is that for including man in the ecosystem?

Aesthetics is strictly personal. I saw the charred snags as exclamations of ugliness that might have perversely been placed on

Yellowstone's hillsides by some installation artist with a crazily flawed artistic vision. Others saw a stark beauty in the juxtaposition of new growth and charred remains.

One forest ranger saw the mature stands of lodgepole pines as something ugly, perhaps. To him they were sterile environments that were never meant to dominate as they did. The unenlightened fire suppression policies of the past had placed them there. After the fires, instead of charged remains, the ranger saw an understory that was expanding the foraging range of elk, something, no doubt, more aesthetically pleasing for him.

Aesthetic quality, being a personal judgment, can never be part of an ecosystem, but fire always is. In its frequency and severity, it is one of those "parameters" of the system. As such, it needs to be managed in one way or another. It is not clear if ecology can point to the direction. Neither of the schools of ecology can. Given a direction, however, appropriate ecosystem studies are more apt to show the way to it.

Although it may be true that aspens have sprouted at Yellowstone where they had not been before, in most cases one-hundred-plus-year-old lodgepoles have been replaced by lodgepole sprouts. The elk probably respond to it in much the same way timber harvesters claim elk and other game animals benefit from the changes engineered by chain saws and yarders.

Fuel reduction through controlled burning in national parks, the currently popular method of managing the fire parameter in park ecosystems, sounds uncomfortably close to the fuel reduction timbering practices that were approved by the Bush (the son) administration for national forests. The only difference, besides the method of reduction, may have been that no one would benefit financially from the Park Service policy.

Alston Chase is correct that ecosystem theory—or the community theory that supplanted it—is being used in the management of natural areas, even if his assessment of its benefits is arguable. But can or should ecosystem or community theory be applied to practical matters, regardless of whether the application might have undesirable consequences, undesirable on some non-ecological basis, such as aesthetics? Does it really work? Does it give us goals that are unambiguous? Misapplication is by no means more desirable than inapplicability per se, but a theory that cannot find a mode of application has no business at all in the practical science of the environment.

None of those attributes of ecosystems described by Odum leap out as being in need of management for our own or the environment's well being. They are not the lists of toxins and eyesores that can be tabulated for many human activities. They just are. They are merely descriptors of a system, parameters, in systems speak. Size, temperature, residence time, and color, for example, are well known descriptors that do not on their own convey value to us, even if we have some knowledge of the object under description. We have to place a value on them ourselves. If I like red, it might be a good color for a shirt, but not if I wish to camouflage myself when fly-fishing. Of course, there is something to be said for giving fish a sporting chance, so the red shirt does turn out to be valuable to me when I fish. On a hot summer night, I would want to minimize the residence time of the air in my house, short of an uncomfortable gale, but in the winter time, I would want the residence time to be as long as it can be without making the air stale. The point is that things that can be measured, as even red is when given the correspondence between the color perceived and the frequency of light reflected by it, indicate nothing to us by themselves. We impose our own personal values on them. Some ecologists have done precisely that. It is their values they impose when they interpret data for the public, not those of their science.

One of the basic attributes of an ecosystem is productivity, primary productivity (that of plants) in particular. In its simplest sense, it is how much new vegetation is produced by an ecosystem. Since the animals that make up ecosystems either eat plants or can eat animals that eat plants, both the animals and the organisms that decompose the detritus of ecosystems can be shown to be supported by plant productivity in almost all examples. A number of ecologists have taken great pains in measuring the amount of primary productivity on earth. One group from Stanford University that included Paul Ehrlich took up the question of how much of that productivity man was consuming. They found the number to be slightly less than half the earth's total. The authors of the study, Peter Vitousek, Ehrlich, Anne H Ehrlich, and PA Matson, in that order, took alarm from their results, as did others. They published them in 1988.

More than a decade later, ecologist Stuart L Pimm, writing for the general public in a book playfully entitled *The World According to Pimm*, led the reader through his own estimate in order to illustrate how a number representing the total productivity of the earth might be acquired. It agreed with the Stanford group's. What is more important is his interpretation of it.

Pimm called the work of the Stanford ecologists "seminal," a "brilliant paper, one of the most important of the twentieth century."

Here was a result of ecosystem (or community to those on the other side of the dichotomy in ecology) study that truly has significance to human affairs. As such, it should presumably provide direction for management, in this case, of the global ecosystem. What they did was to make the concept of productivity, on a global basis, an important management parameter. It has been picked up by others in ecology and is the subject of intense research today. Headlines in news outlets broadcast warnings of similar findings for the future almost monthly from various work groups.

The 40 to 50% of productivity that the earlier studies found that we consume (now reduced by some studies to half the amount) was a dire warning that humans need to constrain their use of the earth's resources. It was new data (with more still coming) for an old argument, one that was well aired on Earth Day. Yet 40 (or 20) to 50% is a valueless statistic until placed into some sort of context. It is of no greater meaning than the statistic of 55 miles per hour offered to you by the kind policeman who pulls you over one day. As a speed, it is a number of little significance by itself. It only becomes significant when you learn the value of the fine that will become your obligation because you were in a school zone. (Not to mention the value the incident will have on your insurance rates.) It would be hideously more significant if, absent the police officer, you unexpectedly needed to brake at that speed for one of those reasons that traveling that fast is prohibited in a school zone.

Ecology has an obligation to interpret that 40 (or 20) to 50% and other such numbers for the public as clearly as the traffic officer does that school speed limit.

Large numbers of scientists on several continents had already been combining their work on productivity decades before the ecologists lauded by Pimm. Their results were published in 1975 in a book entitled *Primary Productivity of the Biosphere*. Scientifically, their feat was as valuable as the exact measurement of the mass of an electron. Remember that what we cannot measure, we cannot understand.

After measurement, perhaps understanding will follow. Measurement of productivity has given us ambiguous insights, however. From a human point of view, maximizing the productivity of agricultural land has obvious benefits. The value of increasing global productivity through advances in agriculture, however, must be balanced by the value of those things that are detrimentally affected by them. This is the lesson that ecology has taught us with help from Rachel Carson.

The maximization of productivity cannot be a goal based on any general ecological principle, however. Neither can maximization of consumption of that productivity.

Closely related to productivity is another ecosystem concept, standing biomass. It carries its own definition. It is the weight (usually dry) of all of the living organisms in an area. Since for terrestrial systems plants outweigh animals by about a factor of 10, like productivity, the standing biomass is essentially determined by the plants in an ecosystem. Think about standing trees in a forest and you quickly get the idea.

Like knowledge of productivity, knowing the amount of standing biomass is also valueless out of context, or from an ecological point of view that is based strictly on natural ecosystems. Is a large amount of biomass to be preferred? That would mean that all deserts are ecological wastelands. The Galapágos Islands, now shrines to Darwinism might better be watered, fertilized, and made to take on a tropical verdure.

Ecosystem productivity and standing biomass do take on an importance in relation to other, more easily valued quantities, however. One is the amount of carbon dioxide in the atmosphere. A considerable portion of the dry weight of trees is carbon. Eat it, burn it, or let it decay and carbon dioxide is produced. The carbon that goes into a tree, however, comes right from the carbon dioxide in the air through the process of photosynthesis. The ecosystem concept of productivity, worked on as strictly basic science by numbers of ecologists, is now a major issue in global negotiations over measures to reduce the amounts of greenhouse gases in the atmosphere. How that 44% that humans might be appropriating effects global warming is a matter of some controversy, as is its effect as climate change.

Nutrient cycling, another of Gene Odum's attributes of ecosystems, can also have some consequences to environmental quality. The biggest impacts from nutrient cycling have been through human induced changes to ecosystems. The major human induced changes have been, quite simply, the addition of nutrients through various forms of pollution. This is what caused Barry Commoner, among others, to declare Lake Erie dead, first in the 1968 *The World Book Year Book*, later in *The Closing Circle*. Closer to the present, ecologists have declared a "Dead Zone" in the Gulf of Mexico resulting from agricultural runoff from Midwestern farms. In a less obvious situation, a consortium of scientists studying a group of small watersheds in New Hampshire successfully applied a very general systems ecology approach to quantify the loss of nutrients through deforestation. It was data in support of the qualitative observations made by George Perkins Marsh more than a century earlier. The series of papers produced by the scientists became known as "The Hubbard Brook Studies," an appellation that merits considerable respect.

The structure of ecosystems basically means how they are put together. It has been characterized by concepts such as species diversity, trophic structure, food web complexity and connectedness. These (and other characteristics) have been quantified and compared across ecosystems and over successional changes in time. To those with concerns about "the ecology," fragility and surprise might be added to connectedness. There is a famous passage from Darwin's *Origin of Species* that is sometimes said to have started ecological thinking. It has to do with the effect of cats on red clover through their suppression of mice.

"Hence it is quite credible that the presence of a feline animal in large numbers in a district might determine, through the intervention first of mice and then of bees, the frequency of certain flowers in the district!" is Darwin's conclusion in his own words.

(Mice disturb the nests of "humble bees," our bumble bees. The bees control the types of flowers in bloom. Fewer bumble bees, less red clover.)

As in my example of predicting the evolutionary changes of Monarch butterflies in the face of climatic disturbances to their overwintering sites, Darwin might also have added, as I did, that they might not. Nonetheless, just that kind of interaction is what much research on community and ecosystem structure is intended to uncover, identify, and warn against. If ecologists can do so, then ecology becomes much more predictable. It can then be a powerful management tool.

MacArthur-style ecology brought to applied ecology the idea of using diversity instead of productivity as a management tool. Diversity was one of Odum's most important attributes of ecosystems, but the theory of diversity, niche theory, was in the Hutchinson-MacArthur camp. Other ecosystem attributes were soon subtended to the concept of species diversity. After all, in one way or another all measures of ecological structure are based on how species get along with one another in an ecosystem.

Trophic structure, to continue this tour of Odum's ecosystem, refers to ways in which animal and plant niches are connected. Its relationship to species diversity is loose, at best, but it exists. Species diversity is a measure of an aspect of trophic structure. Food web complexity and connectedness, which mean exactly what they sound like, are both sophisticated ways of putting simple numbers on the otherwise qualitative (producer, consumer, decomposer) concept of trophic structure. They, too, have diversity at their cores.

Diversity is a relatively simple concept on its surface. It has to do with the variety of different types in an area. Some ecosystems are composed of only a few species. Cornfields (when looked at superficially) and arctic tundra ecosystems are two examples that have few plant and animal species. Cornfields are managed to be that way. Tundra is found in a harsh environment in which few plants can grow; thus there is little diversity. Cornfields and tundra, however, actually have little in common. On an energetic basis, temperate and tropical grasslands and savannas are more similar to agricultural fields. Then, again, it is not clear if cornfields really are less diverse than grasslands when cornfield weeds and pests are included. There are always these complications in ecology.

A tropical forest is composed of incredible numbers of species, many of which we have yet to document. Tropical forest is obviously more diverse than tundra. One need merely add up the species present and compare. But look at the cornfield ecosystem again. As already mentioned, careful study might turn up a number of critters and weeds here and there that belie the concept of a cornfield being pretty much only, well—corn. Our species list might be found to be full of plants and animals that are actually not very common in cornfields, but we looked for them and there they are.

What to do? The power of mathematics can be brought to bear. This takes us into the realm of some truly esoteric ecological exotica. We have the famous Shannon-Weaver (or is it Shannon-Weiner?) diversity index, Preston's canonical distribution, niche partitioning theory, broken sticks, and the multidimensional niche concept. Santa Maria! (Hutchinson, ever the classical scholar, invoked Santa Rosalia to help understand diversity. Santa Rosalia's bones turned out to be those of a goat—not saint—so I invoke a different saint.)

For the time being, the experts seem unable to explain—although there are those who claim they can—why species are found in certain relative abundances to one another. Most commonly, there are only a few species in an area that will predominate in a census, a number of species of intermediate abundance, and a large number of quite rare species, many with only a single specimen per sample. One explanation is that this common distribution, often called Preston's canonical distribution, after its discoverer, an amateur ecologist, is merely a statistical artifact of something called the Central Limit Theorem. In much simpler words, given all of the random factors that could determine the presence and abundance of any given species, one would expect to find exactly the type of abundances one actually finds in nature.

The concept of diversity is very much in the public mind. It is what conservationists are calling upon us to preserve. It is an area on which ecosystem and evolutionary ecologists could have come together. Their basic ideas differed little on it. However, there WAS a difference. Evolutionary explanations for differences in diversity rested on the ability to identify and study the species interacting to produce different levels of diversity. MacArthur's warblers, more attractive to classical ecologists, also became more promising for diversity studies than Odum's disreputable species, the superorganism.

Do Cats Eat Clover?

If the ecosystem concept was found to be wanting and collapsed scientifically, ecologist had to face the loss of their wedge onto the kind of big policy-making decisions that led to big funding. The environmental revolution had opened up many niches for ecological ideas in legislation such as the National Environmental Protection Act and the Endangered Species Act. However, these often admitted only an ecological point of view—nothing more, really than Rachel Carson had called for, care about the connections—rather than the whole hog application of what the science of ecology was making of itself. It was the ecosystem almost without the systems science or the ecology. Or else it was autecology.

Adaptations between organisms and environment, such as those resulting in the habitat requirements of black-footed ferrets or whooping cranes, then and now were best studied by those with specific expertise on each endangered species. They would, of course, need an ecological perspective, but they really had no need of the theory of the niche under development by all of the MacArthurites in their mostly ivy-covered towers.

An understanding of population dynamics is crucial in the study of a species whose dynamic is downward, but this could be accomplished using techniques of game management found in Aldo Leopold's 1933 book on the subject. Applying the mathematical theories of population ecology developed by the MacArthur school was a bit more difficult. They might well apply in general, but policy-makers wanted specifics.

In short, with the exception of the ecosystem approach being discarded, there was no applied ecology. There was, instead, the wildlife and agricultural sciences, forestry, fisheries science, and the rapidly developing parts of physics, chemistry, and health sciences having to do with pollution.

MacArthur and Hutchinson had pointed to a way for theoretical ecology to get back into the game, however. It was through biodiversity. Biodiversity is species diversity and habitat diversity. It is genetic diversity, too. It is essentially all those things that contribute to an animal's ecological and evolutionary niche. All of that variety, according to Elliot Norse, the lobbyist for the Ecological Society of America who coined the term in a report he wrote for the government, needed protection. Biodiversity was so evidently disappearing.

Unusual habitats often hold interesting and unusual species. Sometimes these species turn out to be of incredible importance to human affairs. Certain heart and muscle relaxants, without which major surgery is a much greater ordeal to surgeon and patient, are obtained from little tree frogs found only in the forests of the Amazon. An unusual bacteria that prefers life in the hot sulfur springs of Yellowstone turned out to be the source of the restriction enzymes so important to recombinant DNA technology, the working out of the human genome, and, perhaps, future wonders at which we can only guess. Numerous similar examples are to be found in popular treatments of biodiversity and modern man-caused extinctions and introductory texts on environmental science.

Unlike Preston's canon or niche partitioning theory, with mathematical underpinnings that await our perplexed attention, biodiversity is hardly rocket science. Our lives will be poorer in many ways—some of them unexpected—if we continue to suffer the loss of biodiversity that is currently under way. The sulfur loving bacteria might well have gone extinct if we had allowed the Yellowstone caldera to be used for geothermal energy at the risk of extinction of the geysers, as one proposal would have us do, but would our lives have gone on? No doubt. Might the bacteria not have gone extinct? Possibly. Could we have found another source of restriction enzymes? Most likely. Are our lives now better because we have not gone ahead with development of geothermal energy? That is one I do not know how to answer. Similar questions can be structured around those lovely little Amazonian tree frogs. How can science help here?

One way might be to take current trends to their limits, see how close or how rapidly we are approaching those limits, then sound an alarm if need be. The trick is to get people who can effect changes to hear the alarm. Ecologists have not been as successful at this as they could have been, and it might have been entirely their own fault. Some abdicated any responsibility on the practical issues. Others advocated more than their science could support. In many ways, the science was just not ready when Earth Day called for it.

What is there about biodiversity to concern us? Extinction is proceeding at a rate more rapid than at any time in the past outside of the cataclysmic extinctions found in our fossil record. The fall of the dinosaurs, some 65 million years ago, is generally considered to have resulted from a comet crashing somewhere (now apparently known where) into the earth. Present human-caused extinctions may be approaching numbers to match comet-caused ones.

What does ecology, which has raised the alarm over extinction rates, tell us our concern over loss of diversity should be? The branch of ecology that should be most capable of answering that question would seem to be ecosystem ecology. However, stripped of information on energy and nutrient flows, Odum's ecosystem science becomes MacArthur's community ecology. Both point in the same direction.

Given the concept of the earth composed of ecosystems, defined in their simplest way as sets of interacting organisms and their environment, the loss of a species is the loss of an interaction. If the interactions found in ecosystems are somehow important to their functioning, then we have the possibility of a loss of that function. If also the earth's ecosystems are themselves all interrelated, the loss of a single species could conceptually lead to the whole house of cards crashing down.

Need we concern ourselves about every single species? Ecologists seem to answer yes and no.

Obviously not every single species. Large numbers have already disappeared and are now disappearing without our environmental condition deteriorating as alarmingly as it did from the unrestricted use of chlorinated pesticides or as it might from the accelerating use of fossil fuels. Many extinct and endangered species are obscure creatures found in very small numbers whose interactions with other species are probably unknown and unknowable, but relatively insignificant, especially on a global scale. There are, however, great numbers of species threatened by or driven to extinction by man's action (often deliberately), the consequences of which are obvious. The loss of large predators has clear cascading effects throughout ecosystems. The elimination of wolves over most of the North American continent has been tied to instability in their herbivore prey (deer, elk, moose, mice?) and drastic changes in vegetation. The elimination of American Bison from the Great Plains would have resulted in subsequent changes in that biome had we not already converted it to wheat, corn, and grazing cattle. We are poorer for it, but how? That question is not strictly ecological.

Although the transition from Great Plains to Great Wheat and Cornfields received much ecological attention after the dust storms of the 1930s, purely ecological concerns were less relevant than economic ones. Even the ecologists of the day looked upon that disaster from the same perspective that George Perkins Marsh had warned of the economic losses caused by thoughtless forest and agricultural practices a century earlier. Land use issues almost invariably cause ecological values to butt up against economic values. With only a few modern exceptions, economics always wins in the end. What's more, the disappearance of most of the American prairies did not seem to have resulted in mass extinction of species.

With the exception of nature preserves and parks, decisions to be made on wildlife are economic or emotional, rather than ecological. Most rural areas in the US allow the elimination of property-damaging pests in legally approved ways. In Vermont, bounties on the coyotes that are replacing its long-gone wolves have been reinstated in a sense. Meanwhile, predators that once were used to frighten children in fairy tales have now been rehabilitated and are tolerated even in the face of their most noxious behavior. Alligators formerly in danger of extinction now drag off unsuspecting tourists from boat launches and children from hotel pools. Mountain lions snatch joggers from hiking trails. No amount of property damage or health threat is sufficient to justify the destruction of nature's popular creatures. No amount of science can convince a suburbanite that Bambi spreads disease or risks starvation if deer numbers continue to go unchecked. Not even the threat of Lyme disease is sufficient to take violent-but-effective steps to reduce deer numbers. Neither is there enough science to convince a farmer to accept economic losses from what in his view is a pest. Any science in those issues is muddled with emotion. Yet the defender of biodiversity seeks to bolster his arguments with science.

One idea that ecology has given the defenders of biodiversity is that of the "keystone" species. The presence or abundance of a keystone species directly or indirectly affects the abundance of other species and thus the structure and function of ecosystems. Unlike species that dominate an ecosystem by their numbers, the influence of a keystone species on its community is well out of proportion to its numbers. Wolves, for example, are keystone species.

Robert Paine of the University of Washington first brought attention to the idea of keystone species. Paine, who is on Hutchinson's family tree, found that if he removed a starfish predator, drastic changes happened in the community of mostly mussels and barnacles hanging onto rocks between high and low tides. Diversity declined from 15 to 8 species. Predators, according to Paine, have the ability to regulate the densities of species in lower trophic levels. Other examples have corroborated Paine's findings, but not in all communities.

Loss of keystone species from ecological communities is a concern. They control diversity. The argument that protecting biodiversity in general will also protect cryptic keystone species is a flawed one, however. Keystone species are hardly difficult to identify and threats to their existence, hunting, for instance, are rather direct. Ecology has to look elsewhere for justification.

There are numbers of other arguments for concern over extinctions that can also be set aside, not necessarily for lack of merit, but because they are just not ecological in nature. They are more likely to get in the way of than be of help to ecologists trying to apply their science. The argument of aesthetics is an example.

Endangered pandas are appealing creatures. If we wish to save pandas, then by all means we should save pandas. The world will be an impoverished place without pandas, but it will go on. This much we do know.

Neither is the argument for preserving biodiversity based on there being other species out there like the tree frog that have medicinal properties that might be lost completely convincing. Nor is it ecologically based. Recently, a frog with a promising secretion almost did go extinct. That would have been a shame, but not necessarily a tragedy. Theoretical advances in biomedical technology, such as the use of a technique that might best be described as the artificial natural selection of molecules, trumps the medicine-from-nature argument. Technology in the future should be able to provide a greater array of molecules to search through for valuable properties than nature has. The issue of medicinal value is just not an ecological one. (In fact, mistaken notions of medicinal values have posed the major threats, through poaching, to a number of endangered species. The very real medicinal value of taxol threatened the existence of the yew tree from which it was extracted before chemists managed to synthesize it economically. The wholly bogus value of rhino horn still threatens the existence of that species)

All species have scientific value as objects of study. Their genetic makeup and its expression make up a repository of knowledge. This argument was brought to bear on the preservation of smallpox virus when it was still thought to exist only in single laboratories in the US and Russia. The argument that the two samples should not be destroyed in order to preserve any scientific value they might have for the future recently became moot in a most unfortunate way. Russian scientists, it turns out, made a number of additional cultures whose whereabouts are not entirely known at present.

Ecotourism, another value promoted by ecologists, is also strictly an economic or aesthetic concept. The aforementioned Dan Janzen of the University of Pennsylvania, for example, has committed himself to preserving the Area de Conservacion in Guanacaste, Costa Rica, by making it useful. "The farm," as he calls it, produces a number of products, one of which is ecotourism. It is not a particularly new concept, essentially being travel to exotic places. As an economic value, it differs from other natural resource values, such as timber and game by being less destructive of what is present. It is not clear, however, that ecotourism depends directly on diversity, although Janzen manages his farm with the goal of maintaining biodiversity. What is it people are coming to see? A diverse environment or a spectacular environment? A variety of different little creatures? Or those with fur or feathers or soulful eyes? It is a subject for a different book.

There is, of course, a high ground, morally and ethically, that can be taken in the biodiversity argument. All species have the right to existence. Can ecology be used to justify what is a basically moral argument? A variety of groups that can, with caution, be placed under the umbrella of "radical ecology," but that also include card-carrying certified professional ecologists, think it can. The roots of these groups can be traced to the years around that first Earth Day, although, as always, more ancient roots can be found. Take the philosophies of Jean Jacques Rousseau and Henry David Thoreau and mix them in with the activism of John Muir and you have just created radical ecology. (One could probably also find more modern influences ranging from Gandhi and Martin Luther King to Robert Redford to Betty Friedan and Gloria Steinem to Edward Abbey. This would be a different book if I were to follow those threads. I offer the reader Kirkpatrick Sale's *The Green Revolution* and Michael E Zimmerman's *Contesting Earth's Future* as further reading. They are works I used as my guides through this morass of ideas. The former work is a brief reportorial history; the latter delves much deeper into the philosophical underpinnings of the movements than I wished to follow.)

What is similar about the radical movements is their reliance on a philosophy based on the concept of "ecology" as seen by them. It is Rachel Carson's concept of there being an intrinsic balance of nature. All share the view that current ecological crises arose from man's insensitivity to and willingness to upset nature's delicate balance. For this they blame the anthropocentric view of the dominant Western philosophies toward nature. Historian Lynn White argued in 1967 that Christianity, through its anthropocentrism, "made it possible to exploit nature in a mood of indifference to the feelings of natural objects." Meanwhile, Eugene Odum placed man as "part of, not apart from" his ecosystems. Radical ecologists later added all of the leading ideologies from Marxism to capitalism, anthropocentric humanisms both, to the Christian anthropocentrism that White decried. Ecologists added biodiversity to this intellectual stew.

What Thoreau found in the woods of Walden, Norwegian philosopher Arne Naess found in the fjords of his native country. MacArthur may have found it in the woods of Vermont. None of the quests were physically long or grueling. Thoreau sojourned a mile or so from his home in Concord to live a few months in the "wilderness," with frequent trips into town. Naess found inspiration in the fjords of his native land as a child. Much as Thoreau found Transcendentalism, Naess found Deep Ecology. "Deep" is in contrast to shallow environmentalism, still mired in anthropocentric humanism, whose goal is simply to change certain social practices. Deep Ecology represents a radically different view of the "Self" to include not just other people, but entire ecosystems as part of one's sense of identity.

"'Myself' now includes the rainforest," a follower writes, "it includes clean air and water."

The science behind Deep Ecology is almost nonexistent. Its theories, and those of other radical movements, are little concerned with what is nature. Their concern is how man (or woman) should relate to nature. Michael Zimmerman in the index to his book on radical ecology notes 3 pages devoted to EO Wilson, but 25 to Baruch Spinoza; 2 pages to Gene Odum, 21 to Jacques Derrida. Hegel, Heidegger, and Nietzsche (who has been identified by at least one author as a "deep ecological thinker") overwhelm Ehrlich, Gould, and even Darwin in Zimmerman's text. There are 28 pages indexed to the term "ecosystem." On most of those pages, the word is used, but not the science behind it, with the exception of outdated notions about the stability of ecosystems.

The interconnectedness and stability of ecosystems is used to provide the scientific foundation for their points of view. Ethicist J Baird Callicott, for example, argues that, based on Aldo Leopold's "Land Ethic" and ecosystem ecology, species and bioregions should have precedence over individual organisms.

It is all very muddled, though, the path from Odum to ecotage. Naess, spiritual founder of the Deep Ecology movement, had his epiphany in oblivious ignorance of Eugene Odum. Besides Thoreau and Muir, Deep Ecologists trace their philosophical roots to DH Lawrence (by way of Lady Chatterley's gamekeeper, no doubt), Robinson Jeffers (a mid-20th-century poet), and Aldous Huxley. They take their modern inspiration from Aldo Leopold, Rachel Carson, David Brower, and Paul Ehrlich. What gave one proponent of Deep Ecology the most inspiration was *Silent Spring*'s closing paragraph, which condemns the arrogance of man's belief "that nature exists for the convenience of man."

Naess cites the publication of *Silent Spring* as the event "from which we can date the beginning of the international deep ecology movement."

I venture to guess that the biocentrists would not have run across an explanation of the ecosystem concept in the works of the authors from which they avow to have received their inspiration, not even from ecologist Paul Ehrlich, not even in *Ecoscience*, a comprehensive environmental textbook he co-authored. In the chapter titled, "Populations and Ecosystems," the word *ecosystem* is used only 3 times. Why not? Ehrlich was a population ecologist. Anyway, it is not likely that environmentalists went beyond *The Population Bomb* in their reading.

In short, there is little ecology in Deep Ecology. There may be even less in the other radical movements. Ecology, like Frigidaire and Google, has lost its specific identity. It is no longer a subject and methodology for scientific research. To radical ecologists, it is a way of life. Some of them, remember, are also ecologists.

Then there are the truly radical radicals, PETA, Earth First!, and various underground ecoterrorists and eco-anarchists. All are responses to perceived abuse of the environment. All have as their scientific foundation the concept of a self-regulating ecosystem whose stability depends on all of its working parts, the initial ecosystem concept. All intertwine their science with tight loops of philosophy.

Might ecology be just a philosophy? Might biodiversity just be a premise adopted by it?

David Takacs, in his study of the relationship between biodiversity and ecologists, concludes that biologists find what he calls spiritual values such as those in Deep Ecology because of, rather than despite their science.

They "turned spiritual by biodiversity," he wrote of those he interviewed.

Peter Raven, a champion of biodiversity admitted to Takacs: "As far as we know, we're the only living things in the universe and I think we have a responsibility based on that fact, that spiritual fact, to guard and keep biodiversity."

"It means the inborn affinity human beings have for other forms of life," EO Wilson has written in describing biophilia, "an affiliation evoked, according to circumstance, by pleasure, or a sense of security, or awe, or even fascination blended by revulsion."

And later: "The naturalist's vision is only a specialized product of a biophilic instinct shared by all."

It is a point of view that fits in very nicely with Arne Naess's Deep Ecology.

What seems to be left in the science with which to defend diversity is that old concept of nature in balance that Rachel Carson decried was being disturbed by rampant pesticide use. It is a concept older than ecology. In fact, it is older than the modern science. It can be traced back to pre-biblical times. The idea that nature somehow stays in a balance to which many different—and even lowly—creatures contribute represents a well-worn philosophical assumption. However old it may be, it is not a scientific observation.

We do still find the idea in the science of ecology, however. In its sophisticated form it is called the diversity-stability hypothesis. Robert MacArthur helped give it sophistication with his math. It became the subject of hot research and debate. In its current incarnation, stability has been replaced by function, productivity, in particular, but the idea of that function depending on some sort of natural balance is unchanged.

For the first part of the 20th century it was assumed, whether from pre-biblical historical precedent or otherwise, that diversity and stability were closely tied. By the first Earth Day, Eugene Odum was arguing that ecosystems developed toward the direction of increasing diversity and stability. Speaking to a different audience, Robert MacArthur's voice was right there with Odum's.

"The cause-and-effect relationship is not clear and needs to be investigated," Odum admitted in his landmark 1969 paper in *Science*. "If it can be shown that biotic diversity does indeed enhance stability in the ecosystem," he went on, "or is the result of it, then we would have an important guide for conservation practice. Preservation of hedgerows, woodlots, noneconomic species, noneutrophicated waters, and other biotic variety in man's landscape could then be justified on scientific as well as esthetic grounds."

"Stability increases as the number of links increases," MacArthur wrote in his 1955 paper. Then he added, with a mathematical exactitude that gives no hint of the personal magnetism he was said to have had: "The maximum stability possible for m species would arise when there are m trophic levels with one species on each, eating all species below. Similarly, the minimum stability would arise with one species eating all others—these others being all on the same trophic level."

Diversity-Stability and Other Scams

Early naturalists found that the tropics held more species and its forests were less uniform than temperate forests. I could once walk for miles along Vermont's Long Trail, often through groves containing only hemlock or fir, and see fewer tree species than I could expect to see in a jaunt of only a few hundred yards in the Amazon. I could also have walked for miles from Manaus and see no single-species stand of trees equivalent to the old growth hemlock groves I enjoy in Vermont. (I can still happen upon single-species stands of old-growth hemlock in Vermont by checking maps and asking around. No one, however, can guide me to a single-species stand of trees in Brazil unless it happened to be coffee or some other planted growth.) The science of distribution and abundance of species, as some saw ecology to be in the first half of the last century, had as one of its tasks the explanation of why that should be.

Early ecologists such as Charles Elton also noted that species-impoverished tundra and taiga ecosystems had animals whose populations varied greatly. Lemmings and other northern rodents would irrupt, often at approximately four-year intervals, to reach plague densities, until the creatures' numbers subsided, sometimes in spectacular and seemingly suicidal behavior, such as marching en masse into the sea. Elton found evidence for lemming plagues as far back as the 16th century.

In contrast, naturalists studying tropical flora and fauna found "harmonious" relationships in animal numbers, as Alfred Russell Wallace, Darwin's partner in the theory of natural selection, put it. There was nothing equivalent observed in the tropics to the rodent plagues of northern biomes.

Was the diversity seen in the tropics a result of the stability of its climate? Did the harsh climates of northern biomes result in less diverse food webs, thus causing unstable fluctuations? Both questions have been vigorously investigated by ecological scientists. Notice that the questions are not either-or. One can answer and investigate each independently.

The century between Wallace's observations and the first Earth Day was not enough time to answer either question. Yet reading the ecological and environmental literature of the time one would have thought that diverse systems being more stable was a fact.

Daniel Goodman, who once jokingly confided to me that he had taken some part in almost all of the ecological "scams," meaning ecological theories, took a skeptical look at the diversity-stability con job.

To Goodman the original idea that diverse systems must be more stable "was probably a combined legacy of eighteenth century theories of political economics, aesthetically and perhaps religiously motivated attraction to the belief that the wondrous variety of nature must have some purpose in an orderly world, and ageless folk wisdom regarding eggs and baskets."

In an essay in *The Quarterly Review of Biology* that should have been the final nail on the coffin for the idea, he traced the modern formulation of the diversity-stability hypothesis to MacArthur's influential 1955 paper. MacArthur and his followers built up what Goodman called an "esthetically pleasing body of theory." In his comprehensive review, Goodman showed that the idea that diversity promoted stability really had neither theoretical nor factual support, yet it persisted.

"It has been influential to the extent that it was cited as more or less of a *cause* in much of the literature discussing diversity and was repeated as more or less of a *fact* in textbooks, conservation pamphlets, and the printout of environmental institutes," according to Goodman.

This was the real problem with the whole business.

"The diversity-stability hypothesis has been trotted out time and time again as an argument for various preservationist and environmentalist policies," Goodman lamented. "It seemed to offer an easy way to refute the charge that these policies represent nothing more than the subjective preferences of some minority constituencies."

Goodman had one final scam. After an impressive succession of academic appointments at prestigious institutions where he produced highly theoretical work, he settled down at Montana State University, where he took up working on local environmental issues, particularly those affecting the Montana trout streams he frequented with fly rod in hand.

Goodman's 1975 review should have been the last word on the subject. It had persuasive evidence, both experimental and theoretical, that ecologists probably already knew about, but chose to ignore, showing that, if anything, diverse systems might be less stable. He pointed out the illogical leaps that some ecologists took to support their point. And he rather convincingly showed that a simple equation that Robert MacArthur used to describe stability in a community was anything but that. It was the same equation used to calculate the degree of predictability in communications science and entropy in statistical thermodynamics. In MacArthur's application, it allowed for calculation of the probability of predicting what the next species that might be encountered would be, given information about the relative abundance of each species. Because entropy, or disorder, must always increase in our universe, MacArthur had reasoned, in effect, that maximizing entropy would minimize change. (It is true that the end result of all physical processes over all time will be one of maximum entropy according to the laws of thermodynamics, but an end point infinitely far in the future summed over the entire universe cannot be assigned to dynamic changes taking place on human time scales here on earth. Entropy must increase, but this tendency might as easily destabilize as stabilize a system.) Hutchinson, in his "Homage to Santa Rosalia" paper had gone on to elevate the equation to something akin to a basic theorem in community ecology. Goodman's should have been the last word, but as these things turn out, it was not. I cannot get off so easy as to dismiss that argument here and go on to other things.

Goodman may have hit on more than he knew with his remark that the theory represented subjective preferences. In my own subjective view, those subjective preferences soon led their bearers to pick up some other ecological reason with which to protect diversity when they had to abandon it as something required for stability.

So, back to the main subject. Before the diversity-stability argument can be effectively judged, there is first the question of what is meant by stability. It has several meanings. Stable ecosystems should resist external change (the simplest definition of stability), return rapidly to their stable conditions after a disturbance (usually called resilience), recover from a disturbance to their previous conditions (usually called recovery), or some combination of all three. There are other ways, too, of looking at stability, but stable ecosystems in the sense of meeting those three conditions seem to be exactly what the world needs in these times of assaults on environmental well being.

If more diverse ecosystems are more stable, then enlightened management would aim for greater diversity. Since ecosystems interact with other ecosystems in what might be looked at as one global ecosystem or biosphere, then the biosphere should be maintained at maximum diversity. If loss of diversity from an ecosystem reduced its stability, then the loss of even a single species risks the loss of biosphere stability. To maximize stability, this logic goes, one must maximize diversity.

Ecologists are in complete agreement that species are found in ecosystems, however that term might be defined. Within those ecosystems, species are often found in several different habitats. Migratory songbirds, moreover, are found in ecosystems representing different biomes in summer and winter. Remember that ecologists had pretty much settled on Richard Lewontin's preferred units of selection, genes residing in individual organisms. Genes go extinct with the death of the last individuals carrying them. Species go extinct when the habitats they depend upon are lost through alteration or destruction. Logging, agriculture, urbanization, civic works such as roads, dams, and power lines, wars, and the introduction of exotic animals, all alter habitats in ways that can make them unsuitable for species that depend on them—or, like agriculture, directly replace one set of species with another. If a type of habitat goes out of existence, so will numbers of species beyond those that specifically made it up. This is beyond contention.

When credible ecologists and conservationists call for measures to reduce extinctions, they are calling for habitat protection more or less based on that line of reasoning. Their ultimate rationale for protecting the genes, individuals, and species that are protected when their habitats are protected will be either one of the non-ecological arguments of the previous chapter—or that of local or global stability.

Early ecologists found much evidence that populations were more unstable in simple environments than in complex ones. Population stability may or may not be coincident with ecosystem stability, however. All manner of booms and busts might be happening in the populations of various species making up an ecosystem, yet the ecosystem might go on relatively unchanged in its most important aspects through those changes. Given that possibility, an agreement on a definition of stability was needed.

The classical definition from physics and chemistry is that of neighborhood stability. Robert May defined it very simply as "stability in the vicinity of an equilibrium point in a deterministic system." A grammatically simple phrase, yes, but except for the prepositions and articles in it, every word needs explanation. There is much esoterica hidden in that simple statement, not the least of which is how it applies to ecological systems.

Stability in that classical definition means return to original conditions, the equilibrium condition, after disturbance. That equilibrium condition, if the system actually has one, can be described mathematically. This description consists of parameters and state variables, those quantities from cybernetics, and something called trajectories. A trajectory is the shape of a line obtained when two or more parameters are graphed against each other. It describes the way the system or some part of it changes as parameters change. The whole business is described through mathematical equations that need not trouble us here, but to which we have already been introduced in their most understandable form. This is the logistic equation of the S-shaped population growth so common to population thinking. All that needs to be understood about it right now is that, mathematically, the concept of equilibrium arises from solving equations like it and examining them under certain constraints that are expected at equilibrium. For example, at equilibrium, we expect the values of variables not to be changing. The forces of change should be balanced out. The mathematics describing a system can test for that.

An equilibrium point therefore represents a balancing point. All agents of change balance each other exactly. However, the value of variables at an equilibrium point may be unchanging only at that point. Think about balancing on a tightrope. When balanced, the change in your position is exactly zero. The physical explanation is that your center of gravity is exactly above the tightrope. The actual situation, however, is in fact, a precarious position at which only the most highly talented and trained circus performer can appear to be still. This is where the concept of stability comes in. You on the tightrope are in a situation of unstable equilibrium. If only you could keep those forces balanced! Unlike the trained performer, a single fidget brings you down.

Lying peacefully in bed on a saggy mattress or safely in a hammock, however, your fidgets and shifts of weight merely return you to your position. You may seem to return to your position faster in bed than on the hammock, but both represent conditions of stable equilibrium. The fact that you go through more back and forth changes, oscillations, in the hammock than in the bed may be of much interest for other reasons, but it has no effect on the stability of the equilibrium. Both are stable.

Lying on a flat wooden surface may be thought of as neutral equilibrium or no equilibrium at all. Your fidgeting may eventually take you far from your starting point, but not toward catastrophe, at least until you reach the edge of the surface.

This is where the concept of "in the vicinity" given in Sir Robert's definition comes in. A small move on the tightrope sends you crashing or swinging back and forth wildly until you crash or somehow right yourself. The equilibrium is unstable in the vicinity of your balancing point. The saggy mattress and hammock are stable in the vicinity of your balancing point, but don't roll too far or fidget too much. If you do, a new set of equations takes over. At the edge of the bed your expected trajectory suddenly changes. It may land you on the floor. Your balancing point was only *locally*, not *globally* stable.

What "deterministic system" in May's definition implies is that the mathematics used to describe changes are exactly predictable. Or reasonably so, for actual ecosystems. For the tightrope example, a shift in a certain number of degrees in the line from your center of gravity to the rope results in a turning force of predictable strength that needs to be countered. The circus performer's sense of balance and muscular memory depends on it.

My balance in walking a stream while fly-fishing is thrown completely off kilter when I am weighed down by a heavy pack, or, nowadays, even a canteen. The predictable response my muscles and sense of balance expect is suddenly different. I have to retrain my muscles. Eventually, I am pushing my way through the current or stepping from rock to rock as before. Should the weight of my canteen be changed from moment to moment, as if by some demon set on producing harm, balance becomes an iffy proposition. Changes have become no longer deterministic. They are now at the random whims of the demon. They are now stochastic, or unpredictable. Only a walking stick (with a small input of energy and massive mathematical complexity) can bring back harmony from the chaos.

A pendulum and a spinning top are more common examples with which to examine the classical definition of stability. Consider, for the time being that an undisturbed pendulum is one exhibiting no motion. It is just a weight hanging from a string. The spinning top, while in normal motion, has its axis of rotation motionless. The equilibrium position of the pendulum and the axis of the spinning top are both perpendicular to the ground. When disturbed, both oscillate, but the pendulum does so stably, the top unstably.

The motion of the pendulum will eventually "wind down" to its original, motionless, undisturbed state. We say that it exhibits damped oscillations. (Friction with the air and within the string causes the pendulum to continually lose speed. This is why old "grandfather clocks" have two hanging weights which need to be reset periodically. The weights are balanced so that their pull on the pendulum through some internal gearing to speed it up is exactly equal to the pull that friction exerts to slow it down.)

The top, on the other hand, will wobble more and more, its axis showing more and more precession, until it eventually ends up in a totally different state, its axis parallel with the floor, its spin dissipated, but not before going through a range of motion that is so obviously different that it will fascinate any child or any physicist. Richard Feynman in his autobiography, *You Must Be Joking Mr. Feynman*, tells a story that he claims saved his career. He saw the same kind of motion in a crest embossed on a dish flung through the air during a food fight in a cafeteria at Cornell University. With childish glee, he jumped into the mathematics of the motion, quickly deriving equations never before obtained. Those same equations, produced from nothing more than childlike curiosity and a talent for math, were later instrumental in Feynman's electron chromodynamics theory. They won him his Nobel Prize.

Eliminate friction from a pendulum and there is a third possibility. It goes back and forth forever. This is known as a stable limit cycle. If the oscillations are small enough, we might tend to ignore them as just "noise" in the system and call it a stable equilibrium. This might be our tightrope example with a highly trained and highly talented performer on it. From our seats in the audience we may not even notice the slight shifts in motion needed to keep the performer relatively still on the tightrope.

The homeostatic ecosystem was thought to be much like a damped pendulum. Periodically disturbed away from its static position, it continually tends to return to it. If it oscillates, it oscillates around a stable equilibrium. This is the way we want our ecosystems to behave. Small disturbances, such as an old tree being blown down in a storm are nothing more than small change. The dynamics of the species in the forest will tend to restore it toward its original state. Often the system is said to be at steady state rather than equilibrium.

Technically, there should be no changes in an equilibrium, in which competing forces of change cancel each other out. In the complexity of real ecosystems, we can not expect there to be no changes at all. What stability means for ecosystems is a dynamic equilibrium, or steady state. In a dynamic equilibrium, forces of change do not necessarily cancel each other out, for the system is often not a closed one. Things can enter or leave it. The balance becomes one of compensating gains and losses. Sunlight and carbon dioxide can be seen to be entering the leaves of forest trees, while oxygen is released. The excess energy the leaves pick up leads to growth and increase in tree biomass, or is consumed by herbivores or lost as detritus. At steady state, our ecosystem would exhibit no change in biomass, while the energy and chemical inputs would stay constant, as would energy and chemical losses.

An ecosystem that is *stable* to disturbance will continue with the same biomass and the same relative rates of input and output, in the face of change in parameters. Some truly stable ecosystems like coastal redwood forests might even control parameters such as temperature and humidity that could destabilize them.

A *resilient* ecosystem is one that bounces back to its original state after a disturbance. Prairies returning to their pre-fire vegetation are examples of resilience to disturbance.

Ecosystem *recovery* is an idea that arises on consideration of what variables we may want to remain stable. Usually a bit less well specified than the previous terms for stability, it requires consideration of ecosystem functions, rather than species composition. An ecosystem that has pronounced changes in its species composition after a disturbance yet looks and behaves pretty much as it did before disturbance has exhibited recovery. Conversely, an ecosystem that maintains the same species composition after disturbance but does not function or look the same (stunted pines, for example, rather than fully grown) is one that has not recovered from a disturbance.

The reason for the various definitions of stability is that equilibrium theory is based on relatively simple systems. A pendulum or the chemical processes for which most of stability theory was designed have a small number of types of objects and behaviors. Often the only behavior at issue is a change in the relative proportions of types for a chemical process, or the motion of the parts of an object for a physical process. The objects in the systems are remarkable either in their uniformity or in their small numbers.

In a pendulum, we can neglect the behavior of the individual air molecules or the atoms in the string and bob. In fact, we have the most spectacular of all simplifying assumptions for pendulums. Mathematically, a pendulum exists as a single object (string and bob). The position of its center can be exactly specified with a little math. Its motion depends only on gravitational acceleration and frictional deceleration. It then becomes easy to play with parameters such as temperature, humidity, atmospheric pressure, etc., and gauge their effect on the motion of the pendulum, or to vary the design of the pendulum, itself.

Chemical processes, the generation of electricity in fuel cells by the reaction of hydrogen and oxygen, for example, are more complicated not because we really are dealing with the individual molecules involved, but because we have more possible changes to follow—and the changes can be more complex. The hydrogen molecules can be depended upon to all look and behave exactly alike. This is not quite true on a nuclear level or for certain organic molecules, but those are differences that can usually be ignored. By dealing with average molecular properties, the complications of the individual molecules go away. (Thalidomide is an example, however, of a case in which individual differences in molecules—they turned out to be either left-handed or right-handed in geometry—led to disastrous results when ignored. Testing for safety was done with pure samples of one form only. Treatment consisted of the more-easily-manufactured mix of forms. The untested form resulted in deformities to fetuses.)

In ecosystem analysis, species might be analogous to molecular types. Types of species interactions can be described, much as types of molecular interactions can also be described. This is where the analogy ends, however. We do not get very far into ecosystems before the complications of living organisms separate them from the physical and chemical systems for which much of systems analysis was perfected.

Ecologists can tease out general types of interactions from the complex webs of ecosystems, much as chemists have found general types of reactions, such as oxidation-reduction. Plant-herbivore, producer-consumer, predator-prey, parasite-host, symbiotic relationships, etc., are ecological examples. But chemists work with systems that are purposefully simplified. If hydrogen is to be oxidized in a fuel cell chamber, a chemist will eliminate all other molecules, if he can, that could interfere with the hydrogen-oxygen reaction and effect the performance or study of the fuel cell. Ecologists have no such choice. The systems they study are inherently complex. Interactions of single predator and single prey populations removed from their ecosystems are nothing at all like their interactions within that ecosystem.

Wolves might eat moose, but they will also eat rodents. What is a good place to put a bear in the scheme of things in an ecosystem? Imagine the bear as a top predator leaving a deer carcass to go munch on blueberries. Now the bear is a herbivore.

A leaf-eating sloth has photosynthetic microorganisms, algae, that tint the creature green during the rainy season. In turn, the algae provide food for the larvae of a particular species of moth. Trout fry can be food for larger trout, possibly even their own parents. Where are these creatures in the trophic scheme of things? How do we write equations to describe their interactions? The concept of average properties becomes almost meaningless in light of knowledge about life cycle changes, particularly in metamorphic organisms such as Mayflies, animals exhibiting social structure, mammals and bees, for example, and oddities of behavior, such as the migration of songbirds that allows them to span several ecosystems.

In addition, evolutionary change means that the types under study now are not necessarily the types that existed historically, nor will they be the same as those in the future. Physical scientists can be assured that hydrogen really was hydrogen when the earth was formed and really will remain as hydrogen as the sun slowly burns itself out.

So how does the systems scientist deal with this complexity? For one, evolutionary changes are eliminated by fiat. This is good, because this will appease those who deny evolution above the level of the individual and closes the door on controversy such as group selection. We can think in terms of "ecological time," the amount of time in which changes in distribution and abundance of species can occur, but not "evolutionary time," during which species can change in their very basic properties. Good, yes, but risky. For many organisms changes in gene frequency, microevolution, do not operate on a longer time scale than do the changes in ecosystems that are known as succession. It is doubly risky when an ecologist, in order to explain ecosystem behaviors that should on first principles be identical, but are not, invokes past history in explanation without taking into consideration the possibility of evolutionary change. Studies that use data on the distributions and abundances of tree species through the recent ice ages in order to make some sense of the relationship between climatic variables and forest characteristics are an example. Ten thousand years is some 200 generations or more for eastern tree species. Two hundred generations is plenty of time for all sorts of evolutionary changes. It may not have been only changes in climate that changed the distributions of North American tree species that were subject to the ice ages. Evolutionary adaptations may also have been involved. They are changes that are difficult to model mathematically, however. Evolution is not a cybernetic parameter.

An easier way to reduce natural complexity and possibly rule out evolution is to ignore species entirely. In a true systems approach to ecosystems, the identification of species is technically superfluous. This was one of the original attractions of the ecosystem concept. It does not mean that the ecosystem concept was being misapplied in arguments for diversity. The number of pathways for energy and nutrient flows in an ecosystem might be related to its stability. Those various pathways, of course, are there because of the various species that comprise the ecosystem. More pathways, more species, and vice versa. What could be more clear?

Chaos Comes to Ecology

A totally different way of simplifying ecology is to let the boxes of systems analysis be actual species. But let there be very few of them. Ecosystems as simple as only several species interacting are not found in nature, but that can be put together in a laboratory and treated very deftly with mathematics. The process somehow seems to make the whole enterprise of population ecology more scientific, more like physicists describing the motion of a pendulum. A number of physicists, mathematicians, and even some biologists did exactly that for ecology.

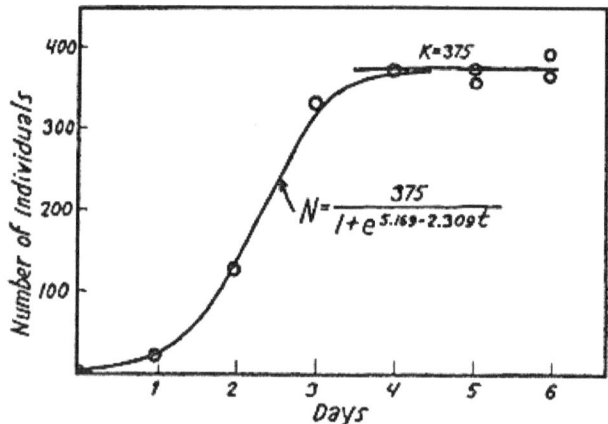

The Russian, Georgii Gause, showed that a simple equation for population growth could fit the laboratory growth of single-species yeast or paramecium populations. Graphs of his data have found their way into almost every textbook of ecology. A number of them are reproduced on this and the following pages. The first graph (above) shows the growth of *Paramecium*, a slipper-shaped, microscopic, one-celled animal found in almost any ponded waters. The mathematical formula on the graph is one form of the logistic equation. Gause filled in its parameters with numbers obtained from the data of the graph. The carrying capacity of his growth vessel he estimated at 375 individuals.

The graph below is for yeast. In this case, it no longer represents the numbers of individual yeast cells, but the volume of the total population found. (Counting yeast cells presents insurmountable difficulties. It is like estimating the exact value of millions of one-dollar bills or trying to pay your tax bill in pennies. Yeast cell numbers must be estimated indirectly.)

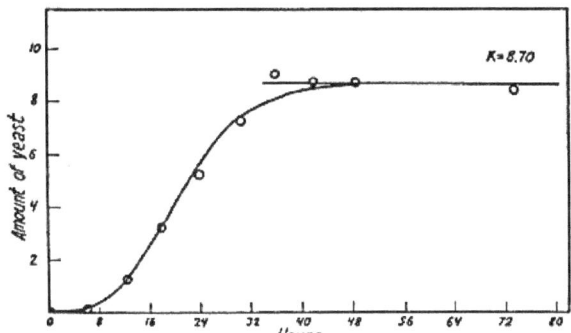

What Gause's data mean is that, yes, the logistic equation is a good descriptor of population growth for simple organisms isolated from all other aspects of the environment, save some limiting resource. Food is usually limiting for microorganisms grown in laboratory cultures, but sometimes lack of space or excess metabolites can limit growth. In the latter case, population growth is limited by its own refuse.

Gause found that mixed species of yeast populations also fit the simple models, but not mixed species of paramecium. The figures on the following page show these results. Experiments in which a predator of paramecium was introduced into the system invariably led to the extinction of both prey and predator. Competition above the level of producer led to one or another species going extinct.

Gause's work, simply interpreted, shows that an increase in complexity results in a decrease in stability. It was not subject to this simple interpretation at the time, for the diversity-stability hypothesis had not yet been given the imprimatur of GE Hutchinson and Robert MacArthur. Neither diversity nor stability had been an issue for Gause; his had been a test of the applicability of mathematical models of population growth.

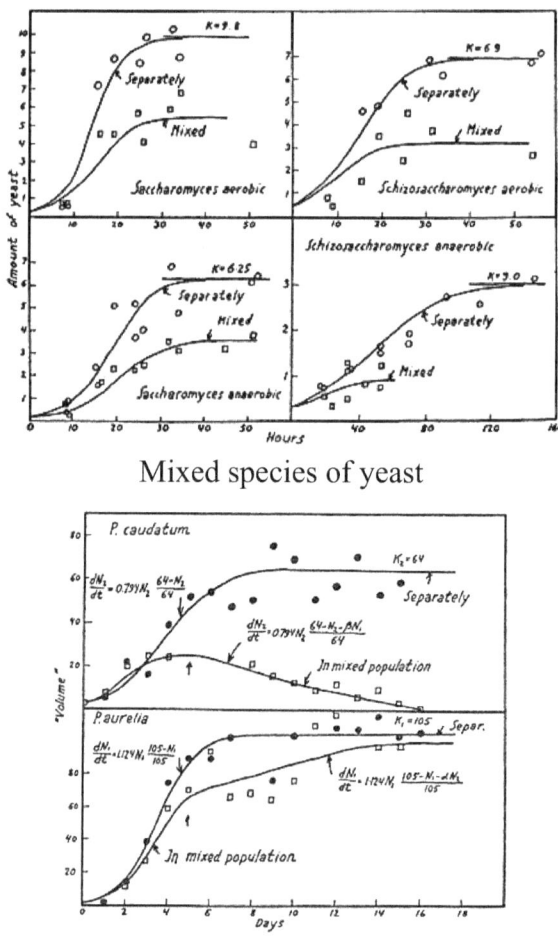

Mixed species of yeast

Mixed species of Paramecium

Gause presented most of his findings before the age of 24, summarizing his research in a book with an aptly Darwinian title in 1934. Gause was an evolutionist, not ecologist. The title he gave his book was *The Struggle for Existence*. His genius was squandered, like many things, by war and politics. Darwinian evolution became a politically unsafe concept to espouse during the era of Lysenkoism in Russia. (Lysenko believed in Lamarckian, rather than Darwinian, evolution. Self-improvement was then big in Russia; inheritance was not.)

Gause's experiments tested the then-current mathematical description of population interactions. This was the "logistic equation" to which I have been slowly introducing the reader. A simple application of Newton's calculus of rates, it has been attributed to various mathematicians all the way back to the early 19th century, but perhaps its most notable investigator was Alfred J Lotka, whose *Elements of Physical Biology*, published in 1925, influenced scientists as varied as demographer Raymond Pearl, cybernetics founder Norbert Weiner, and GE Hutchinson. Lotka's starting point is a rate equation general enough to describe any change in any quantity that I can imagine. I had hauled it out (although not yet in mathematical form) previously to help explain density-dependent selection. Let's rewrite it below in words and without the complexity of the integral form that is on the graphs: The rate of the growth of a population in numbers of its individuals at a given time is equal to the reproduction rate of an individual times the size of the population times the difference between the carrying capacity in numbers of individuals and that population at that time divided by carrying capacity. *AEPPS* has it as: The rate of population growth equals (is given by) the potential increase of the population per unit of time times the degree of realization of the potential increase.

Given all that verbiage, the following little equations might seem paragons of simplicity (which they are, mathematically).

$dN/dt = rN \times (K-N)K$ or $rN \times (1-N/K)$, where

N is the number of individuals at some given time,
dN/dt is the rate of growth of the population,
r is usual called the intrinsic rate of increase, and
K is usually called the carrying capacity of the environment.

A modern hand-held calculator (or phone or other marvel of the 21st century) and a little knowledge of algebra allows the reader to quickly calculate the number of *Paramecium*, *N*, on a given day, *t*, using the equation shown and on the first graph. According to that equation, populations eventually reach a stable level that is known as the environmental carrying capacity, an intuitively attractive concept and one that Gause verified in his laboratory. Tradition has given the carrying capacity the letter "K" for shorthand. In addition, the other parameter of the equation, "r," the intrinsic rate of increase (which, for the sake of confusion, it would appear, was designated by the letter "b," for "birth rate," in *AEPPS* and many other early versions of the equation), was related to the simple demographic mathematical analysis that was under development at the time. The intrinsic rate of increase of the equation can be calculated independently of data on population growth. One need only know a few simple parameters, age at first reproduction, number of offspring per reproductive event, gestation period, and longevity, for example, in order to get a rather firm estimate for it. In Gause's formula, r is hidden in the exponential term. Based on the shape of the curve, the r for that *Paramecium* was 2.309 individuals per day. Gause made no comparison of this estimate to one based on demography; doubling time is usually all the demographic data needed for organisms that reproduce through fission.

That this simple equation could be used to explain actual population growth in a laboratory while also being tied to demographic and ecological concepts (the r and K) is a set of coincidences equal in power to planets being discovered according to Kepler and Newton's Laws. Orbital inconsistencies turned out to be grounds for the prediction of new planets, which were right where they should have been. It was the kind of unexpected serendipity that often marks some deep truth or principle in science. The logistic equation remains a starting point for the study of population dynamics to this day.

The combinations of predator and prey species in their mathematical form are known as the Lotka-Volterra equations. This is the situation for which Lotka used the r and K notation, which Gause then used in his experimental work. It is also the jumping off point for Robert MacArthur's thinking on competition.

"Curiously enough," MacArthur wrote in his final work, possibly having just discovered his 19th century predecessor, "the basic theory of competition was worked out by Volterra, the mathematician, before any of the experiments were carried out."

What MacArthur added to that theory was a new way of thinking about r and K. In particular, he tried to include principles from evolutionary biology. MacArthur liked to think in terms of dichotomies. Specialist versus generalist species represent two ends of a spectrum of types that greatly influenced his thinking. He called the latter a "jack-of-all-trades." Specialists are species found in only certain habitats or feeding on only one type of food, for example. Looking at the logistic equation of growth, MacArthur found another dichotomy. He could subdivide species into "opportunistic" ones, described mainly by the r, or growth term of the equation, and "equilibrium" species, whose population size was described by K, the environmental carrying capacity. Later, in his and Wilson's *Theory of Island Biogeography*, this became "*r selection*" and "*K selection*." Ecologists soon shortened the concept to one of r-species and K-species. Although MacArthur's general goal for ecology was to apply the principles of mathematics to it, what he was doing in a simpler view was looking for explanation of ecological patterns in those simple r and K equations. Their appeal for him must have been two-fold. One is the obvious ease of associating ecological phenomena with their parameters. The other, less obvious, is that they are what are known as differential equations. That is what dN/dt is all about. MacArthur's mathematical expertise, right from the start of his career with his undergraduate Plan of Concentration, was in the analysis of differential equations.

In their simplest form, one predator and one prey, the Lotka-Volterra equations are a pair of coupled equations; meaning one solves one equation in order to have numbers for the other, and vice-versa. They have a disturbing lack of realism. In the absence of predator, prey numbers increase exponentially (no carrying capacity, in the simplest formulation). In the absence of prey, predator numbers decrease exponentially (approach, but do not reach zero predators, at a consistently decreasing rate). Instead of dying immediately when their prey is gone, the predators must resort to cannibalism, it seems, or maybe they can exist, at least for a while, on nothing but air. Lotka himself admitted to a lack of realism to the model, but justified it based on the "considerable mathematical difficulties" in including more realism. Indeed. These are the equations that Gause's experiments with protozoa rejected.

They are also equations that much stability analysis is ultimately based upon. Their attraction in the face of their lack of realism is that mathematicians have little difficulty in going from a two-species system to a many-species system. They are also fairly easy to "linearize." In fact, those equations describing competition, with their r's and K's that MacArthur interpreted, are already linearized. Lotka had taken a very general form of an equation that described rates of change and truncated all non-linear terms. Truncation means that he essentially lopped off the offending terms by ignoring them. (Einstein, in his theory of relativity, was guilty of similar, but even more egregious mathematical legerdemain, which troubled him to the end of his life, but which has now been shown to have apparently been correct.) In their linear form (without the terms Lotka hacked off) they are mathematically interesting because an entire arsenal of mathematical techniques exists for their analysis.

Linearization is a process that is unavoidable in stability analysis. When equations are not linear, they usually are approximated as linear in order to use all of the really "neat" methods of stability analysis. It is also what keeps the results limited to the "vicinity" of equilibrium only (meaning it may not work with the actual system you are analyzing).

A little refresher on high school algebra will help make this clear. Algebra is a method for setting up general relationships between quantities. Take list price, sales tax, and total cost, for example. It is not difficult to set up an algebraic relationship between list price and total cost, given a particular tax rate. Most people who still bother to do math have such a relationship in their heads. For a given tax rate, the relationship between cost and price is linear.

Double the price, double the cost. If the relationship were graphed for a number of prices and costs, it would look something like this:

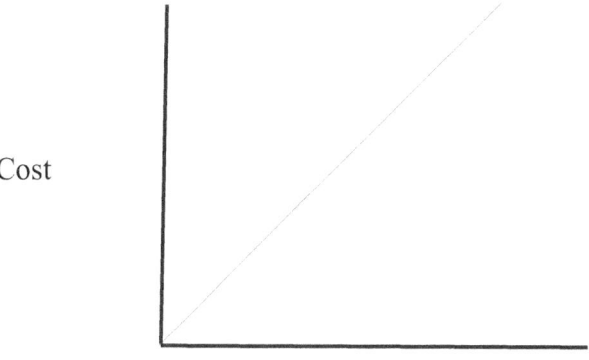

Price

Our cost-price relation turns out to graph as a straight line. (If the line appears to be leaning closer to the cost axis than the price axis, that is because the total cost will always exceed the list price when tax is added. You knew that, didn't you?) This linear graph tells us that the cost to us of an item is determined by its price times a multiplier that includes its tax rate, or Cost = Multiplier X Price. We can simplify by rewriting the algebra using the y's and x's of Cartesian coordinates to $y = kx$, the simplest form of a linear equation. It is linear because the multiplier (the tax) remains the same for any price. I am sure you can remember relationships that were not linear. Quadratics, cubes, exponentials. These are not as easy to interpret and graph. In the old days, before graphing calculators and tablets or whatever (watches?), engineers where taught to use special graph paper (log and semi-log) to make straight lines out of the curves to facilitate their interpretation. The logistic equation, growth rate = $rN(1-N/K)$, is one of those non-linear types of equations.

Mathematicians have mathematical tricks to approximate equations that are not linear with linear ones. After all, a very small segment of any curve approximates a straight line very well. This is why people mistakenly thought that the world was flat. (Some still do, their imagination is so void.)

All of this math has some implications for the real world. If the properties of ecosystems, whatever those are chosen to be, are stable or unstable based on linearized mathematics, they are so only within a small range of their possibilities. This may be OK for the tightrope analogy, but how about the hammock?

Nonetheless, a number of mathematical analyses of the two sets of equations (logistic and Lotka-Volterra) were tried out to see if their mathematical behaviors fit observations and to see what predictions could be made by that math for more complex situations than those that Gause tested. It became something of a cottage industry among the mathematically inclined but ecologically dense. It was not until physicist Robert May turned his mathematical powers to them not too long after the first Earth Day that all of the mathematical complexity was given a thorough interpretation. It was part of what earned him his knighthood.

May gathered together his conclusions about much of mathematical population ecology in a very influential book entitled, *Stability and Complexity in Model Ecosystems*. My copy of its second edition (it came out not much more than a year after the first had in 1973) is well thumb-worn. For quite a while, it and he represented the last word on things mathematical in ecology.

The equations he analyzed had been ridiculed by many biologists as unrealistic for decades. May admitted in a second edition afterthought that ecologists had not devoted enough study to the subject of mutualism and that some models led to "wholly nonsensical" results. May added only analysis, not realism, to them. Still, he turned some ideas on their heads. One of his mathematical findings, for example, was that if complexity in Lotka-Volterra systems is increased by introducing more predator and prey species into the equations, stability was reduced. Having more species of predators and prey, and thus more diversity, made for less stability.

In general, May concluded that "complexity tends to beget instability rather than stability."

Even mutualism, in which two species benefit each other, "tends to have a destabilizing effect on community dynamics," according to Sir Robert.

May's conclusions about complexity (diversity) and stability were couched in terms such as "robustness," "domains," and "parameter space" that mathematicians like to use, but he brought it all together into one simple statement of remarkable significance.

"I would argue," he wrote, "that the complex natural ecosystems currently everywhere under siege, are less able to withstand our battering than are relatively simple temperate and boreal systems."

What the statement implied was: Don't mess with the tropics. It was exactly what ecologists and environmentalists wanted to hear. May had overturned the conventional wisdom then entrenched in biology that diversity begets stability just in time for an entirely new, environmentally conscious ethos on diversity ("Save the Rainforest").

The stability analyses May performed were, of course, based on linearized equations. The exclusion of non-linear behavior by modelers was not lost on some ecologists.

"Where in your models," one asked not of May, but of modelers in general, "does it say that a prey species turns out to grow spines when the numbers of predator species increases?"

Where is natural selection in your models? is a just paraphrase for that question. Evolution tends to be nonlinear. How are your models related to reality? might be another. So, too, is reality nonlinear?

The other thing that the math May analyzed says about the real world is that the quantities being described by the variables in the equations are continuous. This is essentially what is meant by the term, differential equation. Difference equations deal with changes; differential equations deal with very minute changes. Time progresses in imperceptible infinitesimally small increments in differential equations, rather than in the days or seasons or years of difference equations. In addition, a fractional predator can still consume prey in differential equations. Apparently whatever parts the predator is missing are non-essential. Mice are born not in litter sizes of 4, 6, or 11, but 3.76548, if need be. The intrinsic rate of increase of Gause's *Paramecium* can be 2.309, even though there can never be that number of individuals.

I recall Simon Levin, who became an ecological guru at Princeton, as a young professor of applied mathematics at Cornell bringing this up to us graduate students whenever a discussion turned to the lack of realism in some particular mathematical model. It struck me as a rather odd thing for a mathematician to bring up. Real numbers by definition are continuous. Why sweat such a trivial detail? I thought. Truncate the little devil out of existence! Si used the issue of fractional individuals to urge toleration for a model's assumptions as long as that model could give useful insight. The logistic equation is one example. The usual term for such a model is *heuristic*, meaning not very real or even correct, but capable of teaching something.

It was Sir Robert May who dropped the assumption of continuous quantities. It led him to some peculiar conclusions. It also made him famous.

"Some of the simplest nonlinear difference equations describing the growth of biological populations with nonoverlapping generations can exhibit a remarkable spectrum of dynamical behavior," Sir Robert reported having found, "from stable equilibrium points, to stable cyclic oscillations between 2 population points to stable cycles with 4, 8, 16, ... points, through to a chaotic regime in which (depending on the initial population value) cycles of any period, or even totally aperiodic but bounded population fluctuations, can occur."

This was a bombshell in population biology. What Sir Robert was saying was that even the very simplest of the mathematical models for population growth, the beloved logistic equation with its easily interpreted parameters of r and K, had all hell breaking loose with them once time was no longer treated as continuous, but became generational or seasonal.

From out of very deterministic mathematics has come chaos.

By deterministic, remember, we are not allowing chance events. Only those parameters that are in deterministic equations are allowed to cause changes. Deterministic means not random. Differential means changes can be infinitesimally small. The little logistic equation, which seems to demand an equilibrium at the carrying capacity, K, did not behave as expected and as shown in the results from Gause's lab.

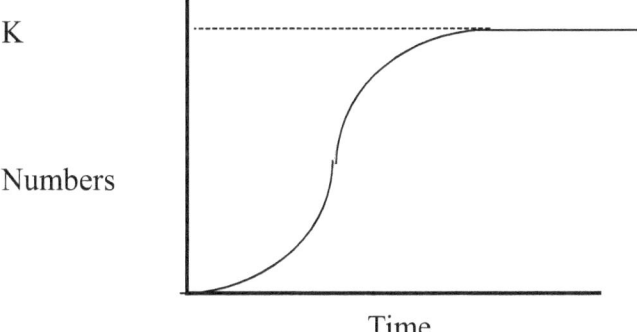

In this picture, once the population reaches carrying capacity, growth rate is expected to be zero. The mathematics requires it. Above carrying capacity, the growth rate becomes negative; numbers must decline back to K or, having overshot, fall to where the growth rate is positive again and return in this way to K. What discrete time intervals suggested might happen is a situation such as below.

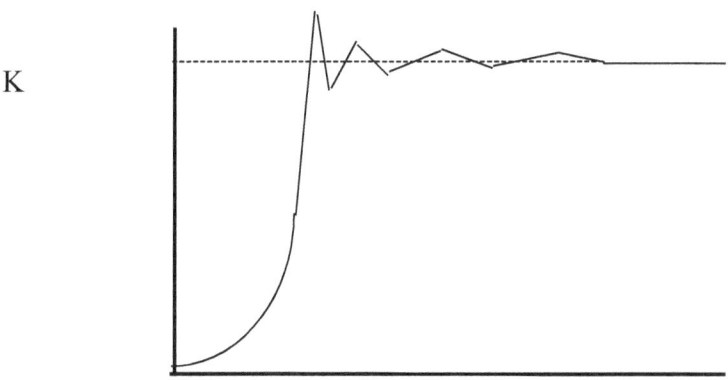

The equation is as deterministic (predictive) as ever, essentially unchanged except for the fact that the time is now measured in days, generations, or seasons. Those are the time quantities of interest to ecologists. For many values of the variables of the logistic equation, the time intervals made little difference. The curve was no longer smooth, but the equation behaved as expected.

For some combinations, all hell broke loose, as described above in the calmer terminology of Sir Robert. The graph looked something like that below.

K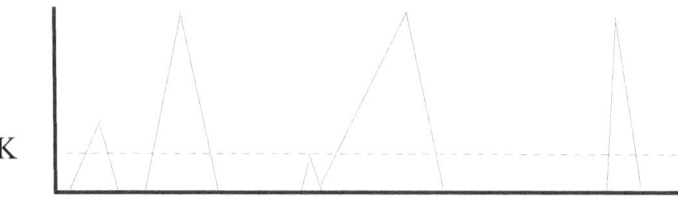

The very model of chaos.

"This rich dynamical structure," May thought, "is a fact of considerable mathematical and ecological interest."

The simple little equations that biologists had hoped would describe the real world but failed to do so had turned out to be chaotic. They behaved as if random, without having any actual random elements. A deterministic equation, which is what the logistic is, is one in which all variables are exactly specified. Deterministic growth rates are exact: 0.1% per year, for example, not on average 0.1% with a standard deviation of 0.02%. That kind of random variation would make the equation no longer deterministic, but stochastic. It would not, however, make it chaotic. Chaos in mathematics means exactly what it does in English: utter confusion. The kind of randomness represented by stochastic equations can be dealt with mathematically to yield predictive results. Meaning can still be found through all of the variation that random events cause. Sir Robert had worked all of that out, too. The only difficulty was when randomness became so great as to swamp out all of the nice algebra. In that case, chaos begot chaos. What Sir Robert had now discovered was something qualitatively different.

May did not inject any randomness into the logistic equation when he discovered the chaotic behavior. All he did was change a differential equation to a difference equation. Difference equations recognize that matter can come in discrete quantities. They can describe the reproduction of rabbits, for example. Litters come in whole numbers, even though averages calculated from them can be continuous.

A difference equation is a perfect descriptor for a seasonally breeding population. For example, the number of Canada Geese found this year around my little neighborhood beaver pond as they return north would depend on how many were born the previous year. In the simplest mathematical case, only those newborns would return. The adults that sired them would die in the interval between breeding seasons. (The reality is that it is more likely for the newborns to perish before their return than the adults.) This single generation situation is exactly the kind of simple model resulting in chaotic behavior. (Letting the adults survive to some degree therefore introduces more realism, but also more mathematical complications.) Interestingly, differential equations, the calculus of Newton and Leibniz, do not exhibit chaotic behavior except in highly specific situations not normally found in nature.

What May had found was chaos out of order. It was very much of interest mathematically. The discovery of chaos theory has revolutionized the way we approach dynamical systems. May's analysis of the population equations used by ecologists did much to advance the revolution.

What it all ends up meaning is that this elegant mathematical analysis leaves ecosystems being about as predictable as the weather. Although May himself collaborated on a paper to show that the parameters necessary for chaotic behavior are rarely found in nature, at least one set of well-documented data, that of lynx populations in Canada as reported through fur-trapping statistics, has been shown to be consistent with chaotic behavior. In May's analysis of the logistic equation, chaotic behavior commenced when r was greater than 2.692, a growth rate of 269%. Few populations in nature could be expected to have such explosive growth rates. Or could they? Gause's *Paramecium* increased at a rate of 230.9%, but this was per day. What might be the proper time interval for *Paramecium* in a pond? Might not an aphid landing on a pea plant and producing offspring that are essentially already pregnant reach and exceed tripling its numbers every generation (a week or so)? What might its rate of increase be in a difference equation? What might it be if we considered its potential for increase over the course of a growing season, rather than a generation? A troubling thought to those who would model aphid numbers based on the fine differential equations of theoretical population biologists.

Actually, the idea that ecosystems can be just as chaotic as the weather can be looked at in two ways. Take James Gleick's fanciful example in his book, *Chaos*, of the butterfly flapping its wings in the Amazon basin, setting up vortices that grow and interact with other vortices and move until they become the current hurricane that descends on Miami. Analogous behavior in ecosystems would give us bleak prospects for safely making even the slightest of changes in an ecosystem without possibly dire consequences. Had Gleick's butterfly been eaten by a bird before taking flight, and had it been a foul-tasting Monarch, Miami may have been spared a hurricane, but the indigestion felt by the butterfly-eating bird might have led to an instability in that trophic level that might have spread and intensified until Salt Lake City was once more faced with a plague of locusts. Try to manage that ecology!

There is another way of looking at weather and ecosystems. Given the chaotic nature of weather, meteorologists have been remarkably successful in predicting changes in the weather. As a boy growing up in New Jersey, for the Weather Service to get the next day's weather right more than half the time was an accomplishment. Had I grown up where I am now in Vermont, I would probably have ignored weather forecasts entirely, preferring the surprise of meeting up with one of the large range of possibilities available in this part of

New England over the disappointment of a wrong prediction. Today, even in Vermont, the next day's weather is predicted with better than the 95% accuracy required of statistical tests for significance in science. Two and three day predictions are accurate enough to pay heed to, and there is talk of preparing accurate two-week predictions in the near future.

What is seemingly impossible to obtain with elegant mathematics apparently can be predicted through brute force. The increase in forecasting accuracy in the face of underlying chaos comes not from the discovery and use of any fundamental principles, but from more and better data, better instruments to collect the data, and better computers to do the necessary number crunching. Accurately knowing in detail what a weather system is doing, the direction it is heading, along with the same information for other weather systems, plus data from years past, can make prediction fairly accurate, if not exactly simple. Perhaps ecology's search for those underlying or unifying principles, the simple clockwork mechanisms that would make ecology like physics, was doomed from the beginning. The almost perverse nature of ecological dynamics place it into the category of phenomena, like the weather, that are relatively well understood in their basic processes, but barely decipherable in their ensemble behaviors. Yet given that every parameter that needs to be described about an ecosystem is quite measurable with today's instrumentation, perhaps ecosystems or entire biomes might, like weather patterns, be modeled in ways so as to make predictions of interest to management. That the time scale of predictions for ecosystems would be somewhat longer than for weather would be an advantage. Unlike weather, there would be time for action to be taken to alter a predicted change, if desired. Isn't this exactly what we all want?

Actually predictive models of ecosystems already exist and have been used for management. The JABOWA forest succession models have been quite accurate in simulating forest ecosystem changes in time (succession) and space. In general, most of these tree-by-tree succession forest models suggest that the very simple steps that are already in use for managing forests can lead to very predictable results.

"The only simple treatments that will have dramatic effects are those practices of seeding, weeding and patch-cutting that enlightened foresters have recommended for centuries," one ecologist has concluded based on his models about ways to manage successional changes in forests.

The JABOWA model is continuously being upgraded, resulting in more accurate predictions. (The acronym is a combination of the initials of its inventors, Janak, Botkin, and Wallis.) JABOWA 4 is today available for purchase.

Interestingly, when JABOWA's main author in terms of the computer commands for the program was at Cornell for a visiting lecture, graduate students such as myself (shamefully) failed to seek him out, categorizing him as a "tool." Tool was a term we used for ecologists who were mainly concerned with methods, such as measurements or computer programs, rather than with the ideas that would generate the need to gather data or simulate nature by computer. The ideas were what attracted us to ecology. Robert MacArthur-style ideas. MacArthur was no tool.

Why aren't models such as JABOWA put to more use? Actually, they are. This is the direction in which ecological management should go. My own model of the Salton Sea in California successfully predicted the demise of its fish populations. The managers (such as they were, for that unfortunate, unnatural body of water) chose not to—or were unable to—heed the model's warnings.

At a relatively recent American Association of Science Meeting, an international team of scientists unveiled a compartmentalized ecosystem computer model for the North Atlantic that, given the collection of adequate data, should do for that fishery what Doppler radar has done for weather forecasting. They expected excitement at a press conference they scheduled. The press never picked it up. The public cared little about it. (Fishermen did, however.)

That kind of detailed, predictive model is often just trouble for the scientists involved. Only part of the trouble is when the predictions turn out to be inaccurate, especially after policy has been based on them. More trouble from such models around the time of the first Earth Day came from a certain group of ecologists and theorists working on ecology looking down on them. The group was easily identifiable. It consisted of those ecologists who had published papers with Robert MacArthur. (As always, there was also a group of wannabes that were unable to do so due to his untimely death, but still claimed descent.)

Robert May was once kind enough to look over the Salton Sea model that I had developed at UCLA. It modeled the various fish populations and their interactions, of course. It also included a baseline of salinity that was predicted to change in a variety of ways due to a number of management options. These had to do with agricultural irrigation, the lining of Colorado Aqueduct canals, local geothermal development, and a rather bizarre solar pond proposed by an Israeli company that was to serve to produce energy, desalinate the Sea, and be a wildlife refuge. Each, of course, had an impact on the salinity of the sea, either directly, as with the solar pond, or indirectly by their effects on flow into the sea. (The solar pond, by the way, turned out to be too good to be true and was never built.)

Struggling with mathematical choices, I scoured the literature of fisheries models. I was not at all surprised that May had the final word on those, too, but some of those words were hidden in rather obscure journals. By mail to his base then in Princeton, I had asked May for a reprint of one of his papers. It was not easily to be found, even at the UCLA libraries. He responded by inviting himself to my basement offices in UCLA's Institute of Geophysics.

This was exciting stuff. For that era, it was a bit like having Christian Barnard consult on an upcoming heart operation.

Fred Turner, a desert herpetologist, with whom I sometimes took refuge from my academic travails in a science that was being invented all the while I was trying to find points for contribution to it. (It was environmental science from an ecological perspective. It managed to eventually invent itself without the benefit of UCLA.) Fred had found a niche doing what appeared to be highly practical work for the tiny little national energy laboratory at UCLA, called simply Warren Hall. Other national energy labs had names such as Oak Ridge National Laboratory or Lawrence-Livermore Laboratory. Warren Hall, also devoted to research related to mostly nuclear energy, may also have had as prestigious a name, but no one seemed to have been aware of it.

UCLA biologists had actually been the first to study the ecological distribution of radioactive fallout. Two years after the Trinity test, they were on site, funded by the Atomic Energy Commission. What Fred did was study the desert reptiles he had a long-standing interest in, somehow relating his work to radiation studies. (Much of his data came from the nuclear weapons testing site in Nevada.)

At least a generation older than me, Fred seemed to me an island of stability in the turbulent seas around me. It must have been at one of our lunches together, which I dearly missed on leaving UCLA, that I mentioned Bob May's upcoming visit.

Fred instantly invited himself to our meeting, except, instead of contributing, all he wanted to do was sit by in a corner and watch the two mathematical minds at work. (Kindly old Fred, he thought one of the mathematical minds in the room was mine.) We met in the basement of the Geology Building, as it was informally called. It was to my office that Sir Robert had come, rather than I to his.

It was at some time during the 1981-2 academic year. Robert May had not yet been knighted or given a peerage. Still, he had a reputation, but I was pleased to learn that it was more imposing than his presence. A clean-shaven, slight man with thinning hair that met high on his forehead in a widow's peak, his appearance was anything but that of a woolly-headed intellectual. (Or radical or non-conformist, of which there had been some in ecology then.) In dress and physical appearance, he seemed more suited to a boardroom than a field site or the front of an equation-filled chalkboard. His suit was dark and perfectly tailored. His shoes were shined. His manner, perfect. Yet there was not a hint of snobbishness to it. He was as gracious as we could wish all of our heroes to be.

I don't remember his accent then except to the extent that it was vaguely British enough to garner whatever cachet that was being accorded at the time. (In ecology, it may have been much.) May had actually been born, raised, and educated entirely in Australia. Like other personalities in this story, May was not then—and never had been—an ecologist. Like myself, he had started out in chemical engineering, except that his PhD was in theoretical physics, and he obtained it at a much earlier age (23) than I earned mine. Theoretical physics then, as now, attracted the brightest of the bright. What theoretical physicists were working out with their exotic instruments and mathematics was no less than the meaning of the universe. Then, as now, the field occasionally attracted far more bright people than it could absorb. This necessitated that theoretical physicists turn their math and computing skills to other areas of science that seemed to need it. Ecology was one such area.

As May tells it, "My transition from theoretical physics to ecology was the result of pure accident. Around 1970, I became involved as one of the founders of the movement for 'Social Responsibility in Science' in Australia. In the course of understanding what I was being responsible about, I encountered Ken Watt's book on *Ecology and Resource Management*, and its exposition of the stability/complexity debate of that era."

May could well have gone on in physics, although his success in that subject would not have been as great as it was in ecology. May admits to it, concluding about his career shift, "In short, it was an accident, but from my point of view a happy one."

He found that parts of the conventional wisdom having to do with stability and complexity were simply wrong and began to redress the failings of ecologists by applying the rather conventional mathematics available to him as a physicist. Much of the math was developed by the Russian AA Liapunov before the turn of the last century. May turned his attention to what happened to stability when Lotka-Volterra type equations were used to describe multi-species systems, publishing his ideas first in the journal, *Mathematical Biosciences*, in 1971 and eventually in his 1973 book.

Having been just recently introduced to ecology, May found himself running into a number of ecologists while he was visiting physics institutes on sabbatical leave from the University of Sydney.

"Most importantly," May remembers, "Robert MacArthur while at the Institute for Advanced Study in Princeton."

The Institute for Advanced Study was once the home of Albert Einstein and, briefly, John Forbes Nash, he of the beautiful mind. MacArthur, a mathematician, himself, should have needed no collaboration with May, but he must have recognized a kindred spirit, for the two did work together on ecological problems, and Lord Robert was soon on his way to knighthood and the presidency of Britain's Royal Society. (One of the most illustrious former presidents of the Society had been May's fellow physicist, Isaac Newton.) May so dazzled the Princetonians that he was lured away from Australia back to Princeton where he spent much of the time covered in this book, until finally settling at Oxford.

May understood the model I had struggled with from just the slightest glance at it. According to him, the results my model was giving were entirely in keeping with the form of the mathematics I had chosen. The numbers were nothing more than could be expected. I had had reams and reams more of computer printouts of predicted fish population changes ready to show him. He waived those aside, however. All he needed was a quick look at my equations.

Of course, he said, that equilibrium is unstable, given those parameters. There is nothing else that can happen with that system other than what your simulations show.

A simple translation of Lord Robert's words is that the model was predictable and boring, although Lord Robert was much too tactful to say anything to that effect.

He did give me some very clear advice, though. I should really abandon the rather "explicit" mathematics I was working on, he had advised, in favor of "strategic" models.

That advice reflected the ideas of Richard Levins, another crony of Robert MacArthur. Levins thought all models must strive for precision, realism, and generality, but at best could only settle for two of those parameters. He divided all mathematical models into three types: (1) those that try to precisely model a real system, but are not very general, (2) those that try to model general systems precisely, but with few realistic details, and (3) those that try to model real systems in a general way, sacrificing any precision in their predictions. It was the third type, because of its generality, that Levins favored. So did Robert MacArthur in making sense of the ecology of his warblers.

By precision, Levins meant that the model made exact, measurable predictions. My Salton Sea Model was precise. Realism and generality are not as easily explained, but refer to opposite poles on a continuum of abstraction. Less abstract, more real. More abstract, more general. My Salton Sea model was clearly not abstract, but how real it was, I still wonder. Of the possible combinations of precision, realism, and generality, Levins found that precision and generality were a logically impossible combination. He decried models combining generality and precision that are sometimes "proposed by those entering biology by way of physics." (Not, we know, Sir Robert, with whom Levins also collaborated.) The only good models, therefore, were those that sacrificed precision to realism and generality. Such as Levins's own.

Sir Robert could see that my model had not an ounce of generality to it.

It was not necessary, Sir Robert further advised me, to exactly specify all of the interactions in an ecosystem. Rather, it would be better to try to tease out generalities from models that can get at the idea of an ecosystem, if not its details. (I think that, actually, he never used the term ecosystem in my presence. Population models were the words he probably used instead.)

Like what Jonathan Roughgarden was doing, he added, giving me an example to emulate. May must have been passing through Los Angeles on his way from Stanford. Although

Roughgarden is about to become a figure in this story, I knew nothing of Roughgarden's work at the time.

I could tell, though, that May was telling me in his own way not to be a tool like Dan Botkin, although, of course, he did not use that term, nor any name. Botkin's JABOWA model also sacrificed generality for realism and precision. Like me, although with more sophistication, he actually was trying to make specific predictions about a particular ecosystem. This was apparently a no-no at that time.

I need to add here that it is only out of narrative convenience that I characterize Dan Botkin as an ecological Bogeyman of sorts. I have the greatest respect for Botkin. His environmental science text is first-rate and his book on the state of ecology at the end of the 20th century, *Discordant Harmonies*, is one of the most sensible things anyone yet has written about where we are going with the modeling of ecosystems. And his model worked.

Nonetheless the advice May gave me was probably excellent career advice. Had I managed to work out some of those strategic models he had pointed me toward, I might have wound up at Princeton, too.

Nutches in Nitches, Atoms in Apples

By now, the reader may be complaining that I had promised a war, but all I am delivering is some characters and some old ideas. It is not Turgenev who seems to be my model, but Tolstoy. And I still seem to be bogged down somewhere in the middle of the "peace" part of *War and Peace*. Nothing but long lists of characters with names and ideas that are difficult to keep straight. When will I get to the war? And what does all of this have to do with practical ecology?

The war is coming. Like 19th century Europe, a war is always on ecology's horizon. And like the prewar countries in Tolstoy's novel, the ecologists were negotiating with each other and gathering together in camps. At stake was prestige. Prestige in a field of science led to research money and research money led to more prestige which—you get the idea. What the Odums and Oak Ridge had shown was that there was lots of money to be had in ecological research. What Robert MacArthur had pointed to was that his brand of theoretical ecology, capable as it was of attracting people like Robert May to it, also had all of the rigor of the physical sciences to it. It too was capable of attracting funding.

Something else was at stake, too. It had to do with the credibility of the quiet science that had been brought to the forefront on the first Earth Day. And that had to do with the particular way that mathematical rigor was to be brought into it.

Mathematical models such as JABOWA or the models turned out by Oak Ridge researchers were intended to be useful for management decisions. In their explicitness, however, the fault May had found in my efforts, they were somehow useless in advancing the knowledge of ecology. Too complicated for use in any situation other than what they were specifically modeling, explicit models, according to one camp of ecologists, could reveal nothing about the functioning of ecosystems.

Unfortunately, the population ecologists could do no better. Unable to predict the exact next state of any real ecological system with their general models, the theorists set about in fierce pursuit of questions they could answer. They were mostly the disciples of Robert MacArthur. Like Richard Levins, smart enough to know that most human minds have trouble truly focusing on more than three things, I group those questions for the reader into three types, also: (1) Why are there so many species? (2) Does spatial structure change the conclusions that can be made from mathematical models? And (3) Can diverse ecosystems be shown experimentally to be somehow preferable to less diverse ones? All three questions are represented by major research efforts. Much of it is mathematical.

The last question is, of course, significant for environmental management. However, science being conducted by people, the first question, seemingly a bit of trivia, garnered the most attention. This question about diversity became the subject of what is called niche theory and niche partitioning and species packing and a variety of other terms having to do with the coexistence of species.

Charles Elton introduced the niche concept in *Animal Ecology* in 1927 as a description of what a species does in an ecosystem, its profession. Earlier, American ecologist Joseph Grinnel had a concept of the niche as a location. This was the habitat niche. The niche was now not just a profession, but a locale for its practice. GE Hutchinson put it into the conceptual mathematical form that still drives current research. Curiously, he presented his milestone idea in an off-handed way in a paper entitled "Concluding Remarks," presented at a symposium, a formal meeting of scientists sharing common interests, rather than publishing it in a scientific journal. Many of ecology's most important ideas, it seems, find their way into the science not by publication in the most widely read and prestigious journals, but as the text of meeting presentations. Curiously, also, what Hutchinson had presented was not a mathematical equation, but a mathematical idea. There was no equation, just a hint of what one might look like, because Hutchinson's multi-dimensional niche could never be exactly specified, nor could data be obtained to fully describe it—there were way too many parameters involved.

Robert May and Robert MacArthur, however, did manage to find a way Hutchinson's niche could be put into mathematical form. They did it using gross simplification. Only parts of the niche were looked at. The rest was glossed over or assumed to be irrelevant. May and MacArthur looked at only one dimension of a niche. Had they been studying humans, for example, they might have limited their analysis to height alone, ignoring all other differences. Their math supported an idea of Hutchinson's having to do with size ratios. Similar species living together seemed to vary in size in regular ways. The magic ratio turned out to be 1.3.

The most sense of the niche, though, may have been made by the famous non-ecologist Theodor Geisel, better known as Dr Seuss. His version, from *On Beyond the Zebra*, is as follows:

> And NUH is the letter I use to spell Nutches
> Who live in small caves, known as Nitches, for hutches.
> These Nutches have troubles, the biggest of which is
> The fact that there are more Nutches than Nitches.
> Each Nutch in a Nitch knows that some other Nutch
> Would like to move into his Nitch very much.
> So each Nutch in a Nitch has to watch that small Nitch
> Or Nutches who haven't got Nitches will snitch.

Dr Seuss got the idea without the math. Community ecology or niche studies (the terms are almost synonymous) are very relevant to the issue of biodiversity. This is ecology applied to global needs in a broad way. Community studies might be able to get a handle on the minimum number of species, the minimum diversity, necessary for healthy biological communities. How many Nitches do Nutches need?

Some ecologists turned away from community studies framed in a practical way, however. They chose instead to adopt a philosophically based stance on preservation issues, only, or, for a small few, no stance at all. Others who did try to make their science useful, found themselves delving very deeply into what it was all about. They began to confront its basis and the basis of science in general. Unlike other scientists, ecologists now found themselves having to be not just mathematicians and modelers or data collectors and analysts, they also had to become philosophers, too. This is an unnerving situation for a science trying to claw its way up to legitimacy. Physics and chemistry got its philosophical issues out of the way long before ecology was born. Physicists and chemists may have doubts about the adequacy of instrumentation or data analysis techniques, but not about the basic premises of their science. What does it say about ecology as a science when modes of thought in the discipline are still subject to question?

Some ecologists were not at all troubled by that question. Whatever ecologists had to do to glean the information they were looking for was fine with Jared Diamond, for example.

"In fact," according to him, "different sciences face different methodological problems, and the successes of physics and molecular biology were made possible by those fields having found appropriate solutions to their particular problems. Ecology must cope with its own distinctive problems. Insofar as ecologists can profitably be guided by other sciences, the best models would be sciences facing problems similar to those of ecology. Molecular biology and some areas of physics are methodologically at opposite poles from ecology and offer poor guidance. Better models are astronomy, geology, medical science, neuroethology, or vulcanology, which share with ecology practical limitations on experimental intervention, concern with historical or evolutionary problems, and conditionality of results. In those fields the recognition that laboratory studies, field studies, and the comparative approach have a place is no longer controversial, as it still is in ecology."

His statement is an excellent snapshot of the situation in ecology and the other sciences mentioned. What is not said, however, is that some of those disciplines have not exactly escaped controversy over whether they are legitimate science. Also, are the methods of astronomy, that part which speculates on the origin of the universe, in particular, relevant to ecology? Conveniently for Diamond, he did not have to answer such embarrassing questions, because post-modernist philosophers began deconstructing all of science at about that time.

Another recap may be in order here. Keep in mind that the superorganism, a concept that, if true, would have been of obvious practical use, died as a result of competition with the more glamorous and more traditional mode of thought in ecology, the evolutionary ecology of Robert MacArthur. Through its mathematics, the new ecology also promised to provide the tools necessary to guide environmental management. At the heart of that new ecology was the question of why there are particular numbers of species in particular places. To ecosystem scientists, it was a question that was totally irrelevant. They aggregated their species into trophic levels. In various transformations, it is still a question whose need for an answer ecologists are using to justify research programs, some of considerable scope. Before that question could be answered, however, ecologists had to face up to questions that essentially indicted the legitimacy of ecology as a science. It was not a pleasant set of events for anyone.

I taught my science students (especially secondary school ones, for whom simplification is very necessary) a somewhat Cartesian method of science. Science is just one of a number of methods of acquiring knowledge, and the basic premises of René Descartes is just one of its underpinnings. In his 1637 work, *Discourse on Method*, and later writings, Descartes established skepticism as one of the underpinnings of what he, Galileo, and Newton launched as modern science. (I suppose Englishman Francis Bacon, its greatest promoter, needs to be noted as a founder of scientific method, even though he never managed to create any part of any science.) Descartes' "*Cogito ergo sum*" ("I think, therefore I am," although what he had actually written was "*Je pense donc je suis*," being French) was reached by a process now known as "Cartesian doubt" in which nothing but the objective observation of one's existence can stand up to credibility. Newton and those on whose shoulders he stood created a science that is founded solely upon observations. What cannot be observed, cannot be studied by science. Given that Descartes was also a mathematician and geometer (the Cartesian coordinates taught in high school math courses, for example, are his invention) the best observations to advance that science are quantitative ones: numbers, measurements.

This idea of measurement as the foundation of thinking is still a very strongly imposed bias in science. It has been stated best by Lord Kelvin, the British physicist known to science students by the Kelvin scale of temperatures.

"When you can measure what you are speaking about," he is quoted in lectures late in his career, some two and one-half centuries after Descartes, "and express it in numbers, you know something about it; but when you cannot measure it, when you cannot express it in numbers, your knowledge is of a meager and unsatisfactory kind: it may be the beginning of knowledge, but you have scarcely, in your thoughts, advanced to the stage of *science*."

Algebra being the way of representing the relationship of many measurements, just as geometry and trigonometry are ways of relating the sides and angles of many shapes, Descartes argued that the best way to express those measurements on which science was to be based was through mathematical relationships, algebraic and geometric, that could be tested. This is the familiar hypothesis-testing-through-experiment part of the Scientific Method. What Francis Bacon did contribute to the methods of science was what is called the inductive method, in which measurements are gathered until a pattern is uncovered. That pattern, given mathematical form, then becomes a Cartesian hypothesis, or prediction, capable of being tested. Skepticism over the results of one's experimental conclusion, although not formally a part of science as conceived by Descartes, is now as important as experiment itself. That skepticism is a direct consequence of Cartesian doubt.

Measurement (observation), mathematical prediction (hypothesis), test (experiment), skepticism (attempts at disproof), the method works. It has given us Newton's Laws, MTV, laser-guided missiles, and the digital revolution. Twentieth-century philosopher Alfred North Whitehead saw the application of science as a "triumphant" activity.

"If success be a guarantee of truth," Whitehead wrote, "no other system of thought has enjoyed a tithe of such success since mankind started on its job of thinking. Within three hundred years it has transformed human life, in its intimate thoughts, its technologies, its social behaviour, and its ambitions."

Trust Whitehead to have also found a flaw in the recipe for science. It has to do with the role of mathematics, which is at once so vital to science and so capable of leading it astray.

"Mathematics can tell you the consequences of your beliefs," Whitehead wrote a few pages later, still on the subject of science in general. "For example, if your apple is composed of a finite number of atoms, mathematics will tell you that the number is odd or even. But you must not ask mathematics to provide you with the apple, the atoms, and the finiteness of their number. There is no valid inference from mere possibility to matter of fact, or, in other words, from mere mathematics to concrete nature."

Some forty years after these words were written, what had become known as theoretical ecology stood accused of not only conjuring up an apple and its atoms, but also of making policy recommendations based on that spurious apple. It did so by ignoring the rest of that recipe for science, the testing of predictions and skepticism toward new results.

It started with some parts of ecology, including Darwin's theory of natural selection and all of the ecological studies founded upon it, standing accused in the 1970s of being non-scientific because they were not testable. This was serious business, for from its inception, Darwin's theory has faced attacks, but usually of a theological nature. Evolutionary scientists have always had to fend these off, but the task becomes uncomfortably more difficult when the scientific basis of Darwinism is questioned.

It was another modern philosopher, Karl Popper, who was invoked by ecology's critics. A valid scientific theory, according to Popper, was not just one that fit the observations at hand, but had to be constructed so as to be "testable and falsifiable." Popper particularly warned about explanations that were tautological.

A tautology is a statement that is true by its logical form alone, as in a circular definition. A trivial example of a tautology is the statement, "All living things are either bees or not bees." If science was a process of testing of predictions, how does one test that little gem? Any possible observations that could be made for that proposition can only serve to support it. It is testable then, but not falsifiable.

There was much *sturm und drang* over Popper's contention that natural selection was a tautological concept. Although it appears to be more customary to trace Popper's ideas back to his 1934, *Logic of Scientific Discovery*, his ideas about knowledge, like all things, also evolved. Here is what he had to say in 1979 about Darwinism at the height of the controversy his earlier ideas had started.

"A central problem of evolutionary theory is the following: according to this theory, animals which are not well adapted to their changing environment perish; consequently those which survive (up to a certain moment) must be well adapted," he wrote. "This formula is little short of tautological, because 'for the moment well adapted' means much the same as 'has those qualities which made it survive so far.' In other words, a considerable part of Darwinism is not of the nature of an empirical theory, but is a *logical truism*."

Popper had no quarrel with Darwin's scientific viewpoint nor with his theory of evolution. In fact, Popper used evolutionary theory to solve Descartes body-mind problem. Popper allowed that Descartes could assume his existence to be true (something Descartes could not prove), because natural selection would have eliminated organs of sense that led their possessors astray in interpreting reality. Still, a tautology is a tautology, while scientific ideas must be falsifiable through testing.

Popper's ideas entered ecology as part of an earlier controversy. Among other things, it had to do (of course!) with density-dependent population regulation. The arguments made by ecologists Nelson G Hairston, Frederick E Smith, and Lawrence B Slobodkin that set it off again have been gone over in so much detail (even to the present day) since their publication in a paper in 1960 that it has been given an acronym, HSS. Its general points are echoed throughout this story. Critics took issue with the deductively obtained hypotheses in HSS. They were not testable. Citing Popper, one wrote, "a widely accepted criterion of a scientific hypothesis is its falsifiability." Testing to show that something might be correct was not enough. The test should be such that the data gathered should also be capable of showing an idea to be incorrect. Popper was in the ecology literature. The year was 1966.

Testing to falsify versus testing to confirm may sound like an argument over angels dancing on the head of a pin, but it is not. An example from the science of geology might make this issue more clear: the now well-known and well-accepted theory of continental drift. Data can be obtained to this day in support of either Wegner's hypothesis of moving continents or alternative explanations. How can that be? It depends on how that data are put together with the desired explanation. Geologists have even come up with a term, "geopoetry," for the process of putting together logical stories based on facts to create an explanation. In his book, *Annals of the Former World*, John MacPhee gives a very good exposition of how "geopoetical" explanations can be produced from plate tectonics theories for observations that may or may not have come about in that way. Ecologists need to borrow that term. The critic could have accused HSS of being "ecopoetical." Their facts may have been true, but their explanation of them need not be.

Ten years later, falsifiability in ecology was raised once more. The arguments now fell on more fertile ground. Young Canadian limnologist, Robert Henry Peters, made them the entire contents of a paper he published. They were ideas he had developed while on a post-doctoral fellowship that he had chosen to spend in Italy.

"I knew the US and UK literature," he later wrote to a friend, "and was not impressed. I posited that different linguistic traditions might have different ways of doing science and decided to learn them."

Knowing very little Italian, he had set out to let the language help him to understand his science. His paper summarizing his ideas about how science should be, yet ecology was not, was printed as the lead paper of the 1976 volume of *The American Naturalist*, which at that time was serving as an intersection point between ecological and evolutionary ideas. Like Samarkand in the Middle Ages, it was a marketplace of ideas from various disciplines, schools, and sects.

"In my research," Peters later wrote to one of his Italian friends, "I have become obsessed with the inability of ecology to provide more than the lore of natural history that I had enjoyed as a child, or even to recognize its deficiencies."

This is the ecologist as little old lady in tennis shoes all over again, the butterfly collectors disparaged by Stanley Auerbach. Peters had apparently expected scientists.

Braver than many at that time, Peters brought his doubts and ideas to the Samarkand of ecology. Other events in other science journals helped to raise the level of concern over his remarks.

The journal, *Science,* had just declared in a news story that ecology was beginning to be a predictive science.

"Ecologists have long been known for their exhaustive field studies in which they describe and catalog species in a given area and search for patterns in the interactions among the species. Although some ecologists still undertake such studies, many are analyzing systems with theoretical models and are using descriptive studies to confirm and extend their models," the news story noted.

The theoreticians identified were Robert MacArthur, who was just recently deceased, Robert May, EO Wilson, and Jared Diamond. There was also a suitable supporting cast mentioned.

"The next two decades," the story concluded, "will see the completion of a revolution in the study of ecosystems."

This statement struck some in ecology as odd in two ways. One is that the Odums' beloved ecosystem concept was never once mentioned in the story, nor were the theoreticians that were reported on ecosystem ecologists. The models of the "new ecology" had nothing at all to do with energy or nutrient flows. The other oddity is that it seemed to indicate that there would soon be only one way of doing ecology. Those doing the exhaustive field studies of the lead paragraph seemed to be getting their walking papers in contrast.

On the heels of the news article in *Science,* two highly respected ecologists called for an end to what they saw was intellectual censorship in their science. Mathematical ecologists (their term for the theoreticians), they claimed, were suppressing the publications of dissenting papers, interfering with the graduate students of others, and maintaining a clique to exclude outside criticism and (the implication went) outsiders in general. They published their complaint in *Ecology,* which had "not yet been seriously affected," according to them.

It was not too long after that *The American Naturalist* (wags had taken to calling it *The American Unnaturalist* at this time due to species of mathematical forms becoming more common than real species on its pages) published Peters' paper.

"I have long had a distrust of contemporary academic ecology," Peters later wrote in explaining his motivation. "I see it as a bog of preconception and logical argument which is largely unable to address the grave environmental problems that confront humanity because it fails as an effective science."

When I began to look at his work carefully, especially his 1991 book, *A Critique for Ecology*, I began to wonder why I had not gotten out of ecology—academic or applied—sooner. It was a gloomy picture that he painted of the science post-Earth Day. It must have been hidden from view in my enclave at Cornell, and I must have refused to see it on leaving Cornell. Reading Peters's book showed me every worm in the can he had opened. Ecology, in his estimation, was a soft, weak science facing dropping course enrollments and cutbacks in staffing. But his most serious accusation was that it was incapable of solving the environmental problems that were its practical justification. Worse still, the science seemed to have always been flawed in the many logical ways Peters documented. Why did I ever go over to it from my engineering studies?

Concerns that Peters had about evolutionary theory and natural selection not being falsifiable were actually allayed fairly quickly by some of the giants of the field. Their response was that it is certainly true that it is easy to frame natural selection in terms that are circular.

"Which are the fittest?"

"Those that survive."

"But which survive?"

"Why the fittest, of course," is a conversation that sums up that argument.

Nonetheless, even if it can be stated in that circular form, the theory does provide powerful explanations and did yield falsifiable predictions. In fact, it was pointed out that Peters' concerns were not new. Evolutionary biologists had become practiced in brushing aside natural selection's circularity.

However, Peters also took on competition theory, Eugene Odum's ecosystem attributes, and theories of diversity. Even given the contentiousness among supporters of those theories, they were still the crown jewels of ecology. In his 1976 paper, he had stepped into the bog of his description at a time when it was about to be roiled up so that all of the smelly stuff at the bottom was brought to the surface.

Peters, up in Canada, was not the only one mulling over the philosophical basis of his science. Farther south, in Tallahassee, another young ecologist was pondering his chosen field.

A Doctor for a Sick Science

Dan Simberloff did not start out as a biologist. He was a math major at Harvard in the early Sixties (pre-Earth Day, but post-*Silent Spring*) when he began to wonder about career choices.

"Towards the end of my junior year," Simberloff recalls, "I began to realize that I did not want to be a mathematician for the rest of my life. I was bemoaning my situation at dinner with classmates, and one of them noted that I seemed to be liking a nonmajors bio course I was taking (with George Wald) an awful lot. I said I loved it. (Actually I'd always loved insects and herps and birds, even as a little boy, but I'd never thought of the possibility of a career in biology.) He told me that bio at Harvard was such that you could get a major in it without any real bio courses as long as you had a lot of other science courses. (I did.) I checked it out the next day."

Wald was soon to be a Nobel Prize winner. His enthusiasm was infectious. (Only at Harvard might you never even see some of the illustrious scholars the institution employs, yet have your freshman science course taught by a Nobel Prize winner.)

Inspired by Wald, Simberloff wound up seeking advice from Frank Carpenter, who was EO Wilson's mentor at the time.

"Carpenter was very encouraging," Simberloff continues. "He said of course, I could get a bio major even at this late date by just taking a couple of bio courses and organic chemistry (which I did in the summer), and that many areas of ecology and evolution were exciting fields where people with math backgrounds could do interesting work and were badly needed."

Having taken so few of Harvard's offerings in biology as an undergraduate became the basis for advice for staying on at Harvard for graduate school. Wilson, then involved with Robert MacArthur in their theory of island biogeography, became his advisor. For his thesis, Simberloff tested the predictions of the theory with Wilson. It probably suited his mathematical bent to do so. Wilson and Simberloff's experiment won a prize from the Ecological Society of America. It also may have made him wonder about other theories in ecology. How were they tested?

Ideas about the nature of scientific theories that were very similar to those of Peters began to incubate in Simberloff's mind. The circularity of the idea of a "balance of nature" was one of them.

An address to the British Ecological Society a few years after Earth Day by its president introduced the ideas of Popper and Kuhn to Simberloff and set him to thinking about methodology in his science.

Amyan MacFadyen, whose address was published in 1975 in the *Journal of Animal Ecology*, admitted that "the assumptions, methods and activities of ecologists have from time to time been subjected to some quite strong criticism and I do not think these criticisms should be quietly forgotten in the way that seems to happen so often."

He continued, "There are those who argue that ecology, like human history, is concerned with unique events and that these are not supposed to be open to the 'scientific method.' Is this true and does 'scientific method' referred to in this context differ from its meaning in other sciences?"

The lack of introspection in ecology over its methods troubled him. Ecologists seemed to have developed too thick a skin. Criticism of their science was just so much water rolling off a duck's back.

"Many colleagues have little time for philosophy of science," MacFadyen wrote, "but I believe this to be an arrogant attitude in a subject which has been split by semantic schisms and methodological muddles, an attitude which we cannot afford."

He also found that Kuhn's contention that a science in its early stages lacks paradigms, causing every publication to start from first principles, "to be uncomfortably close to home."

Anyone who could gather an audience of an editor and sympathetic reviewers could set about defining basic ecological principles. Although MacFadyen identified four outstanding men (one of whom was Hutchinson) who successfully used the hypothetico-deductive system, he feared that their example encouraged "lesser mortals" to overly hasty generalizations.

"These usually come to the scientist in the form of 'bright ideas,'" he wrote, "and they are an essential and exciting element in scientific work. But they can become destructive tyrants also. Apart from psychological difficulties deriving from protective attitudes to people's 'own' bright ideas, they set limits to the kind of questions which can be asked. There are plenty of examples from other fields of biology of the tyranny of bright ideas."

"I began to think," recalls Simberloff, "is it really this bad? And concluded that it was. I didn't see at that time a nefarious plot in all this."

Ecology was a "sick science, awash with all manner of untested (and often untestable) models, most claiming to be heuristic" is what Simberloff eventually concluded about much of the theoretical work in his field. It was right around Earth Day 1980.

The models were someone's bright ideas. Simberloff's use of "heuristic" was sarcastic. The dictionary definition is "providing aid or direction in the solution of a problem but otherwise unjustified or incapable of justification." A heuristic model is very much what Si Levin had thought of the logistic equation. It was what Robert May wanted me to make of the Salton Sea. (Actually, the part of the model that had to do with the salt content of the sea drove the biology in ways that could have been predicted without all the math that bored May. Maybe that was the heuristic part.)

The mathematical models of the day were of little, if any, use in the task of environmental impact analysis that NEPA had set before the science. These were the mathematical models mostly of the ecosystem theorists, but also those of MacArthurites, some of whom were beginning to tout their work as practical.

Oak Ridge, in particular, trumpeted the usefulness of ecosystem modeling. Stanley Auerbach was the lead trumpet. Its scientists were surprised, therefore, when they worked on an impact assessment for a proposed nuclear power plant on the Hudson River. The surprise was that they found themselves studying the autecology of striped bass, rather than the ecosystem in which it was found. The questions that needed answering about changes caused by the power plant required detailed knowledge about striped bass populations that could be obtained with little concern for the place the fish had in the Hudson River ecosystem. Only three years would elapse between an article promoting the use of ecosystem analysis for impact assessment by Stanley Auerbach and one proclaiming the futility of the concept by one of his subordinates. The evidence must have been quite compelling.

"When I set out from graduate school I saw the National Environmental Policy Act environmental impact assessment process as an opportunity to combine effectively modern ecological thought with political activism," Glen Suter's lament began. He was that subordinate at Oak Ridge. "Five years and over 20 NEPA documents down the road, it is apparent that part of my baggage of modern ecological thought (the ecosystem coveralls) has never been unpacked."

Suter found that the predictions given by ecosystem theories were not testable, not useful, or both. His diatribe was shown around to others at Oak Ridge.

"None of them agree with everything that I have said, but some of them will defend to the point of personal discomfort my right to say it," was the support his colleagues gave.

Suter's criticism was not so much over methods, but results. He had been sold a practical gadget, ecosystem analysis, and was annoyed that it was not working. Rebuttals to his complaint pretty much said that things weren't really that bad. Ecosystem analysis was not the only questionable bill of goods being sold by ecology at the time, however. The MacArthur camp had its own special inventory. Criticisms of their methods in the science, such as made by Peters, were more summarily dismissed by ecologists, by dismissing Peters instead.

"There is no one way (especially not that of Peters) of doing science," was one of the simplest retorts. It was a recurrent theme in ecology.

"Reading the ecology literature my first graduate year," Simberloff recalls about his introduction to ecology, "I recognized that a lot of the quantitative work was basically models that more or less fit some data, without considering other models. I am probabilistically inclined, and I quickly began to think of what various 'random' patterns might look like. After I read Popper, soon thereafter, I learned the right terminology."

Ecology graduate students are drawn into philosophy much more so than students in other sciences. My dissertation had what on hindsight is a totally unnecessary section on Baconian induction. Sadly (to me), it was what most interested one of my examining committee members. (Worse would have been had I read Popper. The dissertation might never have been written.) Perhaps ecologists delve into the philosophy of science more than other scientists do because their interests are broader. Perhaps it is because the field is more of a philosophical premise than a science. Or had been that way. Or still remained so to some of its practitioners.

When he joined the Florida State University faculty after completing his doctoral requirements, Simberloff recruited likeminded ecologist Donald R Strong, Jr. What Strong and Simberloff began to look at was not just the underlying philosophical basis of ecology, but something more fundamental and easier to understand. They looked at the experiment part of the scientific method as it was being done in ecology.

Experimentation in science looks for a cause-and-effect relationship. It is easier to tease it out in some sciences than in others.

For example, apply a force (a push) to an object and you get a change in its motion. Push harder and you get a bigger change. Measure the forces and changes precisely and you get Newton's Second Law of Motion, which can be applied to a variety of situations, each a separate test. If a push does not result in the expected change of motion, some other factor had to have been overlooked. This factor can be identified and isolated and studied by itself. Such is the progress of science.

However, take a houseplant, give it lots sunlight and water, and it still may die, rather than grow as predicted. Something was overlooked. The possibilities for unknown factors interfering with our expected outcome in the biological example are notably greater than in the physical example. Let that houseplant instead be outdoors in the vagaries of nature and the factors multiply. Include such phenomena as genetics, that are inherently random, and the further vagaries of soil, weather, and diseases, that might as well be random, and a cause-and-effect relationship is not so easily determined. Whatever changes might be found could have resulted from the chance occurrence of other factors.

There is help, of course. It comes from mathematics. The science of statistics was invented for biologists to tease out real effects from random events.

Experiments in ecology, however, were notable by being exceptions, rather than standard procedure. Although laboratory experiments such as Gause's were well known and influential, they were few and far between. For ecologists, the laboratory environment was a confounding factor in trying to understand natural relationships. Any graduate student faced with the prospect of controlling all of the variables in nature that could confound an experiment, while somehow keeping things natural, is soon tempted to forego the whole idea.

Those voles that I trapped and brought into the lab turned out not to perform the most rudimentary parental tasks on which the existence of their species depended. Caged voles would not become pregnant. PREGNANT voles would abort or kill their offspring. Some voles in outdoor enclosures would go about their business in a way that would let me tally up the intended experimental observations, while others, the Steve McQueens of the rodent world, seemed to see escape as their sole purpose in life.

Field experiments in ecology, when they were performed. were influential but were few. Yet these lab and field investigators showed that experimentation in ecology can be done and is done. Why didn't more ecologists do it then?

One difficulty with field experiments becomes obvious. The time required to study ecological changes does not match the time for a degree. One is measured in decades, the other hopefully maxes out at four years.

Ecologists, like geologists and cosmologists, have found a way around the time limitations of a graduate career—or human lifetime. They let nature do their experimentation for them. Find two natural areas with a single major underlying difference and you have a natural experiment. Not easy, but it can be done. Easier is to have a theory make a prediction, then test it against the natural world. Is nature organized the way the theory predicts? Even easier is to find some sort of pattern in nature, concoct a theory to explain it, then test that theory against the pattern found. This is the circularity that troubled Peters so much. It was at the heart of the hypothetico-deductive methods of Robert MacArthur.

Strong and Simberloff turned their attention to how we determine that a pattern found in nature really does support a theory. What Simberloff had discovered early on in his career was that ecologists were not really using statistical inference when testing theoretical findings. Indeed, modelers concerned themselves not at all with testing their models with formal statistics, for that would take precise data, while Richard Levins had argued for generality and realism, only, in models that were to help in understanding the complexity of the ecological world.

It was an important criticism brought up by the Florida State researchers. It also exposed underlying issues that were more than just scientific. Some were political, and emotional.

The controversy their criticism generated has not been dealt with adequately in literature meant for the general public. Jonathan Weiner's Pulitzer-Prize-winning *The Beak of the Finch*, for example, is a splendid introduction to some of the later work having to do with the evolution of Darwin's finches, as they are commonly known. The question then, as in Darwin's time, was whether the differences seen in the birds evolved through competition. The matter of how one goes about testing for competition is a serious one.

Simberloff, Weiner wrote, "argued that Lack's famous patterns might be nothing more than the faces we see in the moon, in clouds, in Rorschach inkblots."

In fact, it was Simberloff's influence, along with others, that led the Grants to the very carefully determined results that Weiner went on to describe. Trying to rid themselves of an annoyance, they discovered a treasure.

David Quammen in his very popular *Song of the Dodo* treated most of the participants in the controversy and some of the science at length, but reduced the very serious scientific arguments to flippant little paragraphs. I don't think he understood the science involved. Simberloff, however, who thought of Quammen as a friend, disagreed with me.

"He's a very bright guy," Simberloff tossed aside my suggestion.

To read Quammen's account, the entire matter was little more than childhood bickering. Apparently, he decided that the reader did not want a detailed account of it, nor of the closely related controversy of applying island biogeography to conservation. Yet island biogeography was the subject of Quammen's book.

"He loses me occasionally with his brisk, concise explanations of intricate scientific ideas," Quammen said of his conversation with Simberloff, "and at one point he shows impatience with my obtuseness."

Somehow other personalities in the story seemed not to notice the obtuseness that kept Quammen from treating the issue that Simberloff had brought up with the same detail he treated other science in the book.

"Like Robert MacArthur, Dan Simberloff is clearly a very smart guy," Quammen noted, having never met MacArthur, nor tried to assess his intelligence.

Based on Quammen's account, though, Simberloff seemed not to have that part of MacArthur's intelligence that made the other person with him seem to think more clearly.

Quammen went on to say about Simberloff that "he doesn't squander his charm on strangers." His intelligence, Quammen called "dangerous." Besides being "curt and unemotional," "blunt," and caring "zip about what most people think of him," Quammen went on to describe him using words more appropriate for an ax murderer.

"He would gladly cut the throat of an ill-conceived idea and swat down an innocent misstatement like a fly," he wrote.

A few pages earlier, Ted Case, on the other side of the controversy, was described as "the swash-buckling herpetologist."

Michael Gilpin's "laugh, sounded often, is a full-hearted, high-pitched bray," Quammen told us about another. "He reads widely, plays hard, thinks fast but deeply, and his world remains larger than science, though his scientific world is large."

These two relatively minor characters in the controversy are presented in juxtaposition to Simberloff. Swashbuckling, fun-loving guys take on Hannibal the Cannibal would have been a fitting subtitle for Quammen's chapter. Reality is a bit different. Quammen seemed to find no need to assess the dour countenance of Jared Diamond, much more key to the controversy than Case or Gilpin, their leader, in fact, nor did the sunny personality of Don Strong, Simberloff's closest collaborator, grace the pages of his book.

Simberloff is far from unfriendly. The closest thing to excessive reserve in him that I can recall is from a time we were thrown together in St Louis. He was at Washington University (the one in St Louis) to give a seminar on his classic work with mangrove islands. Like two Englishmen recognizing each other abroad in an uncivilized land, we were two ecologists on a campus then going totally molecular. Professionally, at least, we were kindred spirits. Simberloff approached me as such, with a greeting that was genuinely friendly. Then I made it a point to make certain he understood that I was only in St Louis temporarily while on my way to UCLA. He suddenly seemed to withdraw. Or at least so it seemed in my memory. UCLA was then a den of MacArthurites, Jared Diamond foremost among them. This was in 1978. Maybe it was because Simberloff was wary of where I, as someone hired by

UCLA, might have stood on the controversies which then centered around him. I think it was, and he was waiting for me to bring it up. Or maybe he just had nothing more to say. Maybe he was thinking about his seminar.

From my experience, Simberloff is as quick to smile as any ecologist I know, although less so than department chairmen looking for someone to perform an unpleasant duty, or salespeople and insurance agents. Judging from his published works, Simberloff's reading has to be as wide as any of the ecologists with whom Quammen became so enamored. But no, Simberloff never surfed, nor was he ever a director of a Zen center, as was Case.

And no, personality was not the direct cause of the counterrevolution begun by Simberloff and Strong, nor its vituperativeness. But then, yes, personalities, going all the way back to MacArthur, were a part of the messiness that resulted. So, too, might have been accidents in geography.

The pinnacle of the controversy might have been a special issue of *The American Naturalist* in 1983. The editor, George W Salt gave it the title, "A Round Table on Research in Ecology and Evolutionary Biology." He took it upon himself to provide the final, authoritative words on the methods behind the science. I will begin with them.

That title seemed to refer to two sciences. Although, then as now, "ecology and evolutionary biology" was used by many to refer to a single body of knowledge, recall that there was a branch of ecology in which evolution seemed to play no role: that of the ecosystem. That the contributions Salt gathered together could pretend to speak on all research in ecology shows that the outcome of the competition between Odum-style ecology and MacArthur-style ecology was by then not much in doubt. Remember, also, that it was Salt who lent the prestige of his editorship of the *Naturalist* to sweeping away any ideas that ecosystems could have emergent properties and thus were unworthy of attention from those who called themselves "evolutionary ecologists." (These semantic distinctions can best be understood in reference to the Monty Python crew's movie, *The Life of Brian*. Recall the People's Liberation Front of Judea and the Judean People's Liberation Front, or whatever the exact, but meaningless differences were.)

For the first half of the 20th century, according to Salt, ecology's methods were mainly the inductive science of Bacon: "the accumulation, collation, and recording of observations." The appearance of high-speed computers, with their capabilities of massive data analysis, did not lead to any major conclusions, Salt concluded without mention that the computers had mainly been put to use to elaborate the ecosystem concept. A more deductive approach that cast hypotheses into mathematical form, according to Salt, had a "revivifying effect on ecology and evolutionary biology." This hypothetico-deductive approach, quite common to science, strongly focused ecological research, he went on, but the method turned out to be too simplistic in comparison with the real world. The result was that what Salt to be the fraternity of ecologists becoming splintered into "empiricists" and "theoreticians."

Salt's characterization of the division being one of data versus abstract mathematics is only partially correct. There was the much more fundamental issue of hypothesis testing. The true experimental method requires a hypothesis that is testable in the Popperian sense and through proper experimentation. To Salt, however, the need for falsifiable hypotheses did not apply to ecology and evolutionary biology. In the view of some ecologists, the one Salt gave the lead article to in the issue, for example, common sense is a better guide than scientific method.

The controversy Salt tried to resolve had surfaced in 1976 with two important papers besides the one by Peters. They were far removed from philosophy. One was by Hal Caswell, who otherwise has little part in this story. It was based on work he did for his doctoral dissertation at Michigan State University. The other was by Simberloff and his Florida State colleague, Lawrence Abele. Caswell examined what he called a "neutral model" of community structure, rather than a null model, possibly to distance himself from the controversy that was brewing over random models. The structure he looked at was essentially what the number of species and their relative abundances in a community might be, those pages and pages in the ecological literature listing what species were found where, some of which even had data on the densities of their populations.

"I have used the term 'neutral' rather than 'random' to describe these models because randomness is not the only way to achieve neutrality," he had explained.

Caswell saw his models (he had tried out several) as ways of eliminating factors, such as competition, that might be causing the patterns found. In actuality, he accomplished that by having populations of a species mathematically appear, grow in numbers or disappear, based on certain probability distributions. Colonization and extinction were neutral to biological interactions, but free from random change, they were not.

Competition, by way of the Competitive Exclusion Principle, had been the biological mechanism then thought to keep very similar species from occurring together. Having a basis in the mathematics of the Lotka-Volterra equations and Gause's experiments, the principle was given its most concise and easily remembered form by Garrett Hardin. In his words, the principle is its own mnemonic device. "Complete competitors cannot coexist." It is also an untestable tautology, according to Peters.

So why is it that Area A has eight species of warblers, while Area B, only six? Competitive exclusion. In analogy to Gause's experiments, Area B has two fewer different culture vessels.

How do we know it is competitive exclusion? Why, look at the data: eight in one, six in the other.

Why is this little warbler found, here, in this particular forest, but not there, where another little warbler species seems to have taken its place? Competitive exclusion.

Why is one little warbler so abundant, while some other species seem to be barely hanging on in comparison? Competitive exclusion.

There were also other questions the principle answered. MacArthur's Three Influential Papers and the work that followed, his and Wilson's *Theory of Island Biogeography*, most influentially, put the tautology into mathematical terms that seemingly could be tested. The reason that Area A has eight species and one is more abundant was exactly due to competition, but now there were mathematical equations, rather than a highly alliterative tautology in explanation.

Caswell did not specifically set out to do so, but his work was a way out of the circular reasoning of the Competitive Exclusion Principle. Caswell thought he could remove the biology from community data in the same way that models on the spread of neutral alleles removed selection from population genetics data.

What Caswell discovered through his neutral models was that patterns found in biological communities having to do with the distribution and abundance of species could be generated by entirely nonbiological processes.

"At any rate," Caswell concluded, "the general role of biological interactions in generating community structure clearly needs to be rethought. The results of this study indicate that the diversity of natural communities may be maintained in spite of, rather than because of such interactions."

Caswell's approach was no different than that LaMont Cole had brought to the analysis of population cycles a generation earlier. These are the well-known booms and busts in numbers of hares and lynx accumulated by Canadian fur trappers starting from the middle of the 19th century. The march of lemmings to the sea is another example of the phenomenon. Other species, many of them non-mammals, were also found to have population peaks that recurred at regular intervals. Three- to four-year intervals, often overlain by an approximate ten-year cycle of especially high peaks, were most commonly found. Explanations ranged from predator-prey dynamics to sunspots. What Cole brought to the argument was to first compare the cycles seemingly found to those that might be expected if the population sizes varied entirely at random. If the cycles could be generated at random, then no theoretical explanation was necessary, according to Cole. It was very much like the null hypothesis of randomly generated results that is tested against a proposed hypothesis in experimental work. Cole found there to be three- and nine-year cycles in series of random numbers. His suggestion that the same kind of randomness might explain the observed population cycles was not appreciated. His lectures on the subject included the bravura acknowledgment that Canadian wildlife biologists were known to "use my name to frighten babies."

Donald Strong has pointed out that there had also been a history of null model use in community ecology that started in 1947. Contemporaneous with Cole's skeptical look at wildlife population cycles, it had been a quieter controversy. It had had to do with a pattern first found by Charles Elton having to do with the ratio of species per genus in isolated communities when compared to the ratio found in the larger, general area in which the communities were found. The isolated communities had fewer species of the same genus than was expected based on the number of species of that genus in the region at large. For example, there are more than twenty species of warblers of the genus *Dendroica* found north of Mexico. (The exact number depends on the vagaries of ever-changing taxonomic studies.) I can only expect to find some dozen in New England. If I staked out a particular forest on a mountaintop in Vermont and conscientiously kept track of warblers found there, the number would be less. So my species-to-genus ratio goes from 20-plus to 12 to 6, or even fewer, in going from most of a continent to eastern forest (isolated by the Great Plains from western forests) to a particular isolate of northern coniferous forest. Elton interpreted similar findings as being a result of competitive exclusion. His discovery kept being rediscovered. Each time it was, a null hypothesis was found that showed the pattern to be no different (less so, even) than could be expected by a random sample of the regional fauna. Each time, the null results were ignored. Peter Grant and Robert MacArthur both rediscovered the relationship for island birds. In 1970, Dan Simberloff was among the last to try to put that relationship to rest by showing, once more, that one could essentially shuffle a deck and pick cards from it to get the pattern found.

Were the natural patterns random? The question misses the point. Up to Caswell's paper, the community patterns based on total numbers of species on islands and the presence or absence of particular species had unquestioningly been used to test various mathematical models. The models were based upon certain assumptions about competitive interactions between species, mostly traceable to early papers of Robert MacArthur. If the model fit the pattern, then the model was seen to be correct. This was exactly the sort of thing that had been troubling Simberloff ever since he had set off to become an ecologist.

It did not take long before Simberloff and his colleagues at Florida State began to flood the ecological literature with their dissent. Before a data set could be used to test a theoretical model, they argued, it should first be shown not to have been a possible result of random interactions. Using a similar philosophy, but different math than Caswell, they went on to show that, for a number of the data sets used to test theoretical models, the patterns identified in data could as well have developed through chance processes as by the methods proposed by the theories. How well the predictions of a theoretical model fit real data, they argued, should be compared to how well a random model that eliminated the biological interactions important to the theoretical model fit the same data. If the theory fit better than the random model, then investigators could begin speculating on the biological causes of the model. If not, other evidence should be looked for.

The idea can be illustrated with games of cards. Did the couple that went home from an evening of bridge 10,000 points ahead of their hosts do so because of superior card playing skills or from sheer luck? This is the identical question that null models asked about data on diversity: was competition the answer for differences? In the card-playing example, a duplicate bridge tournament serves to eliminate chance (although not entirely) by having all contestants play the same hands by the end of the night. What ecological null models did was eliminate skill, in essence, to see if the winners were still the same.

These random models came to be known as a "null models." They were analogous to the null hypothesis in statistics, without which trying to tease meaning out of experimental data can be futile. On the surface, the requirement of a null hypothesis is a perfectly reasonable one. It is already an essential part of hypothesis testing in situations where chance factors can affect the interpretation of results, as in drug trials, for example. The biological peculiarities of the individuals chosen for the trial, if not taken into account, could lead to misleadingly dangerous results.

However, we are also on some rather delicate personal ground here.

There was another paper that Simberloff published, this one in 1976, having little directly to do with the null hypothesis, but having much to do with the brou-ha-ha that was building. He and Lawrence Abele published a short note in *Science* in which they tried to correct an overly hasty application of island biogeography theory to the design of refuges. This entire and separate controversy (although like all things ecological, not quite separate) will be taken up in detail later. Suffice to say now that it did not endear Simberloff to a number of ecologists who were staking their careers on doing just what he was warning against. It was Cole and wildlife cycles all over again, except that no one was frightening babies by this time. What is more relevant is that some of those people were members of the clique that had grown up around the ghost of Robert MacArthur. And that clique had grown to the point that warnings about intellectual censorship previously mentioned had to be published in the open literature.

Quammen glibly called one side "rational Unitarians" and the other "devoutly agnostic." He was closer to the mark in his earlier analogy of MacArthur to Lenin. The rift was in a revolution, rather than religion. It was Lenin and Stalin, those the revolution had placed into power, versus the Trotskyites.

In truth, by the time the controversy came to a head within the *Naturalist*, Simberloff had actually had to confront some real Trotskyites. Inspired by reading a paper by population geneticist

Richard C Lewontin to finally write down ideas he had been mulling over for a number of years, he sent his own paper off to a journal little known to most biologists. The journal was *Synthese*, and Simberloff's paper was an intellectual ramble through areas as diverse as Darwinism, theoretical physics, Greek philosophy, and of course, ecology. It was exactly the sort of paper that a scientist having broad reading habits and interests could be expected to write about his field of study and science in general. Usually they do so toward the close of their careers. The state of ecology apparently kept Simberloff from waiting until then. Looking at it through the philosophical lenses he had donned for the paper, Simberloff had many critical things to say. Barry Commoner and Eugene Odum's ecosystem took the biggest knocks, but the work of MacArthur, Robert May, and Richard Levins, was also found wanting.

Simberloff's effort so excited the editors of the journal that they decided to make a special theme issue based on it. Until then, the journal's editors had had no special interest or expertise in the subject. The outcome of their enthusiasm was two issues and eventually a book. Robert Peters wrote a chapter recapitulating his arguments on the need for falsifiable hypotheses in ecology. Simberloff's ally Don Strong also contributed, writing a chapter on null models. Robert McIntosh, an ecologist and historian of ecology, wrote a useful chapter on the history of ecological ideas that he later expanded into a book. Others also used Simberloff's ideas as a jumping off point for their own chapters.

Then came the ambush. In advance of publication, Esa Saarinen, a young (24), newly minted Finnish PhD in philosophy, who was serving as an editor for *Synthese*, sent Simberloff's paper to Marjorie Grene, a philosopher on the journal's editorial board, and to Richard Lewontin, among others. Grene had been a natural choice for Saarinen; Lewontin had been picked based on Simberloff's suggestion. That began the fireworks.

Lewontin recalled having been invited by Saarinen to write "a critique and discussion of Simberloff's ideas."

"Having then read Simberloff's paper it seemed to me appropriate that a joint paper with Levins would be a good idea and we then collaborated on a reply to Simberloff that appeared in the special issues," he remembered for me.

"The paper is Levins and Lewontin because our joint work always uses the alphabetical order of names. I have no recollection of how much each contributed to it."

The exchanges that ensued were biting. One reviewer, remarking that the dialog in *Synthese* was something the discipline needed, nonetheless found "images of hand-to-hand combat or a barroom brawl" coming to mind. He found the following exchange between philosopher (Grene) and ecologist (Simberloff) to be particularly spirited.

"Many biologists, when they turn to philosophical (epistemological or ontological) questions, abandon the standards of accuracy that, at least in the layman's view, ought to govern discourse as scientists. Simberloff's argument forms an unusually flagrant example of this practice," had been one shot.

"Many philosophers, when they turn to biological questions, abandon in favor of captious logomachy the quest for epistemological or ontological enlightenment that, at least in the layman's view, ought to govern their discourse. Grene's argument forms an unusually flagrant example of this practice," had been the reply.

What the philosopher was saying was that Simberloff failed to precisely define his terms. What the biologist (Grene somehow could not identify him as an ecologist; biologists in other fields have little need these days to brood over how we know the things that we know) was saying was, "Give me a break." Simberloff, not meaning to make a strictly philosophical contribution, had forgotten perhaps the precision with which philosophers use their terms. I gave up some 200 pages into one of Immanuel Kant's great works when it suddenly dawned on me that he had spent all of those pages defining existence and still had not done so.

The sharpness of Simberloff's reply to Grene matched the sharpness of her criticism but had also come from a feeling of having been set up, particularly by Levins and Lewontin.

"I was a bit taken aback by the evident vitriol and somewhat dismissive tone adopted by them in that paper, particularly since some of what they said was junk," he told me.

"Simberloff's essay seems to us to embody the false debate based on three fundamental confusions," they had brushed away his ideas with.

"As a result of these confusions, Simberloff, in his attempt to escape from the obscurantist holism of Clements' 'superorganism,' falls into the pit of obscurantist stochasticity and indeterminism," they had summed up his approach to science, before going on in the rest of the paper "to develop implicitly a Marxist approach to the questions that have been raised."

Self-avowed social revolutionaries, Lewontin and Levins wore their Marxist philosophies aggressively. Soon after this episode, they published a book together, *The Dialectic Biologist*, that was "the result of a long-standing intellectual and political comradeship," to use their own words. In it is also reprinted the rather sophomoric prank Lewontin and Levins pulled off to belittle EO Wilson. It had come in the form of a letter to *Nature* from a fictitious Isadore Nabi, who turned out to be the two Trotskyites plus a friend, Leigh Van Valen. (It also turns out that Robert MacArthur helped to create Nabi in the days before Wilson's fond memories of a "Marlboro Circle.")

One can judge their abilities as Marxist scholars from the following rationale presented for their final chapter: "After collecting these essays, we were dissatisfied. The assembled work *illustrated* the dialectic method, but it did not explain what dialectics is. ... We then set about to write a chapter on dialectics—only to discover that in twenty-five years of collaboration we had never discussed our views systematically!"

The Hegelian dialectic, as explained to me, is a philosophical concept that occupied the minds of 19th century European romantics and anarchists. Serious readers of 19th century Russian literature, among whom I count myself, cannot avoid having to learn at least the basics of this philosophy. Those "superfluous" men might well have been Hegelians. According to Hegel, an early 19th century German philosopher, every idea or concept (thesis, as he called it) had within it its own negation (antithesis). Eventually, the original idea is destroyed by its internal negation to create a new idea (synthesis), which starts the whole process again. It is infectious in its simplicity. Freedom, for example, is meaningless without the concept of repression, yet to guard against freedom might require repression. There you have it. Hegelians soon found it difficult to think of anything at all without immediately recognizing its negation.

Although I am being free with my opinions on philosophical issues, I am hardly an authority on them. Neither do I want to become one. So I have borrowed an opinion on Hegel from someone who was a philosopher, Bertrand Russell.

In his *History of Western Philosophy*, Russell had the following to say: "Hegel's philosophy is very difficult—he is, I should say, the hardest to understand of all the great philosophers."

The words Hegel used for definitions, according to Russell are "very obscure" in translation and "even more difficult" in the original German. When they are understood, his ideas are "self-contradictory."

See, and I thought it (the dialectic) was simple.

Of Hegel's historical ideas, Russell wrote: "It is an interesting thesis, giving unity and meaning to the revolutions of human affairs. Like other historical theories, it required, if it was to be made plausible, some distortion of facts and considerable ignorance."

In summary, Russell concluded, "almost all of Hegel's doctrines are false," yet his philosophy "illustrates an important truth, namely, that the worse your logic, the more interesting the consequences to which it gives rise."

Marx and Engles appropriated the Hegelian dialectic to prove that Capitalism was doomed. Levins and Lewontin, apparently assuming that this background was well known to the biologists for whom their book was intended, applied it to biology. Other than a remark about self-negation, they attempted no explanation of the underlying philosophy for their ideas. Mostly, their concern was over the need for a holistic view of biology. That is their dialectic biology. Russell sums up in the following paragraph what Levins and Lewontin failed to manage in almost 300 pages.

"The view of Hegel," according to Russell, "and many other philosophers, is that the character of any portion of the universe is so profoundly affected by its relations to other parts and to the whole, that no true statement can be made about any part except to assign its place in the whole. Since its place in the whole depends upon all other parts, a true statement about its place in the whole will at the same time assign the place of every other part in the whole. Thus there can be only one true statement; there is no truth except the whole truth. And similarly nothing is quite real except the whole, for any part, when isolated, is changed in character by being isolated, and therefore no longer appears quite what it truly is. On the other hand, when a part is viewed in relation to the whole, as it should be, it is seen to be not self-subsistent, and to be incapable of existing except as part of just that whole which alone is truly real. This is the metaphysical doctrine."

This is the holism at the heart of dialectical biology.

Levins, recall, had made his reputation arguing for holistic models (such as his own) that sacrifice precision to realism and generality. Lewontin could be both a holist and reductionist, given the Hegelian concept that, if the parts were not understandable without the whole, neither was the whole understandable without the parts.

"To understand the world," Lewontin explained his view to me at the time of my first writing this book, "there are different ways to cut it up for different purposes. The hand, the fingers, the joints of those fingers, the tissues, the cells, the molecules of the cells are all appropriate units of function and levels of dissection for different questions. It is simply not true that everything is effectively connected to everything."

Levin's three criteria made his models untestable by any null model, the subject of the controversy that was then brewing around Simberloff, although Simberloff never once made mention of null models in his *Synthese* paper. Lewontin's beef with Simberloff was perhaps more personal.

Both Levins and Lewontin were in the MacArthur camp of ecology. Both had collaborated on papers with Robert MacArthur and had been part of the "Marlboro Circle," as EO Wilson came to call it, collected in Vermont by MacArthur, although Wilson apparently came later to the circle than Levins and Lewontin. As the Nabi prank and his resignation in protest from the National Academy of Sciences showed, Lewontin took his science lightly and his politics seriously. Even though Wilson had been instrumental in bringing Lewontin to Harvard, despite his Marxism, Lewontin felt little obligation to him. In 1975, he led the Sociobiology Study Group, which included Levins, in a campaign to make Wilson's life unpleasant. The group was formed to expose Wilson as a "counterrevolutionary adventurist," in Wilson's words, leading him to suffer various unpleasantries, culminating in 1978 with Wilson having a pitcher of water dumped on his head while speaking at a AAAS symposium. Wilson's sin had been to speculate that much (or at least an important part) of human behavior might be genetically based.

A more serious attack on Wilson by Harvard's Marxists was a letter to the *New York Review of Books*. Wilson felt blindsided. The resulting cold shoulder he received from his Harvard colleagues made him feel "like an atheist in a monastery."

I wondered if Simberloff, being one of Wilson's most notable students, had himself become a target of the bile directed at his mentor.

"You know, until you just said this, it never occurred to me that the tone, at least, might be partly due to their associating me with Ed Wilson!" he had replied to me.

Lewontin had little to say about the incident.

"I have no recollection that Simberloff was Wilson's student. Whether I knew that at the time I couldn't say, but I wouldn't attack someone just because he was Ed's student," Lewontin had insisted to me. "As you can see from our paper we thought Simberloff had made a number of errors and confusions. The paper really speaks for itself."

Although Lewontin has to be taken at his word, someone capable of concocting the Nabi prank to annoy Wilson might also have been capable of putting a little extra sting into his criticism of Wilson's student, if only subconsciously. When they reprinted their *Synthese* article in *The Dialectic Biologist*, Levins and Lewontin found it necessary to edit out the parts that made it "a polemic against" Simberloff.

"We have edited it to remove the flavor of *Anti-Duhring* and to tie the discussion less to a specific disagreement," was how they had put it.

This brings us back to the sequela of the argument in *Synthese*, to the issue of the *Naturalist* with which I started this chapter. The side in opposition to Simberloff and Strong was not represented by those who had argued in print against their interpretations, but by the person Robert May had held out at that time to be an example worthy of my emulation, Jonathan Roughgarden.

This also brings up a somewhat delicate issue for me. By the time I first saw Roughgarden he was an attractive, energetic, middle-aged blonde called Joan. I offer the information in the same spirit that I identified Dan Goodman as a fly fisherman. The reader can make of both pieces of information what they will.

Both Roughgardens were (are still?) based in Stanford. Jonathan had success in creating theoretical mathematical models with great heuristic value, as already mentioned. Joan seems to have used her mathematical skills to save the world's endangered fisheries.

It Wasn't a Barroom Brawl. It Was an Athletic Contest

I want not to be flippant about the arguments made over null models. The issue is much too serious to treat the way David Quammen did. Yet now that I have come to writing about it, I find myself being highly sympathetic to Quammen. How do you make the matter palatable to the general reader?

Forgive me reader, but my only inspiration is to translate the arguments into the simple terms of a sporting event.

In that vein, Roughgarden can be said to have made the following complaints about criticisms raised by the Florida State group: (1) Damn that referee! (2) My star player is better than yours, (3) That rule doesn't make any sense, anyway, and (4) My team has a better record than yours. Of course, there were issues lurking in the background that never really got into print.

"I am not aware of a single finding that emerges from what Simberloff and his colleagues have written," Roughgarden was quoted as saying. "But I am aware of a lot of bitterness it has caused."

Why should there be bitterness? It has to do with all of those things that motivate scientists who really did not go into their sciences simply out of love for their subject. Roughgarden had summed it up quite nicely in his paper for the *Naturalist*.

"The place of ecology within the sciences depends on the perception by scientists from other fields and by the public that progress is being achieved in ecology," he wrote. "This perception becomes the basis for science policy decisions concerning faculty billets and research support. The advance of theoretical ecology during the 1970s is an important event in contemporary science: it has begun to attract serious attention from other areas of science and from mathematics; and it has contributed to enhancing the stature of ecology as a result."

These are some of the same concerns that led Auerbach to jump on the ecosystem bandwagon at Oak Ridge. The "theoretical ecology" advancing during the 1970s, according to Roughgarden, was MacArthur-school ecology.

Roughgarden marginalized the field of statistics, something in general neglected by the MacArthur school, when he expressed the fear that "the extreme antagonism in the rhetoric about theory doesn't reinforce the inherent disinclination people have to learn all that math that is so necessary to the study of ecology."

I don't remember one-time math major Simberloff having been disinclined to learn math. And I thought statistics *was* a mathematical science.

In his criticisms of the Florida State group and their "tone of righteous indignation" and "antitheoretical rhetoric," that I have broken down into those four sports-related analogies, with (1) Roughgarden denied that scientists use formal rules (the scientific method). This is the only interpretation that can be put on the following statements.

"As scientists," Roughgarden wrote, "we rarely abide by formal rules establishing scientific facts."

True, we know that scientists do not go about their work with the dispassionate objectivity we would hope them to have. TS Kuhn had shown that scientists sometimes hold on to their pet ideas to their death, even in the face of overwhelming evidence, but nowhere had he or others suggested that abandoning proper controls in doing a study was anything but fraud or charlatanism. Roughgarden cited F Suppé's *The Structure of Scientific Theories* to support the statement quoted.

A short time later, Jared Diamond, for whom Roughgarden was acting as something of a stand-in, also used Suppé to try to discredit Popper.

"Finally," Diamond wrote, "while Popper's philosophy in general and the falsification criterion in particular were formerly considered to be among the significant views within the philosophy of science, most professional philosophers other than Popper's disciples abandoned these views by about 20 years ago. For example, in the comprehensive presentation of major modern philosophies of science edited by Suppé, the falsificationist criterion is cited only briefly in a few places to explain why particular modern philosophers discarded it."

Suppé, of course, was not an ecologist. He wasn't any kind of scientist. He was a philosopher. By the time he wrote his book, philosophers other than Popper were no longer concerned with an objective scientific method. They were trying to make sense of reality through a haze of text deconstruction, semiotics, and moral relativism. Science to them was no different than any other human activity, poetry, for example.

For (2), Roughgarden invoked the philosopher, Ludwig Wittgenstein. This was a strange choice. Wittgenstein may have been a "giant in twentieth century philosophy," as Roughgarden described him, but Wittgenstein never made any serious inquiries into scientific method. Wittgenstein and Popper were on opposite ends of a philosophical spectrum. Wittgenstein's was the end that was totally bereft of modern science. To Wittgenstein, language, in particular puzzles posed by language, not science or mathematics, was the only material suited for philosophical studies. His opinions on science were therefore immaterial. Besides Roughgarden's choice of Wittgenstein as an authority on method in science having been an odd one, so had a quotation he had chosen from Wittgenstein. It had everything to do with language and nothing with science.

Science has at times progressed through flights of fancy. Kekule's dream of a snake biting its tail leading to his solving the mystery of the structure of benzene is a commonly used example. Such conceptual leaps are rare, however, and almost always launched from a solid platform of data. Only a few flights of fancy turn out to take permanent wing or even see the light of day. Fortunately, the nightmares suffered by scores of other chemists all over the world were left to remain dreams. Those of ecologists flooded the pages of *The American Naturalist* and other journals.

Complaint (3) was that null hypotheses are not easily and reliably developed. Fabricating random communities against which to test theory itself had procedures that had not been tested. Unlike null hypotheses used in controlled experiments, a process that has been perfected through generations of scientific use, null models were a new concept. There was little to guide an investigator in the nuances of choosing a null model. Given those difficulties, why use them?

There was another point of view about what to do when the going gets tough.

"The challenge of community ecology is to understand the structure and function of some of the most complex natural systems," a group of ecologists who had great success using experimental techniques has written. "Those who accept this challenge have no excuse for complaining about its difficulty."

The Florida State ecologists had devised a check on data not obtained experimentally that was being used to test those bright ideas so common to ecology about which MacFadyen had warned. And it was the most basic of tests. Instead of comparing patterns found in nature to what some hypothesized cause might have produced, the patterns were compared to what could have arisen through chance alone. The MacArthur school, at least in the person of Jonathan Roughgarden, was unwilling to concede the need for such testing.

Roughgarden must have been confused by the frictionless pulley he invoked in defense of theoretical ecology. This was something, no doubt, he had borrowed from Robert May, who spoke of theoretical ecologists and the "perfect crystals" that physics started with to achieve a basic understanding before it could tackle real crystals. Had May used a pendulum, his analogy would have broken down. Physicists can ignore friction in many situations, as Roughgarden pointed out, but only if the motion of a pulley or pendulum, for example, with friction is not drastically different from one without. Simplifications made by ecological theorists, however, are often more like doing away with the pendulum bob because it is too difficult to model, rather than ignoring friction. True, without friction, the pendulum would need no bob—a rigid rod, free to swing can exhibit pendulum motion—but in the real world, no pendulum motion will be observed in a string lacking the inertia to push through the air. Friction becomes critical. Perhaps Roughgarden could have still argued for a modeling success, as some had in the past for models that did not fit reality, for the bob-less pendulum demonstrated the necessity of a bob for pendulum motion. Without the model, we would not have known!

In fairness to the theoreticians, however, simple models with seemingly trivial results can sometimes be useful. Mathematical analysis can show an action to lead to irreversible and undesirable consequences. It can signal a need for caution. The pendulum bob, in analogy to ecological systems, could be a predator. Loss of the predator could result in loss of function for the ecosystem, much as loss of the bob resulted in a non-working pendulum. Most often, however, ecological warnings coming from mathematical models are heeded only if they match our intuition. Surprising results just bring down suspicion on the legitimacy of the model.

Roughgarden's fourth complaint was not really a complaint, but a boast. His (4) alludes to the successes of ecological theoreticians, such as May's discovery of one such surprise, chaos, in the Lotka-Volterra equations. Roughgarden saw the advance of theoretical ecology as "enhancing the stature of ecology as a result."

Simberloff responded that the enhanced stature might exist only in the impression that ecologists were speaking the language of mathematics, even if ill applied, rather than Latin.

"Physics-envy is the curse of biology," he had once before quoted another ecologist, Joel Cohen, as saying.

Physics envy is an affliction not limited to ecology. Neither should it be a curse. Newton may not have been so remarkably more intelligent than Darwin. The physicists just solved all of the easy problems first, the ones that actually were suited to their Cartesian methods. Ecologists have harder problems to solve.

Today, the use of null models is an established technique in community ecology, necessitating a textbook devoted to it, *Null Models in Ecology*, by Nicholas Gotelli, a former student of Simberloff's, and Gary R Graves, who also was trained at Florida State. The most appropriate manner of creating the models for given situations, of course, is no simpler to determine today than it was when Strong and Simberloff were pioneering their use. But at least there are guidelines for avoiding certain pitfalls.

Why was it then such an issue? Why the resistance to it from Roughgarden, May and—always in the background—Diamond?

For the answers, remember that a clique had been formed around Robert MacArthur's ghost that by the 1970s was dominating opinion on what should be the proper subject for ecological research. It turned out to be those areas that had been investigated by the mathematics of Robert MacArthur: competition theory, the subject of his Three Influential Papers, and island biogeography, a natural outgrowth of those ideas about competition. What MacArthur wanted to show, and tried to show with his simple mathematical models, was that the patterns having to do with the number of species and their commonness or rarity seen in communities were a result of competition between members of the community. His warblers avoided competition in the various ways he showed in his doctoral dissertation. They occupied slightly different niches. One of his Three Influential Papers described a model that came to be known as the "broken stick" that he thought would explain a particular pattern of the abundances of competing species. The pattern is that, in almost any environment, a few species are so abundant that we consider them to be common. Most are less so, others are rare.

All this was probably well known to naturalists of Darwin and Wallace's time, but it became a focus of ecological research after Frank W Preston published his analysis of it. In 1948, he devised a way of pictorially presenting data on the relative abundances of species in an area. The problem Preston solved was a highly practical one: how to graph the data. Graphing, remember, is, of course, a first step in discerning how two variables might be related. It puts us on the road to theory. It shows us the form of the equation.

In any community, the most abundant of species have many more specimens—Preston gave examples of tens and hundreds of thousands—than those less common, which might number in the dozens or less, or the single specimen by which many species are represented in a sample. Finding a useful scale for a graph of such data is impossible. Engineers of that era turned to specially ruled graph paper, semi-log or log-log, that compressed the larger numbers and expanded the smaller ones. Not only could all of the data now be successfully shown on a single page, but often it also fell on a straight line. The same effect could be accomplished by first transforming the data in a non-linear way, such as taking logarithms of all the data, then graphing. Preston saved ecologists from purchasing graph paper at engineering supply stores by organizing abundance data into octaves. An octave essentially represents a doubling. The eight notes that form an octave when middle C on the musical scale is 256 Hz (vibrations per second), have twice their frequencies on the next octave, at which C is 512 Hz. Increases by a constant proportion are known as geometric or logarithmic. As with the musical scale, Preston chose his octaves to represent doublings in abundance.

"Commonness is a *relative* matter," Preston wrote. "One species, we say, is twice as common as another …"

When plotted as octaves (which progressed geometrically, octave 5 representing 8 times as many individuals as octave 2, for example), the picture Preston obtained looked remarkably like the Gaussian, or "normal," curve of statisticians. A number of the species-abundance curves Preston obtained using data for birds and moths are reproduced on the following pages.

Preston's biography may strike some by-now familiar notes. First of all, he was British, born and educated in England. Second, his career was carried on in the US. Finally, his PhD was in physics. Like many Englishmen, his interests varied widely. Although for most of his life he did research and consulting on glass technology, he was an ardent bird-watcher and amateur geologist. His interest in birds led him to describe the patterns of abundances found for birds and moths mathematically, while his geological interests led to the creation of Moraine State Park in Pennsylvania. The pattern that Preston uncovered came to be known as log-normal (since the log transformation resulted in a normal curve) and as Preston's canonical distribution, a term he gave it himself. It has been the subject of much speculation in ecology. MacArthur's proposition that diversity increased the stability of a community had a basis in it. The canonical distribution was also key to MacArthur and Wilson's theory of island biogeography.

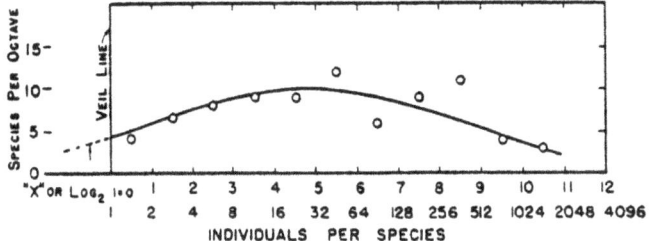

FIG. 2. Saunders' breeding birds. The octaves are definitely not equally filled, and the curve looks as if it is humped up in the middle.

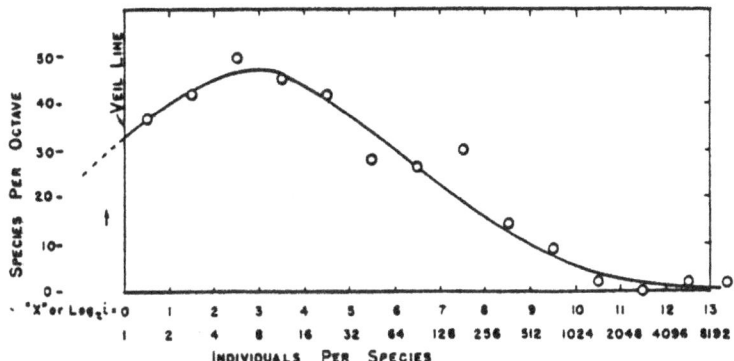

FIG. 3. Dirks' moths of Orono, Maine. The octaves are not equally filled; the curve is humped; we see the right-hand tail, but the left-hand one is hidden behind a veil.

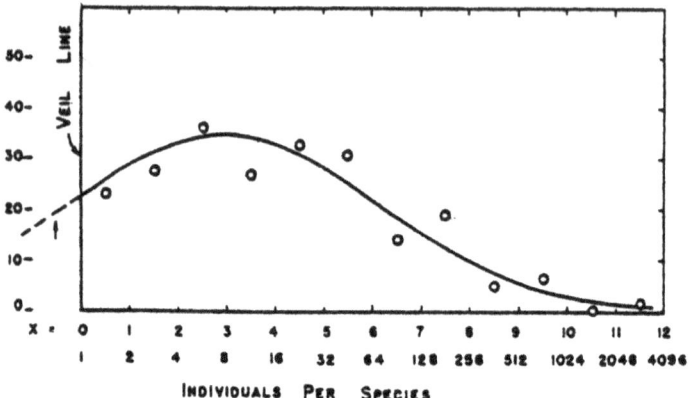

FIG. 5. Williams' moths. This resembles figure 3.

Except for island biogeography theory, it is all ground we have already gone over. It had to do with why there are so many species. Put species-abundance distributions together with GE Hutchinson's mathematically-exact-but-impossible-to-specify niche idea and you have a theory of community structure. In its simplest formulation, community theory says that species in a community are organized by competition into the niches in which they are found. If it sounds tautological, as it has been accused, it is, but it can be tested. The testing is carried out in the very best Cartesian tradition. Hypotheses are posed in mathematical form. That's the model. Predictions of the model are then tested against observation. This should mean experiment, but we have already discussed how difficult experiment is for the world of ecology. It is easier for confirmation to compare the theoretical world produced by the model to whatever parts of the real world seem appropriate.

Simberloff preferred that real experiments be done to test theories, as in his dissertation, rather than having to depend on poorly documented "natural experiments." Failing an experimental test, then at least make a statistical comparison of the fit between data and model with the fit between data and chance circumstances, Simberloff, joined by Strong and others, argued.

Nonsense, said others, Jared Diamond in particular.

Diamond, like many in the MacArthur camp, also did not start out as an ecologist. In fact, he is to this day not an ecologist. He is also, however, a best-selling and prize-winning author, a giant in the field of conservation, and a MacArthur Foundation (different MacArthur) designated genius. For much of his adult life, Diamond was a professor of physiology in the UCLA medical school. His specialty was kidney function. It is what he did his graduate work on at Cambridge. He is now a professor of geography at UCLA.

My personal impression of Diamond is more severe than Quammen's impression of Simberloff. Diamond was hardly Mister Personality when he came at my invitation to speak to UCLA's Environmental Science and Engineering students. Perhaps he was wary of any connection I might have had to Simberloff and Strong, since this had been at the height of the null model debate of this chapter and the refuge design debate of later chapters. I had recently published a paper having to do with species-area relationships that had attained a certain notoriety. Species-area relationships were exactly at the heart of the debate over refuge design. Diamond may have been waiting me out to reveal myself as being in one camp or another—or he may never have heard of my paper, did not know me from Adam, and had no reason to ingratiate himself to me in any way. I can report that he did look like a miniature version of Gregory Peck as Captain Ahab. There was the same ridiculous beard that followed only the chin line, the same dour look on his face, as if he, too, carried a painful burden with him always. Never a smile. You don't smile when you are looking for a white whale.

What was he doing as a leading ecological theorist? I have no idea. He liked birds, like many in the MacArthur camp. He liked to go on expeditions to the South Pacific. Perhaps his medical school salary subsidized that travel initially. He fancied himself to be an evolutionary biologist and biogeographer. Perhaps having followed bad vocational advice until the present, he was late in finding his true calling: he is now an anthropologist by interest in the Department of Geography at UCLA. Someone else, I imagine, is teaching medical students about kidney function. At any rate, Diamond's special area of interest seems to be the spread of human cultures during prehistory. It is an area in which experimentation is nigh impossible and statistical analysis not always relevant. Yet it is an area that lets his fertile mind run rampant. His very readable *Guns, Germs, and Steel* won him a Pulitzer Prize. It also caused a furor that, fueled by his as-popular and as-controversial later books, such as *Collapse* and *Upheaval*, continues on the Internet today.

Diamond's contribution to the debate at issue here was an almost book-length chapter in a symposium volume. In it he tried to explain the pattern of distribution of birds on islands in the South Pacific, for which he had good data. Out of that effort came a new theoretical construct: assembly rules for species in communities. The distribution patterns at which he looked had mainly to do with the presence and absence of certain species on certain islands. In short, why here, but not there? In what Diamond clearly must have then seen as his most important contribution to ecology, he developed a set of rules based on the biologies of the species available to the islands to explain why here, but not there. The rules were remarkably unmathematical and appealed greatly to common sense. Their goal was to provide a definitive answer to that question about the number of species.

"The working hypothesis is that, through diffuse competition, the component species of a community are selected, and coadjusted in their niches and abundances, so as to fit with each other and resist invaders," he had explained in the paper.

"Much of the explanation for assembly rules has to do with competition for resources and with harvesting of resources by permitted combinations so as to minimize the unutilized resources available to support potential invaders," he had added in explanation of the rules, which took up the next hundred pages.

The eight rules were remarkably simple to understand, no more complicated than the rules for chess. They were common sense, really, given the properties of the species involved, mixed in with a little island biogeography theory. Some species had excellent colonizing abilities. Diamond called them "supertramps." Certain species were able to colonize an island only if a particular other species was not already on the island. The rules were things like that. In short, the rules had to do with the ability to colonize islands and compatibilities between species. It was Diamond, though, the expert on the birds at issue, who determined which rules fit which bird in which situation.

The papers in the symposium, held in 1973, did not make it into print until two years later, leading one reviewer to remark that the book had the "dubious distinction of probably being the most cited while-in-press book in ecology." How did papers that had yet to see the scientific light of day get cited in affirmation by those who had not presumably seen them? Remember the clique that had formed around MacArthur? It was at its height. The symposium at issue was one organized shortly after his death to eulogize MacArthur's contributions to ecology.

Being in some ways akin to testimonials given at an Irish wake, the papers had never been refereed before going to press, but they spread influence on their science over those two years before publication. What is more, without facing critical review, their ideas became firmly established as exemplars to be followed. Hutchinson, a participant, pointed to the volume as one of "the best places to get an idea of what is being done at the moment." He called Diamond's chapter "a very impressive work."

Edward O Wilson, writing with Edwin O Willis in the memorial volume, implied that Diamond's assembly rules put it "within the power of science not merely to hold down the rate of species extinction but to reverse it."

It was the "species packing problem," as it became known, on which Wilson and Willis based their hopes, with Diamond's assembly rules being one of its "more sophisticated" developments. They did caution that the state of the art still needed development, but "species packing is one of the techniques of applied biogeography that seems likely to become practicable within the next several decades, on the basis of current and projected research."

For a policy maker willing to grasp at straws, it was one of the few visible straws hanging down from theoretical ecology—or ecology in general. And all this, recall, is out of respects paid to Robert MacArthur.

Once the volume was actually published, critics were skeptical, but impressed with Diamond's assembly rules.

"One of the best studies of bird community structure," a reviewer praised. The assembly rules "approach is new and deserves wider application," he urged.

"It is a monograph in its own right and will be of lasting interest to ecologists and evolutionary biologists," another praised, but went on to say: "Not all of Diamond's arguments convince, and in places they seem inconsistent."

A third wrote that while Diamond's "hypotheses are intriguing, they need to be tested thoroughly."

Who was there better to test Diamond's ideas than Dan Simberloff? Simberloff had been troubled by the lack of hypothesis testing in ecology since graduate school. He had just published a critique of a paper that used data generated by his own experiment with Wilson on recolonization of islands. He was skeptically testing his own hypothesis with his own data. That is what scientists are supposed to do before bringing a discovery out in public. They first consider all of the ways they might have gotten it wrong. What caused Simberloff to reanalyze his data after the fact was a paper, one of whose authors was Richard Levins, that found evidence for competitive interactions between species in Simberloff's data. No, Simberloff wrote, the changes found were no different than what might have been expected from random occurrences. It was his first use of a null hypothesis. It followed quickly on the heels of Hal Caswell's paper using the mathematics of neutral alleles to look at communities.

By this time, Simberloff was already in a tiff with Diamond over the application of island biogeography theory to the design of nature reserves. Simberloff, who should have been the foremost expert on island biogeography theory, given MacArthur's death and Wilson's immersion in his new science of sociobiology, thought the application premature.

I can't say that the Florida State group went after Diamond's research. It just seemed to have happened that way. What they took on instead with their first of the set of null model papers was an old chestnut in ecology: how the differences observed by Darwin in the finches of the Galapágos Islands could have arisen. That it was through natural selection was not, without evidence for it, a sufficient answer to modern ecologists. Extensive data obtained by British scientist David Lack on beak dimensions, wing sizes, and distribution among islands and later by Canadian (formerly British, now Princetonian) Peter Grant on their behavior, together told a very pretty story of natural selection acting to reduce competition. In sum, the finch communities on the Galapágos were structured through competition. That is the old chestnut. Evolution works through natural selection to make closely related species different when they are found together. It is either that or one goes extinct. Or one goes extinct and another arrives from another island that is sufficiently different not to breed with the first. Or one winds up on another island where the same is true. New species form just by that mechanism. But how do we know it's true?

Other evidence that competition structured communities was something called "character displacement," a term coined by EO Wilson and ant expert William L Brown, Jr, when both were at Harvard. Species that never coexist, being found in slightly different environments, are never entirely identical. Brown and Wilson thought that competition and natural selection made them different in areas in which they did coexist. That similar species are not found together could, of course, also be the result of a multitude of other reasons. Proof of competitive exclusion, therefore, required subtle methods.

When two similar species are found together, went Brown and Wilson's reasoning, natural selection favored offspring that differed from the other species in a way that lessened competition. Character displacement, then, could have driven the changes that resulted in the differences in bill sizes that David Lack had reported for Darwin's finches. This was an important idea to evolutionists. Lack's findings had come at just the right time to support what had restored Darwinian theory to its current position in biology. Before that, there had been a sort of Dark Age in which mutation, Lamarckism, and "hopeful monsters" abounded in explanations for evolutionary changes. The New Evolutionary Synthesis gave evolutionists a more complete story to tell about how those finches became different. Yet Lack's data was only suggestive, rather than confirmatory of competition-driven changes. What he showed was a pattern that was consistent with a theory, not a mechanism and an experimental test. Brown and Wilson's character displacement added logic, but not proof, to the story.

Scientific proof needs data. Hutchinson thought he had found some in a mathematical rule to show competition between closely related species. He observed that the sizes of feeding-related measurements, such as beak length, varied by a ratio that seemed to approach 1.3 for closely related species in a community. Each was some 30% larger than the next. He interpreted his finding as resulting from character displacement. His inspiration had come from examining two species of water boatmen coexisting in an artificial pond located below a cave in Italy that served as a sanctuary for the bones of Santa Rosalia. For his water bugs, body size was a food-related measurement all by itself. Ever the misplaced classicist, Hutchinson could not fail to also report what little was known about this patron saint of Palermo, whose bones they were.

His evidence for character displacement was meager, and better evidence was either not forthcoming or contradictory. Ratios were reported in support of Hutchinson's contention that were perilously close to 1.0, essentially no difference in size of feeding apparatus in coexisting species. Not much support. Papers actually concluding that there was no support for competition-driven differences in size ratios found their way into the literature only rarely, though. The concept had appeal and was held to. The "1.3 rule" became dogma. Progression in sizes among co-occurring species in a community became indicative of competitive interactions. Again, it was not proof, however.

Peter Grant and his colleagues tried to dispel any lingering doubts about the importance of competition to the structure of the communities of Darwin's finches in the Galapágos Islands in a data-filled paper in 1976. A previous study, in 1961, had looked at Lack's data and at the vegetation found on the islands. That study found that the differences in finch beaks could be explained just as well by the differences in the plants found on specific islands, and thus the food available to the birds, as by competition. The birds may have become adapted by their beak sizes to the particular habitat available to them on an island—in the absence of competing species. The differences then remained when two species came to cohabit the same island. It was disproof, in a sense.

Using a state-of-the-art linear regression analysis, Grant and the Abbotts, a husband and wife team, concluded that it was both food supply and competition that resulted in the pattern of species diversity found in Darwin's finches. Vegetation differences may well have started the changes in beak sizes, but competition put the finishing touches on their divergence. (Regression analysis is a sophisticated way of comparing a pattern to that expected from what are singled out to be its causes.)

Not so fast, said Simberloff and his graduate student, Edward F Connor, in a paper just as full of data and only a year later. Connor and Simberloff did not specifically write the paper in criticism of the study by the Abbotts and Grant; theirs had just been the latest in a series using the Galapágos and other island fauna to support ideas about competition. None of those papers had the statistical niceties Simberloff wanted to see. Simberloff knew from his own experiments how much randomness there was in determining which particular species reached and persisted on which particular island. How then had the Galapágos and other island studies incorporated chance events?

They had not.

You should have used a null hypothesis, Connor and Simberloff told the current experts on Darwin's finches. They provided and tested two null hypotheses. Both suggested that a large part of the differences found in measurements of birds on various islands might just as easily have resulted from chance occurrences having to do with whether species get to an island and survive there. In their model, those chances were unaffected by other species present on the island. (That is, unless the island had already been dealt the number actually found on it. Colonization stopped, in a sense, in their model when an island known to have 3 species of finch, as an example, attained that number of randomly chosen finches. (Any additional species, it might be argued, were then kept off by competition, an example of one of the annoying difficulties in building a null model.))

Connor and Simberloff also took issue with the data used in showing competition on island communities. Trying to figure out what species is known to be where is always fraught with difficulties when using data originally collected by others, as was true of a number of the island studies.

Connor and Simberloff also said a few things about Jared Diamond's work. It was critical, but not overly so. He had used evidence "based on weak inference," they asserted. He had assumed that the habitat for birds on the West Indies remained unchanged for 10,000 years. Things like that. They also admitted that the "occurrence distributions" they used for one of their null models was similar to the "incidence functions" Diamond used for his assembly rules. Diamond's "supertramps" behaved in the same way for Connor and Simberloff, for example.

Diamond did not come under criticism directly until a year later. It was then that Simberloff published the paper that took a second look at his own data on the colonization of mangrove islands. Coincidentally, it came out in the same issue of the *Naturalist* in which Robert Henry Peters published a defense of his critique of the philosophical underpinnings of much of ecology.

Simberloff's experiment had consisted of killing all the arthropods on a number of mangrove islands by enclosing and fumigating them. Mangroves are shrub-like trees of the tropics that throw out prop roots into the ocean. Capable of anchoring themselves farther and farther off the ocean shore, if they could be attributed with intent, it would be to fill in the earth's major bodies of water with mangrove forests. In the process, they form islands close to shore. Like Krakatoa after the eruption, the islands that Simberloff fumigated were then available for whatever creatures could get to and survive on them. Unlike the situation with Krakatoa, however, the mangroves survived the fumigation, so that only animals where removed, instead of an entire ecosystem. Simberloff then recorded what turned up, what disappeared, and what stayed on.

What would happen, Simberloff asked of his data, if you just put all of the species found in total on the islands in a bag and blindly removed them from the bag for each island until every island had the right number of species? This was one null model. What would happen, he found out, is that actual islands shared more species than expected by the random draws. Competition, however, would be expected to reduce the number of similar species on any given island. There should have been fewer shared species between islands than from random draws. His data suggested no competitive interactions between species. Organisms seemed to arrive at an island or not, succeed there or not, independently of what other species were doing. This was a result also found for plants on the Galapágos in a study Simberloff cited in his paper. Yet Diamond's assembly rules required competition between bird colonists.

Perhaps it was not a very realistic model, Simberloff, had reasoned, given that various species have varying dispersal abilities. Why not, then, allow the species to have differences in dispersal abilities for the null model? Some species are naturally better fliers, or have their seeds drift farther, or are just tougher than others and survive better to gain a foothold on an island. However, getting *a priori* data on those kinds of things was not easy and could well have led to circular reasoning. One might decide, for example, that species found on more islands are better dispersers, whether or not they really are. The model fits the data, which then fits the model. Simberloff had a way around this, though. One was by using his own data on defaunated mangroves, for which information on how rapidly various species reached islands was easily available. The other was to go ahead and assume that the number of islands on which a species is found could be used as a proxy for its dispersal ability. Rather than fitting the data observed, the second null model also predicted fewer species on each island than in reality. Where then was competition?

Diamond should do his homework, Simberloff implied, and do the work of comparing his bird data to expectations from a null hypothesis. Otherwise one has to conclude that insects and plants, for which there have been constructed null models, must differ in their biology from birds. A possibility.

The same year, Donald Strong added his voice to Simberloff's on the issue of null hypotheses. This time, along with graduate student Lee Ann Szyska, they specifically addressed some of the work with which Peter Grant had made his career: his evidence for competition between Darwin's finches. Their criticism was now predictable. In order to infer that patterns in body or beak size are the result of competition, one had to first compare that pattern to what would be expected from a biologically null model. What they did, once more, was to compare the size ratios predicted by Hutchinson (what they called community-wide character displacement) to ratios that might be expected at random from the species of interest (their null hypothesis).

The details of that argument go something like this. The species found together on islands vary in a regular way in particular features that can be given a linear measurement. (That 1.3 ratio.) The features of interest, such as body or beak size, are those that can be related to food gathering, and thus competition. Not all species are found on all islands, even though habitat seems to be available for them. The conclusion then is that competition, as mediated by island size and other factors, caused species to be found having the size ratios observed. Originally similar species either diverged to meet the size ratio found on a particular island, or intermediate species were excluded from islands due to competition. The size ratios prove it. If that sounds tautological, the Florida State group pointed it out, invoking both Peters and Popper in support. They called any theory that predicted both character displacement and character convergence due to competition a "catholicon," a term that must have made many reach for their Funk and Wagnall's. (The convergence part of competition theory was all quite logical, but need not divert us here.)

What would happen, the Florida State group asked, as Simberloff had before, if you just put all of the birds found in total on the islands in a bag and blindly removed birds for each island until you obtained the right number of species for each island? How would the size ratios turn out? That was their null model.

The ratios predicted by the null model for particular islands turned out to be statistically indistinguishable from the ratios actually found on the islands. Does that prove there is no competition? No, but it should make you less excited about explaining the patterns found as the result of some sort of competition. They might be, but then again, they might not be.

As it was, they found no evidence for character displacement, but they did find evidence for convergence. Some species on some islands had size ratios closer than expected, which was what might be expected from natural selection working on both species to adapt them to the local conditions. This supported that 1961 study that Grant had been trying to disprove for years.

"Although many imaginative theories have been invented that propose how competition could structure multispecies nature," the group wrote, throwing down the gauntlet, "such theories have rarely been treated as alternative hypotheses and, thus, have not been tested critically. Instead, research is often directed at finding particular circumstances that are consistent with competition theory, while ignoring contradictory evidence."

These were fighting words. Competition is as important to Darwinian theory as Darwinian theory is to ecology, at least that part of ecology not having to do with ecosystems.

It Wasn't a Barroom Brawl. It Was an Athletic Contest

The testing of community theories, much of it with data gleaned from the literature, had to that point been done only to see if competition could be rejected as a possible explanation. Based on their results, the Florida State group saw the need for a different approach. It was very much based on what passes as standard experimentation, in which it must be shown that a result cannot be explained by chance alone. If a result could have been observed by chance alone, it was not a result, in essence, and should not be used to infer any cause-and-effect relationship. True, the replication and variety of controls that are possible for experiments are unavailable for data that had already been collected, some a century before, but the Florida State group saw a way around that.

"We propose another possibility with logical primacy over other hypotheses," they wrote, "that other hypotheses must first be tested against, but that is rarely considered at all by ecologists. This is the null hypothesis that community characteristics are apparently random."

Their method of random species lists was the way to frame that hypothesis that must first be eliminated. By arguing for logical primacy of the null hypothesis, they ratcheted up the stakes in the controversy. Simberloff had been very careful to state in his publications that he was not rejecting competition, he was just rejecting the evidence for it in the cases under inspection. Yes, your evidence could be interpreted as resulting from competition, but it might just as well have resulted by chance, he seemed to say. The two hypotheses had been in competition on an equal footing, in a sense. Now, taking a cue from the science of statistics, they were saying that chance factors needed to be disproved first.

The tone was now one of outright rejection of competition theory: "... not statistically distinct ... no tendencies ... cannot detect ... no evidence."

Although they admitted that their "results do not disprove interspecific competition," they showed that it "is not as common in nature as believed."

The term, Florida State group, is not of my own devising. They were also called the Tallahassee group and the Tallahassee school. Simberloff had taken the position at Florida State to be close to the mangroves he was studying. No doubt, at the time, he could have written his own ticket at almost any university. He had lured Don Strong to Tallahassee in order to build up what is often referred to as a critical mass in an area of research. The Florida State group included Larry Abele and lots of graduate students. It was a logical way to refer to them, by their academic institution or the city in which it is located, but in a science that was still working out whether it was to follow the lead of engineers with rolled up sleeves or British country gentlemen with binoculars, the Florida location carried a sociological connotation.

"We are known as the Tallahassee Mafia," Strong had remarked at the time.

Their situation was not without professional peril.

Strong, who is now at the University of California at Davis, a premier institution for ecology research and theory, and had been Editor-in-Chief of *Ecology* for a number of terms, was quoted to have remarked at the time: "The devout MacArthurians are all in powerful positions in powerful universities."

Eugene Odum was at Georgia at the time, recall, and Howard T, at the University of Florida. Theirs was a concept apparently on its way out in ecology. Hutchinson was at Yale, MacArthur had been at Penn and Princeton, May at Princeton, Richard Levins at Harvard, and Jared Diamond at UCLA, along with MacArthur disciples Martin Cody and Henry Hespenheide. Jonathan Roughgarden was at Stanford. Peter Grant, at the time, was at the University of Michigan, but on his way to Princeton. Theirs was a concept in ascendancy. Put it together yourself.

I once asked Nick Gotelli whether as a graduate student under Abele at Florida State he had felt that it was Ivy League versus them, a southern university. He had arrived at Florida State in the middle of the controversy.

"Absolutely," he said. "That's exactly how we saw it."

He had come from Berkeley to work with Larry Abele in marine ecology. He eventually got his degree under Simberloff.

"When I visited Florida State, I had never even heard of Dan Simberloff," Gotelli told me. "He was not mentioned in any Berkeley classes."

This had been despite the fact that one of Gotelli's professors, Rob Colwell, was busily at work at the time to try to find a way to bring down the whole idea of null models.

"Yeah," Gotelli had finally been told by the person who had suggested he go to Florida State, "there is this island biogeography nut out of Florida, Dan Simberloff, and he's an interesting person. So, when I got to Florida State, I had not seen or heard of null models, or any of this work."

He remembered having walked into an exciting situation.

"Every new issue of a journal that would come out, people would be peering through it," he recalled. "Lots of drinking in those days with arguments at bars late at night over this kind of stuff. You sort of immersed yourself in it.

"People who were reading Popper and Lakatos, We would soak up all that stuff."

There was a down side, too.

Of the *Naturalist* issue, he remembered, "People would walk around the halls holding up the issue 'Can you believe this bullshit! Look what just got published? How can they say this?'

"There was also the sense that Florida felt very different and distinct from all the other American universities. In fact, it seemed at the time that the people who thought the most the way Florida State did were actually all out of the US."

There had been a drawing up of wagons and there was retribution. There was the issue of funding his dissertation work.

"Larry tried to get NSF funding a couple of times for that," Gotelli told me, "and got vicious, vicious grant reviews back. I mean some really nasty, unbelievable stuff. In fact, after a couple of years, he kind of decided to retool and he basically left community ecology and focused all of his effort on systematics."

NSF is the National Science Foundation, a preferred source of funding for ecologists.

"I wanted to keep doing community ecology," he continued, "so that's when I switched to Simberloff for my PhD. If we had had money, I was originally going to work on the epifauna that occurred on mangrove prop roots in the tropics. There is a big community there of sponges and other invertebrates, but we didn't really have the money to do that, so I had to work closer to home. I ended up working on the population dynamics of a couple of subtidal marine invertebrates in the northern Gulf of Mexico."

There was disappointment apparent in his voice. Neither had Gotelli's professional path to his current position and success at the University of Vermont been without bumps and setbacks.

Grant and Abbott put together a very reasoned response to the Tallahassee Mafia's conclusions, taking issue with their methods. There were choices, according to them, that the Florida State group made in interpreting existing data on birds that made their analysis problematic. Using slightly different and what they thought were more realistic choices, they failed to duplicate the results of the Florida Staters. However, even though they thought that null models should have no more logical primacy than any other models, they did not reject the approach.

"Our criticisms have not been directed at the need for appropriate null models," Grant and Abbott wrote, "with which we firmly agree."

They even ended their paper with some kind things about Simberloff's mangrove experiments. Grant's subsequent work on the Galápagos finches, aided by his wife and a changing cast of graduate students, is capably detailed in *The Beak of the Finch*. The 30-year study may not have been as powerful a substantiation of Darwinian evolution in action, nor would its surprises have come to light, had he not taken to heart, perhaps, Simberloff's reminder to ecologists that there is a method to science. In the end, the best results Grant got had little to do with analysis of community-wide patterns, but were based on the extensive data provided by the natural experiments possible in a long term study. I recommend them to the reader. Grant's *Ecology and Evolution of Darwin's Finches* is my first recommendation, but the husband-and-wife team's technical paper in *Science*, "Unpredictable Evolution in a 30-year Study of Darwin's Finches," can be understood by any reader who has slogged through my text to this point.

Less reasoned was an analysis by a scientist at the Academy of Natural Science of Philadelphia. I have no idea what motivated him. I can only speculate that Ruth Patrick, the grand old lady of the academy, whose career skyrocketed through her use of MacArthur-school theories in her studies of polluted streams, may have had something to do with it.

That scientist, John Hendrickson, looked at the same data as the Florida Staters and had conclusions that were "strikingly different from theirs." He also found errors "at various stages in data assembly, computation or transcription."

I bring this paper up, even though it is just a minor skirmish in a greater war, for two reasons. The first is that the "errors in data" attack is always a good one to the desperate. By claiming errors in a study, even though those errors have no effect on results, one is disparaging the competence of the workers making the study and, indirectly, any conclusions they may have made from it. Any study using large amounts of data that was not generated by the investigators doing the study and that is available for perusal, as it should be, will have room for differences in interpretation—and even outright mistakes. They happen in any activity. The question is whether they are mistakes that are fatal to the study. It might still be a perfectly valid study even in the face of computational errors. Someone not familiar with the research might easily conclude, however, that a study identified by others as having errors is too flawed to believe.

It's a cheap shot, in other words, that might better not been printed, or even written.

The second reason that the paper is interesting is the response given to it by Strong and Simberloff. The kid gloves were off. There is a tone of arrogance to it right from the title, "Straining at Gnats and Swallowing Ratios: Character Displacement."

"Our only disagreement is with his conclusions," they wrote. "He strains at statistical significance, finds little, and thus substantially reinforces our original inferences."

Noting that out of 15 tests Hendrickson made with his now error-free data only two pointed significantly to competition, they took issue with his "obsessive seeking of interspecific competitive explanation," while ignoring simpler alternatives. In principle, this alone should have kept the paper from ever being published. It is standard editorial practice to instruct referees to reject a paper that presents conclusions that are not supported by its data. It is clear that Hendrickson had produced just such a paper. That editorial misjudgment, however, may have been what caused Strong and Simberloff to instruct the ecological community on some basic principles of science.

At any rate, they were not going to be quiet in the way they made their points.

"General alternative hypotheses such as character displacement," they explained, "are not proven by finding one significance here, another there, each with different fauna, morphological feature, and combination of tests; the haphazard and sparse catch of Hendrickson's usually empty nets is only a red herring."

The reference to fishing has to do with a clever discussion of "data dredging" that Strong and Simberloff borrowed from two statisticians. Data dredging is a non-pejorative term for seeking patterns in the available data. To ecologists of that era, there sometimes appeared to be an intractable morass of data to wade through. Data dredging would be the weak statistical techniques analogous to a fisherman, having no idea what he expects to catch, lowering his nets out of hope. The hope is that that first catch might then lead him to change to techniques more suitable to the fish found.

The scientist-fisherman, in Simberloff's view, would have hypothesized a catch, based on whatever theory he was testing, then checked his nets for that fish. The data-dredger makes his hypothesis after looking into the nets. Worse yet, he uses the same catch to test his hypothesis. At the same time, in another paper, Simberloff was also accusing Hutchinson and MacArthur, both for data on the "1.3 Rule", and Diamond of dredging for data in another of his papers. Diamond, for one, may not have seen the characterization as non-pejorative.

The character ratios with which ecologists had become so enamored were just the result of such data dredging, according to Strong and Simberloff. It led them to another analogy.

"As evidence," they wrote, "these infrequently significant ratios are analogous to rare, large impressions, reminiscent of footprints, in the Himalayan snow. Both are flimsy; neither can shed much light on cause."

Their target was broader than just Hendrickson. They indicted an entire school of ecology. The cast of characters soon reached unmanageable proportions.

David Quammen punted his responsibility to the reader on this issue in his *Song of the Dodo* by giving up short of where I do. Instead, he wrote that the journal papers "pile up into a longish list: Gilpin and Diamond (1976), Abele and Patton (1976), Brown and Kodric-Brown (1977), Simberloff (1978), Diamond (1978), Abele and Connor (1979), Connor and Simberloff (1979), Gilpin and Diamond (1980), Simberloff and Connor (1981), and dozens more. Each of those papers was full of elaborate logic and conviction. Many were besmeared with math. Some also contained facts."

Other than arrange the dates in order, Quammen did little to inform the reader. His excuse seemed to be: "The specifics were intricate and tendentious. I recommend them as a cure for insomnia."

One of my own concessions against insomnia has been to reduce the cast of characters. Often, when I attribute views to Strong or Simberloff, I neglect to add the others, not all part of the Tallahassee Mafia, who held or expressed those same views.

Getting back to main characters, gloves were also off when Simberloff Connor took on Jared Diamond's work specifically.

"We challenge Diamond's idea that island species distributions are determined predominantly by competition canonized by his 'assembly rules.' We show that every assembly rule is either tautological, trivial, or a pattern expected were species distributed at random," they wrote, then they set about to dismantle each of Diamond's rules either logically or through the use of null models.

The quotation given makes up the bulk of their abstract. (As an aside, Connor and Simberloff present Diamond's rules more clearly in that paper than Diamond did in his monograph.)

By this time, Simberloff and Diamond were no longer communicating directly with each other, an unusual breach of scientific collegiality.

"The reason for the debate is primarily for protecting and building reputations," Simberloff has said. "A lot of people want to be famous, especially in a young science, like ecology, where it is still possible to jockey for prominence. A lot of people would like to be viewed as the heir to Robert MacArthur."

Apparently, MacArthur had designated Diamond as his heir apparent, or so at least one story went.

And it was not just that the MacArthur school did not use null hypotheses that it seemed to have been targeted by the Tallahassee school, but that the MacArthurites were afflicted with a general sloppiness of approach compared to other sciences. We are not quite back in the butterfly-collector territory that was such anathema to Stanley Auerbach, but there may be some Little-Old-Lady-in-Tennis-Shoes birdwatchers in view.

The butterfly collectors, by the way, try as they might, had difficulty fitting their data to predictions of the MacArthur school. Oh, the species abundances of butterflies and moths made very nice log-normal patterns, but what was underlying those patterns was not competition. Simberloff's frequent colleague back then, Don Strong, demonstrated that quite unambiguously through his research on insect communities.

Br'er Rabbit and the Tar Baby

The Diamond and Simberloff controversy came to a head—or it's lowest point, from another way of looking at it—in a symposium of papers, sometimes referred to as the Wakulla conference, after its Wakulla Springs location, not far from Tallahassee, collected by the Florida State group. It was an "anti-Cody and Diamond" (the editors of the symposium volume devoted to the works and influence of Robert MacArthur).

The set of papers that are the most quoted from the Wakulla conference were never presented there, it turns out. By that time, Diamond was refusing even to be in the presence of Simberloff, perhaps in fear of a barroom brawl breaking out. Instead, he carried on something like a long-distance school-cafeteria food fight.

Here is how Simberloff remembered it: "The general idea was to have a bunch of people talk about some of the issues in community ecology. Diamond did not attend, but Peter Grant, Tom Schoener, Jim Brown, Rob Colwell, and Stuart Pimm, among people we generally did not agree with, did attend and gave papers. My paper was about birds in the Tres Marias and Galapágos, because Grant was there. I can't remember if Gilpin attended, but he might have."

How things that were never said at the conference got into conference proceedings is explained as follows: "Bob May, who was there, wanted us to publish the book with Princeton University Press; we had originally been thinking of Harvard University Press, but he made a good pitch. However, he said he couldn't consider it if Diamond didn't contribute a paper. So we invited him, and he gave the paper (senior-authored by Gilpin) reaming us out."

By "gave the paper" Simberloff presumably meant through the mails.

"It was just atrocious," Simberloff continued his reminiscences, "many errors, and vicious. So Ed Connor, who hadn't contributed a paper, and I, as the people attacked, wrote a response, then they wrote a response to our response, and we wrote a response to their response. All are in the book."

By the time the book came out, it was clear that the Florida State group had already made the larger point: community research was no to be longer done without null hypotheses. For example, a book edited by Ted Case and Martin Cody, dyed-in-the-wool MacArthurites, was soon criticized for use of an inappropriate null hypothesis.

There had been holdouts, however.

"Finally, the last thing ecology, or any other science for that matter," wrote a scion of the field, "needs is to have a few self-appointed guardians of *the* scientific method (note that the editors use the singular with great emphasis) tell others in the field how to do science."

Those young whippersnappers! Imagine, trying to tell us how to do science!

But wait! Is there an echo here of the lament just a decade before that there was a clique around Robert MacArthur trying to control what research ecologists do? This was not a minor skirmish in a larger war. This truly was a counterrevolution.

It also echoed an older controversy, carried on at another resort, Cold Spring Harbor, on density-dependent population regulation. Bird populations seemed to be density-dependent, insects not. This time, in more southern climes, it was established that insect communities were not structured by competition. Birds, of course, were, although one bird ecologist did express his doubts.

"I began my own studies of communities fervently embracing the existing views of competitively structured equilibrium communities," he wrote. "But I have become skeptical of much of this dogma, and believe now that we know far less about the patterns and processes of communities than we think we do."

To that turncoat, John Wiens, the evidence was just not convincing. What was needed, he had decided, was "manipulative field experiments." Short of that, "the temptation to generalize prematurely should be assiduously avoided."

Besides some more vituperation, what the Wakulla Symposium volume had to add to the controversy was firm agreement that producing an appropriate null model was not a simple process. The Florida State group had, like Hal Caswell, posed the simplest null models they could devise, with absolutely no biology whatever to them, if possible. No, said demurrers, you have to add to your null model all of those things so dear to biologists, with the exception of that one being tested. Hmm. For example, instead of communities, study guilds, only, said Diamond and Gilpin. Then the null model can be rejected. Besides switching the argument to one over guild structure, Diamond and Gilpin also had the following to say.

"In summary, the 'null hypothesis' by Connor and Simberloff is characterized by hidden structure, inefficiency (reliance on tedious, expensive simulations), lack of common sense (diluting the effect they seek to demonstrate), imprudence and statistical weakness (failure to note the breakdown of their procedure in two limiting cases: checkerboards and nested distributions), and ultimately by scandalous disregard for the result of their own procedure (claiming acceptance of their 'null model' when it was actually rejected at levels as extreme as $p < 10^{-8}$)."

(Gilpin, by the way, was at the University of California at San Diego. There was a sort of symmetry to the argument, as Quammen noted, but it was not two against two, but Southwest versus Southeast. You might want to look up in old *US News and World Reports* the relative status then of the institutions involved.)

It was a remarkable diatribe, crafted in a way that knowledgeable—even some expert—readers would take from it the impression that Connor and Simberloff's arguments had to have been, indeed, flawed. Diamond even refused to use the term, null, without having it set off in parentheses, "lest the reader be lulled into accepting its implication of primacy for their particular hypotheses." Passages like that are just not found in the scientific literature. Where there is smoke, there must be fire is the usual suspicion about them.

There was a lot of smoke. Statistics can be applied arbitrarily and selectively—and inappropriately. Those low probabilities of random effects calculated by Diamond and Gilpin using "results obtained by Connor and Simberloff" did not invalidate their conclusion. Statistics are meant to keep investigators from foolish interpretations of results; there is no guarantee that the same statistics might not take off into foolishness, themselves, when followed blindly. Statistical tests will often produce a very low probability of randomness for a minuscule change. It usually happens in cases were an awful lot of data, perhaps way too much data, has been collected. It's up to scientists knowledgeable about what is under study to decide whether what is significant is also important. Connor and Simberloff showed graphs that, according to statistics, were different, but looked remarkably similar to the eye. Those graphs remained unchanged by Diamond and Gilpin's calculation of a low probability of similarity.

The argument about guilds was more smoke. A guild is a concept invented at a young stage in the career of an ecologist who was desperate for immortality. The ecologist was Dick Root and the guild concept came out of his dissertation work having to do with competition in a "guild" of five species of insect-eating birds. It was sort of a West Coast version of MacArthur's study of warblers. Root really knew his gnat-catchers, but at Cornell, he spent a career studying the ecology of insects. Like MacArthur, Root also had a personal magnetism. He was beloved by his students in ways no other professor was then at Cornell. "An immortal rock ballad, 'In Dr Root's Garden,'" was recorded in homage to him by the group Chrysalis. Another of his students led a spoof of the null model controversy by finding no evidence of music in the prelude to Bach's First Cello Suite. The spoof was itself spoofed. Nonetheless, Diamond and Gilpin used it as one more point in their favor in the null model debate. The whole musical affair warrants much less attention even than I am giving it here. Be that as it may, Root has to his credit the guild concept. It is now in all of the undergraduate textbooks as a useful ecological concept. Immortality, after all.

The use of guilds in place of communities is a narrowing of Diamond's assembly rules. In addition, guilds are difficult to define unless one sticks to species that are closely related. Competition, however, can be much more diffuse. Is a caterpillar munching on a birch leaf competing with the yellow-bellied sapsucker using his woodpecker beak to draw sap from its bark? In combination they will destroy their resource much more surely than if each was exploiting the tree independently. Are they in the same guild? Are they in the same community?

Diamond's original "working hypothesis" was that communities were structured through "diffuse competition." Diffuse competition, one can conclude, would be like the example of the caterpillar and the sapsucker. In separating smoke from fire, bringing up guilds was changing the argument. The shifting target strategy used by Diamond and Gilpin is an effective one for debate, but not science. It may win a point, but it will not lead to understanding.

This brings us full circle back to the dialectics of the *Synthese* volume and the *Naturalist* issue that were devoted to the philosophy of ecology. There are also some parting shots of interest. Diamond and Gilpin took the Florida State group to task for a number of flaws in Connor and Simberloff's rather harsh criticism of assembly rules. They did so by invoking the ghosts of Darwin and MacArthur, in whose ideas competition played a primary role. In so doing, they were the first of whom I am aware to formally identify themselves with a "school of ecologists" aligned with Robert MacArthur. They were proud of it.

Oh yes, and they also accused Connor and Simberloff of ignoring their evidence. Not only that, but they accused the Florida State team of questioning their morality.

"They now recognize this issue and proceed to befog it by applying the term 'degeneracy,' with its mathematically and morally pejorative connotations," are their exact words.

"Degenerate" is (or was—maybe in a world of PC, it is no longer allowed) a value-free term used in mathematics to describe the simplest, often trivially so, special case of a broader category. A circle, for example, is a degenerate ellipse; a straight line, a degenerate curve; a point, a degenerate line. Gilpin and Diamond's words seem to accuse Connor and Simberloff of inventing the term to use against them.

There was no intended insult, there were no flaws, Connor and Simberloff replied, and there was no published evidence from them to be examined by others.

Oh yes there was, Gilpin and Diamond replied. Just look at our two 1982 papers.

Those papers were published in the journal *Oecologia*. Recall the weight given to papers based on rigor of review: *Science* first, then *Nature*, then journals published by scientific societies, then those of commercial presses. The null model debate was played out in *The American Naturalist*, *Ecology*, *Ecological Monographs*, and *Evolution*. Diamond's rejoinder came only in collaboration with Michael Gilpin and in *Oecologia*, published by a commercial press. Whether they first submitted to one of the other journals is unknown, but the delay between the appearance of papers critical of Diamond's assembly rules and the *Oecologia* papers suggests it. (Not that *Oecologia* publishes rubbish—the problems with commercially published, open-access journals appeared a generation later and mostly in the biomedical field—but it is beneath the other journals mentioned in this paragraph in its standards.)

Diamond's original publication outlets were a symposium edited by himself, in which his paper would have received scant—if any—peer review, and the *Proceedings of the National Academy of Sciences*, one of MacArthur's favorite publication outlets, where papers submitted by way of members are accepted without question. Like MacArthur, Diamond may have been speeding his way to fame by bypassing the process of review.

By the time the Florida State group put out their own unrefereed volume, Connor and Simberloff began to show evidence of weariness with the whole argument.

"We are beginning to feel like Br'er Rabbit enmeshed in the Tar Baby," they wrote. "Each of our attempts to elucidate the statistics of species combinations on islands seems only to elicit a mass of goo."

"With no briar patch in sight," Connor and Simberloff referred readers back to their previous work.

This was not "a hot little battle fought out parenthetically to a larger war," as readers of *Song of the Dodo* have been informed. It was a major conflict having to do with how a science is to be conducted.

The conclusion of the conflict? Ecology has changed in the direction Simberloff wished it to go. Perhaps null models really have no logical primacy, but at least their use admitted that the theories proposed to explain patterns found in nature needed more rigorous tests than the coincidence of similarity. Still, there are ways around them. To ecologist Brad Hawkins, who showed that a specific null model for geographic changes in diversity was untenable, the difference between null models and neutral models remains that "one was useless and the other pointless."

So does competition structure communities? We still do not know, although it is very likely to be true. That had not been the point of the Tallahassee Mafia. Experimental evidence showed that it did not structure some communities. Experimental evidence was needed to show that it did structure other communities. They were not denying competition; they were denying some of the evidence for it. The bird communities that Diamond examined, and some others, probably are structured by competition for resources; insect and marine communities probably are not. Theoretical edifices such as those put up as general truths by MacArthur in his studies of birds should be used only with care. Bird ecology might be different from insect or plant ecology. The organisms are certainly different.

What on the surface was a neutral panel of ecologists, including Sir Robert and two British scientists surveyed the status of the null controversy in 1983. Their conclusion?

"The recent emphasis on the need to evaluate perceived patterns in community structure against null hypotheses is—as seen with hindsight—overdue."

This was around the same time at which Sir Robert was being quoted elsewhere that "it is paradoxical that some of those who are most sensitively aware of the need to keep sight of alternative explanations for observed patterns in community structure seem, at the same time, occasionally to accept that there is only one True Way to do science."

The revolution—and its counterrevolution—was ending. In proper Kuhnian fashion, however, it will not end until it outlives its participants.

Oh, by the way. That evidence in the *Oecologia* papers? It wasn't there. It had to do with actual evidence of competition, not just its inference from the distribution of species on islands. In case I missed it somehow, so did Sir Robert.

Life is Simpler on an Island

Had Darwin not stopped in the isolated Galapágos Archipelago, he would not have seen so clearly the evidence for what he later termed natural selection. Almost half a century after Darwin, David Lack published a book appropriately titled *Darwin's Finches* that filled in some of the details of the adaptive radiation alluded to by Darwin. Lack's 1947 book spurred much of the work on competition and evolutionary ecology that has been the subject of the last few chapters. Much of that latter work, like Lack's and Darwin's, was based on single visits to the islands.

At about the same time that Lack had been making sense of his birds, Eugene and Howard Odum were measuring the structural and functional characteristics of another, even simpler island, the Eniwetok Atoll ecosystem. Their study showed a path for ecology totally different from that set out by Darwin and Lack.

Much evolutionary biology has been learned on islands. Some of it is very scary. Island species are particularly sensitive to extinction. The loss of species in Hawaii is a well-documented tragedy. Of course, the nonexistence of the Hawaiian Islands throughout the vast majority of the history of life on earth would argue against the necessity of those species. Still, they were there and now no longer are. Does that raise ecological concerns? It does to many ecologists.

Recent lessons learned from islands have been quickly applied to practical matters. The ecology of islands is simple enough to be easily untangled. There are fewer species, fewer numbers, less variation in general, and—maybe more importantly—the living things on them pretty much stay put. Yet I doubt that those conveniences are the only reasons for the attraction of islands for ecologists. Pond and reef ecosystems are as amenable to analysis as island ecosystems, but bibliographies on community and evolutionary studies based on them do not come close to matching in extent the studies of just the birds on the Galapágos. If you stacked all of the studies on each into separate piles, I would bet that the ones on island birds would tower over those for ponds and reefs. Not surprisingly, the names of those involved in the studies are nonoverlapping. Ponds and islands represented habitats for ecologists having separate traditions.

Is it that the Galapágos Islands were so important in helping Darwin to formulate his principles of evolution that leads ecologists to continue trooping down to their barren landscapes to the present day in search of answers to ecology's riddles? It obviously was for the Grants. Needless to say, the ecologists that flock to islands are now exclusively evolutionary ecologists rather than ecosystem ecologists.

There is a well-known, but cautionary example on how lessons learned on islands might apply to practical matters, however. This is that of the dodo (the bird that things are as dead as) on the island of Mauritius. The dodo is more than dead, it is extinct. The way that extinction might now be threatening a local tree with its own extinction is often used to illustrate the importance of keystone species. Unfortunately, it turns out to have been just some ecopoetry. Ecology, especially evolutionary ecology, mathematical evolutionary ecology, in particular, was positively waxing ecopoetical in the 1970s.

In 1977, Stanley Temple, a wildlife ecologist with the University of Wisconsin, published a rather speculative paper on the history of the dodo. Its title, "Plant-Animal Mutualism: Coevolution with Dodo Leads to Near Extinction of Plant," explains itself very nicely. Surprisingly, his speculation landed in *Science*. The dodo, his story went, disappeared from the island some 300 years before, mostly as the result of the need for sailors to obtain fresh meat. Although not particularly appetizing, based on descriptions of its taste that have come down to us from the sailors, the dodo was, at least, available and easily captured. Unafraid of humans, it could be walked up to and clubbed to death.

Located in the Indian Ocean, Mauritius had been a convenient refueling stop for early ocean voyagers. *Calvaria major*, or the "Dodo Tree," as it has become known in the environmental literature, also found only on Mauritius, also coincidentally stopped germinating 300 years before. Three hundred years is an ancient age for Dodo Trees, which, according to Temple were now threatened with extinction because the dodo was no longer present to have seeds pass through its digestive tract, making them suitable for germination in the process. No doubt the paper got through *Science's* rigorous review because, like Ali Baba in the fable, Temple had found and used the magic words that were then the *open sesame* to favorable review. They were mutualism, coevolution, and extinction. Like the dodo's stomach acting on seeds to prepare them for germination, those words acted on reviewers to prepare the paper for acceptance. A good story, that of the dodo; it is used to argue that care should be used to maintain biodiversity, for its inadvertent loss might just bring down an ecosystem.

Many other species depend on the *Calvaria* tree for food and shelter. That is the potential for ecosystem collapse. The story of the dodo, however, turned out to be mostly apocryphal. Parrots may have been more important to the *Calvaria* tree than dodos and the tree may not have been heading for extinction, anyway, according to researchers who spend all of their time on, rather than just visit, the island. The story does not make a good case to fear the unnoticed loss of a species, anyway. The dodo was hard to miss. It could have been easily managed away from extinction just by keeping sailors from eating the birds, had that era been as enlightened environmentally as today's.

Still, island examples are useful. Phenomena that can be perplexingly complicated in mainland ecosystems become clear in the simpler worlds of islands. A complete census of species is possible. In fact, for some species on some islands, a complete population count can be obtained without superhuman effort. Variations between islands, such as size, isolation, and climate can be easily quantified. In addition, since all natural habitats are insular in one degree or another, the principles derived from the remotest of archipelagos should be applicable in some degree to all habitats.

In 1963, Robert MacArthur, collaborating with EO Wilson, continued to give ecologists reasons against the ecosystem-as-superorganism way of doing ecology. What MacArthur and Wilson did was take some patterns found for the numbers of species on islands and fashion a theory in explanation. Out of it came a mathematical model, $S = CA^z$, that on first view appears to be every bit as elegant as $F = ma$ and $E = mc^2$. Like the two famous equations from physics, the species-area equation, as it is known, was developed from basic underlying principles, had data in support, and could be shown to have testable consequences. It has been called the "most seminal branch of ecological theory" and "one of community ecology's few genuine laws." With that equation as ammunition, MacArthur and Wilson had taken aim at creating a general theory to explain a number of the things that ecologists had been puzzling over.

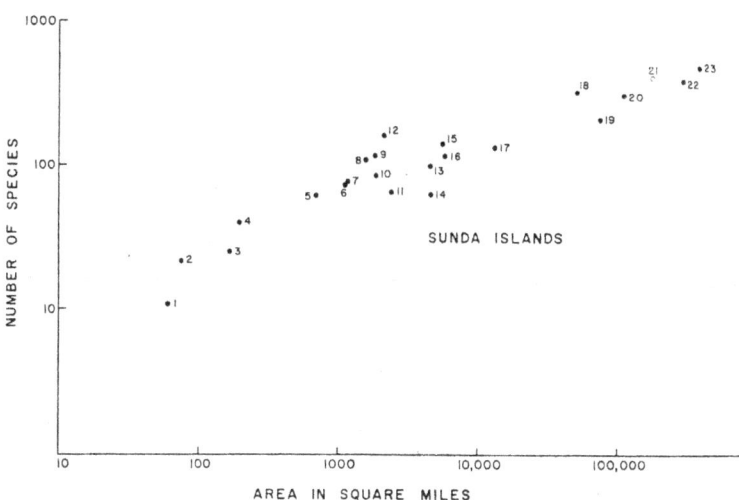

FIG. 1. The numbers of land and freshwater bird species on various islands of the Sunda group, together with the Philippines and New Guinea. The islands are grouped close to one another and to the Asian continent and Greater Sunda group, where most of the species live; and the distance effect is not apparent. (1) Christmas, (2) Bawean, (3) Engano, (4) Savu, (5) Simalur, (6) Alors, (7) Wetar, (8) Nias, (9) Lombok, (10) Billiton, (11) Mentawei, (12) Bali, (13) Sumba, (14) Bangka, (15) Flores, (16) Sumbawa, (17) Timor, (18) Java, (19) Celebes, (20) Philippines, (21) Sumatra, (22) Borneo, (23) New Guinea. Based on data from Delacour and Mayr (1946), Mayr (1940, 1944), Rensch (1936), and Stresemann (1934, 1939).

The theory itself is as simple as its mathematical summary. It predicts that the number of species will be greater on larger islands.

That relationship was already well known when MacArthur and Wilson trotted out their equation. Surveys of similar islands having different sizes invariably found that the number of species, birds, butterflies, whatever, present varied in reasonably predictive ways with island size. It was the underlying reasons for the pattern that had not been identified to complete satisfaction before.

On the page opposite, I have reproduced the species-area graph that MacArthur and Wilson presented in their paper, scanning it from the musty pages of the library copy of the journal in which they first published it for the benefit of the reader. By now you should be able to discern that there is a linear trend and remember that where there is a graph, there is a corresponding equation. $S=CA^z$ is it. Note that the axes are not linear. The scale is a logarithmic one.

MacArthur and Wilson's insight was to see the situation in the graph as the outcome of balanced factors. According to them, the number of species on islands is determined by rates of colonization by species happening upon the islands and the extinction of those already there. When as many species are lost from an island as colonize it, when those forces of change are balanced, the equilibrium number for that island is established. Using simple logic, some differential equations, and graphical representations that look suspiciously like supply and demand curves from economics, MacArthur and Wilson then tied the rates of colonization and extinction to the size of the islands and their distances from a mainland source of colonizers. That fit the species patterns observed for birds on island archipelagos in the Pacific.

I have also reproduced one of their theoretical graphs from that same musty-smelling journal on the following page. The equilibrium number of species for an island is read off by dropping a line down from the point at which two curves (representing distance to mainland and size) intersect. There are multitudes of lines on the graphs, representing the multitude of possibilities for each curve. Taking any one of the curves labeled "near" to "far" on the left and tracing it through the set of curves labeled "small" to "large" on the right, one finds that the intersection point for it is at a larger species number on larger islands. Larger islands should therefore hold more species than smaller, as is true in a number of studies used in their analysis. Islands that are further away from the mainland should have fewer colonists reaching them and thus the equilibrium number of species should be less. This can be seen by tracing a single line for a particular island size through the curves representing distance from mainland, the mirror image of the previous process. The data in support of this are less impressive, either anecdotal or derivative, rather than direct. Nature seemed not to provide sets of similar islands in a progression of distances from a mainland.

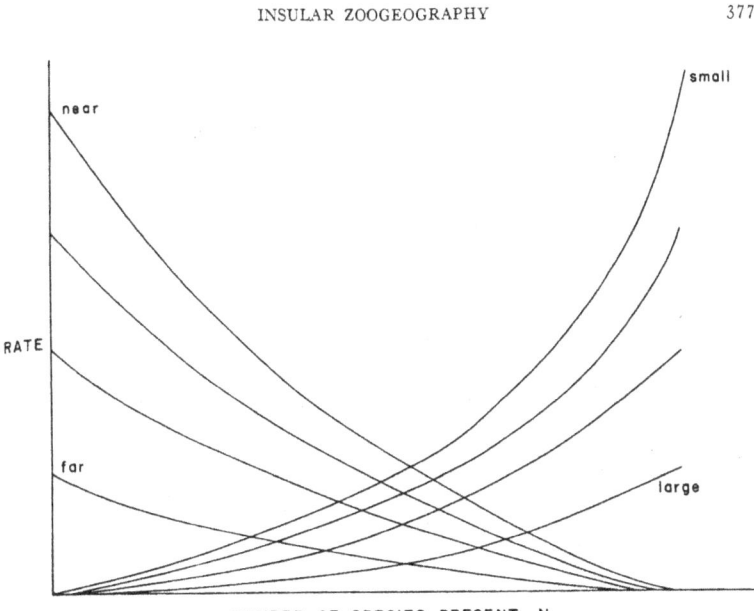

FIG. 5. Equilibrium model of faunas of several islands of varying distances from the source area and varying size. Note that the effect shown by the data of fig. 2, of faunas of far islands increasing with size more rapidly than those of near islands, is predicted by this model. Further explanation in text.

Nonetheless, the theoretical distance effect is what distinguished their ideas from those of Frank Preston, he of the lognormal distribution and the glass technology laboratory. Preston had published a paper that went on and on about every possible aspect of species-area curves. It was so long that it had to be published in two parts in separate issues of *Ecology*. It came out in 1962, however, beating MacArthur and Wilson's paper by a year. After all of his analysis, Preston managed to blunder through a statement that MacArthur and Wilson interpreted as expressing the idea that island faunas were at equilibrium, as they were proposing also. They must have sensed the importance of their result, for they went to pains to establish their priority over that of Preston. Although he did discuss the remoteness and isolation of islands from each other and the mainland, Preston had missed the chance to specifically put a distance effect into his mathematical model, $N=KA^z$. Distance, Preston had as an unspecified complication of "isolation" in the exponent, z.

Mathematically, MacArthur and Wilson's model incorporated Preston's log-normal distribution. That had been Preston's *entrée* into species-area relations, through his canonical distribution. If one starts exhaustively collecting organisms in an area, most of one's sample would consist of a few dominant species, but at the same time, most of the species would be relatively uncommon. Preston's species-area equation was identical to MacArthur and Wilson's, and even the underlying causes Preston gave for it were almost the same. Preston emphasized an analogy between sampling and colonization and extinction. Basically, larger samples have more species; larger islands have more species.

This makes sense based on larger areas holding more habitats, with niches resulting for more species.

Increases in habitat types had been the earlier explanation for the area effect, but there seemed to be something more than that going on in the island examples. An increase in samples, therefore, more individuals and more species, were expected. To MacArthur and Wilson, however, the islands were not just areas being sampled, they were colonization points, locales at which the forces of biology placed and removed species. They (no doubt it was MacArthur's idea) presented a differential equation in explanation. I have been mostly shielding you, reader, from mathematics until now, but it is time that you set eyes on a differential equation. It is part of what biologist Joel Cohen, who introduced us to "physics envy" in a previous chapter, has called "the strongest language I know, apart form poetry."

"In fact," he goes on, "its terseness, immense analogical power and frequent difficulty make mathematics the poetry of the sciences. If you haven't read some mathematics or some poetry lately, you're not having as much fun in life as you could be."

The equation that follows should add to the fun in your life, then.

$$dM(t)/dt = \lambda_{M(t)} - \mu_{M(t)}$$

The use of lower case d's is one way mathematicians have of signifying a difference, in this case an infinitesimally small difference, thus making it a differential equation. The left side of the equation is the rate of change of M, the number of species on an island. The t's in parenthesis, (t), signify that all of the quantities can change with time. The Greek letters on the right of the equation are typical ways of representing birth rates and death rates, in this case, for species on the island, rather than individuals. The rate of change of species on an island is therefore the difference between the colonization (λ) and extinction (μ) rates for the island. When they are equal, the rate of change is zero and the island is at equilibrium. There is more to MacArthur's math than this, of course, but it gives a taste for what it was like. There is nothing new to the analysis, other than new identities for the variables in the equation given. It has all been worked out before in terms of population growth. For individuals, when the birth and death rates are independent of population size, the equation could represent the intrinsic rate of increase and fit very nicely into the equation from before of unlimited population growth. And like the equation for population growth, MacArthur's differential equation can be shown to result in an exponential relationship: the species area curve for islands. Let either the birth rate or the death rate change with population size and *voila*! You have the logistic growth equation. There is nothing here that MacArthur could not have whipped off as an undergraduate at Marlboro College.

The other two quantities in the species-area equation, $S=CA^z$, are not easily assigned meaning, however.

"C is a parameter that depends on taxon and biogeographic region," MacArthur and Wilson informed. "z is also a parameter, but one that changes very little," was all the information they thought the reader needed to know about that constant.

Preston's explanation of z seemed to satisfy them. In his canonical distribution, many uncommon species are actually expected to be found anyway in a sample, even if they are very uncommon; doubling the area one samples only increases the number of species found by about 20%. That yields a z of 0.263, good enough for MacArthur and Wilson. Preston worked out the z's predicted from his canonical distribution to be 0.262 and from available data, 0.27. He did it numerically, meaning it was he who crunched the numbers, rather than some high-speed laptop.

However, by looking at islands as much more than sampling devices, MacArthur and Wilson's theory claimed to have brought principles of ecology and genetics to bear on biogeography, taking it out of its natural history stage. They did so by considering the effects of loss of genetic diversity in small colonizing populations (propagules they are called, consisting of as few as one individual, hopefully pregnant) and the known high rates of increase of good colonizing species. They then contrasted the adaptations of good colonizing species to those a species needs to persist when populations are at their carrying capacities. MacArthur and Wilson deemed the rapid colonizers to be r-selected and the persistent ones, K-selected, species. They also offered two "prospects" that were almost challenges. One was to experimentally manipulate island biotas, the other to treat islands and continents together with their new theory. They clearly saw embedded in the equilibrium theory of island biogeography the greater issue of biodiversity itself.

Wilson and Dan Simberloff met the first challenge almost immediately. Mangrove islands varying in size and distance from the mainland were fumigated to remove the entire fauna of arthropods (insects and spiders, mainly), then monitored as they were recolonized. It sounds simple, and is. But it required overcoming some logistical challenges—How do you wrap a tree that is off shore?—plodding through swamps, and no doubt getting wet. Earlier, Ruth Patrick of the Philadelphia Academy of Sciences found a way to test the theory by ingeniously creating artificial islands for diatoms from glass slides, which she then placed in streams. Other studies followed quickly. The data gave the theory resounding support.

Further study dampened the enthusiasm, however. Immigration rates did not necessarily decrease in a simple way as the number of species on an island increased. The immigration rate for plants, for example, apparently increased with increasing diversity. This should be obvious right from the most basic theory of ecological succession. Early colonizing species of plants can modify an environment (soil, in particular) to improve the chances of success for other species. Very rigorously obtained data—Peter Grant was involved—for bird species on islands near Australia and New Zealand supported neither a distance effect nor an equilibrium number of species.

Along with the less than resounding support that the theory obtained when put to the close scrutiny of actual data, there also appeared to be a theoretical fly in the theoretical ointment. Mathematical guru Robert May placed the whole species-area relationship back onto the level of being a mere sampling artifact, where Preston had had it. He showed that, in the event that the relative abundances of species derives from a variety of interacting factors, statistical theory, namely something called the Central Limit Theorem, requires a species-abundance pattern that is essentially that of Preston's canonical distribution. In overly simple terms, throw a lot of unrelated population data together and out pops a species-area curve. In simple communities, where only a few factors affect the relative abundance of species, other distribution patterns are predicted, according to May. These also lead to species-area curves of the form predicted.

Given the assumption that the larger an area one samples, the more individuals of all types one finds (which May called of "doubtful validity"), the species-area curve is a logical outcome of generally accepted species-abundance distributions. What this means is that for situations other than actual islands in which the total number of species can be directly censused, resulting species-area curves have no more biology to them than the null models of the previous chapter.

A consideration of the ecologically-related sport of bird-watching might help to show why that is so. In starting out on one's life list, a little effort will add a lot of species, but as your list increases, more and more effort is required to add new species. Meanwhile, the number of birds in your region is unchanged by your effort. That is the biology. Your tallying of sightings is just sampling. According to May, if bird-watching effort is equated to area (whether you actually increase your area of observation or just do more observations in a limited area) your life list would fit very nicely into the species-area (or species-effort) patterns found in nature and predicted by the theory of island biogeography. Should you manage a complete census of species in a region, as is possible on islands, and compare different regions, the species-area pattern you will find will be biologically based, but on a multiplicity of independent biological factors. Incomplete censuses, such as for the "virtual" islands that MacArthur and Wilson thought might exist on mainland areas, lead to statistical artifacts that only appear to be equilibrium species-area curves.

In an example of statistical artifact, my own woefully meager life list suddenly turned up species without any seeming increase in effort from me. It happened when I found out that my wife was adding birds that only she had sighted—one of which was the endangered whooping crane, that I had never set eyes upon—to my life list.

"What do you mean 'your' life list?" she had asked me accusingly when I pointed out that she should be keeping her own list.

"Since we're married, it should be 'our' life list," she had announced.

It's a case of communal property that I am not certain how to go about fitting into the Central Limit Theorem. It does, however, make my list "artifactual." There can be no equilibrium to my life list if my wife keeps slipping in endangered species that she stumbles upon.

Once May had shown the artifactual nature of most species-area curves with his mathematical analysis, there seemed to be no biology left to island biogeography theory. In a review and exhaustive statistical analysis of published species-area data, Edward Connor and Earl McCoy of the Tallahassee Mafia ratcheted up May's conclusions. They could find no way to discriminate between an equilibrium as postulated by MacArthur and Wilson, a relationship between area and habitat diversity, or what is essentially the null model of statistical artifact.

Connor and McCoy were, of course, stepping right into the null model controversy. The lines along which disputants formed their arguments were similar but for one considerable difference. Whereas in that argument it was agreed that the biology must make the mathematics so, in this case one side reversed that logic. Mathematics must make the biology so. It is a bit of esoterica that I take up here because it helps me to make a point and provides an entrée into some of Robert MacArthur's ideas that, for simplicity, I have been avoiding until now.

George Sugihara, a colleague of May's at Princeton, in a paper that I to this day do not understand, took up the challenge to defend the biological reality of the species-area curve. There is not any single thing in it that I do not understand—I can follow all of the parts of it—it is the transitions that lose me. Sugihara starts from the clue that the most interesting of the species-abundance distributions "seem to apply only to taxonomic collections having some measure of ecological homogeneity." This is an echo of Jared Diamond's plaint that his assembly rules applied only to guilds. It also takes into account May's analysis that communities having a "broken-stick" pattern of species abundances happen to fit the relationship $S=CA^z$ perfectly. The broken-stick model is MacArthur's famously *heuristic* attempt to explain the pattern of species abundance curves, published in one of his Three Influential Papers. It did not fit data from very many biological communities, and MacArthur eventually gave up on it.

Essentially, MacArthur's "broken stick" model compared the way a stick might be randomly broken into pieces with the way competing species might subdivide a resource. The number of stick pieces after breakage would represent the number of species sharing a resource. There would be large pieces, of course, and many smaller pieces. These would be analogous to the relative amount of the resource each species used. Given some simplifying assumptions—roughly identical size for all species is one—the relative amount of resource each species used would be proportional to its relative abundance. The largest piece would be the dominant species in a sample. A good idea, beguiling in its simplicity.

Sugihara resurrected the broken stick by deriving a formula for what he called "minimal community structure." It fit very well with Robert May's analysis of the species-area curve. Sugihara's model essentially represents the way two or three species might subdivide a habitat. From his basis of MacArthur's broken stick, Sugihara generates both the most common form of species-abundance relationship (Preston's canonical distribution) and the species-area curve. Sugihara's stick had to be broken sequentially, instead of all at once, as in MacArthur's original formulation, so as to better fit the species-abundance patterns of communities with only a few species.

More specifically and more technically, Sugihara later argued that the exponent for the mathematical form of the species-area relationship, z, cannot be ¼, the value Connor and McCoy assign it in their statistical analysis. Again in my own overly simple terms,

Connor and McCoy saw the z-value as being a possible combination of two parameters each of whose values could vary between 0 and 1, with biology ignored. The parameters were multiplicative. Since ½ is midway between 0 and 1, z should be ½ of ½. They could not find any statistical difference between this null value, 0.25, and data used in support of MacArthur and Wilson's theory that ranged from 0.2 to 0.4. Preston's tedious calculations were just so much wasted effort, while MacArthur was a lovable, but false icon.

Sugihara's rebuttal is subtle. He also gets ¼ as the expected value of z when randomly generated, a straight-forward result of Connor and McCoy's assumptions, but argues that the probability distribution about the ¼ mean (another word for average, but preferred in statistics) is too different from the data Connor and McCoy present. This difference in distributions between Connor and McCoy's randomly generated z's and actual data must lead to the rejection of their argument at the 99% probability level, according to Sugihara.

Rejection of Connor and McCoy's simple statistics in this way, according to Sugihara, points in the direction of biological meaning, both for the species-area relationship and the exponent. There are technical questions raised by the whole thing that have no answers. Could a more sophisticated null model than Connor and McCoy's have been as easily rejected by Sugihara? More importantly, is the variance around a value really as important as the value itself?

An analogy to illustrate the subtlety of Sugihara's argument has to do with women's skirt lengths. I don't think anyone in the world knows why they go up and down, aside from a suspicion that they are at the random whim of designers. Suppose we are given actual data on skirt lengths and data from a model that has the changes being random from year to year. Suppose further that in both cases the average change is identical. Connor and McCoy interpret this to mean that the skirt lengths changed randomly. Sugihara, looking closer, sees that, although the changes average out to be the same, the actual year-to-year changes were sometimes greater and sometimes smaller than random. From this, Sugihara implies in my analogy that women are consciously choosing their skirt length. Possible? Possible. In my view, though, Sugihara used a statistical cannon (his Jacobean matrix) to fend off an observational ant.

There is a parallel to Sugihara's finding evidence by careful sifting through a statistical distribution. Jonathan Roughgarden, she of the null controversy, was capable of equal subtlety. Roughgarden combined with a mathematically inclined population geneticist to look at differences in the shapes of species abundance distributions. Distributions with thick tails, according to them, permitted more species to coexist than distributions with thin tails. They did "as if the shape could be empirically demonstrated," as Lawrence Slobodkin admonished.

Sugihara's paper, which really proves nothing, was given much credence in the theoretical part of the academic community of ecologists. Ruth Patrick included it in her collection of the most important ecological papers having to do with diversity. Null model papers were coyly omitted from the collection. In a similar series of classic studies in ecology under the prestigious imprimatur of the University of Chicago Press, Sugihara is given the last word for finding biological meaning in Preston's canonical distribution and in the species-area exponent. It is done on the same page on which the inappropriateness of MacArthur's broken-stick model is discussed. (Oh, by the way, they also left out any mention of null models. Perhaps being mathematicians, rather than experimentalists, they saw null models not as theoretical advances, but as retrenchments.)

One prestigious ecologist characterized the broken-stick model "as dead as the dodo," even after Sugihara's paper. The dodo is perhaps a more apt analogy in many ways than the ecologist might have intended. Meanwhile, Robert May and colleagues found that the very narrow range of communities to which Sugihara's model might apply served "as a corrective for the vagueness of earlier models."

One senses out of all this that people's reputations were at stake.

This was also about the time that another prominent ecologist wrote, "Beginning around the time of Robert MacArthur there began such a population explosion of theorists that it seemed for a while as if no biology department was complete without its resident theoretical ecologist. Recently, however, the market seems to have crashed, and biology departments no longer feel incomplete without an indigenous theoretician."

The relationship between applied mathematicians and the science they purportedly serve is mutualistic. Mathematicians are like cats on a farmstead. The farmers feed the cats because they know that they might eventually bring them a mouse. In ecology at that time, however, the cats may have taken over the farm. The farmers then realized how much keeping the cats was costing them.

SLOSS: Islands in the Wilderness

The second of MacArthur and Wilson's "prospects" for application of their theory of island biogeography created a cottage industry of studies that tried to apply the theory to what came to be known as "virtual islands." Initial focus was on island-like areas of the mainland. These are generally isolated pieces of habitat, such as forest "islands," for example, whose isolation is due to something other than being surrounding by water. Imagine the landscape that is the eastern megalopolis of the United States and the significance of such forest "islands" becomes obvious. There was evidence of "faunal collapse," a loss of species, when areas were cut off from similar habitats to become islands. The collapse occurs because the equilibrium number of species on an island, according to the theory, depends on the island's area. Reduce the area;, reduce the number of species in consequence. Preston had warned in his 1962 paper of the inability of preserving all species originally in a park, due to such an area effect.

MacArthur and Wilson's species-equilibrium theory was also applied to truly "virtual islands," such as individual trees for the insects found upon them. I was involved in one of these "virtual" efforts. The islands were very virtual: each was represented by the area in which a particular species of rodent dwelled and the mite species reported to be parasitic on it. It turned out that the larger the area of the rodent's range in the United States, the more species of mites it was reported to have. When plotted, the data made a reasonably good species-area graph. The z value, 0.37, was on the high end for island faunas, but lower than values reported for insects on trees and higher than expected for continental areas, based on data then available. The habitat diversity hypothesis seemed not to fit these creatures that spent all of their lives either under a host's fur or in its nest, except, of course, that for mites with larger rodent host ranges there were greater chances of interactions with other rodents. The area effect could have been a consequence of that.

It was the paper that LaMont Cole had—or, actually, would have, had I given him the chance—belittled so as to cause me to heap revenge on him. Actually, it was criticized not just by Cole, but also by two ecologists from Santa Barbara. Their criticism mostly had to do with our use of published species lists. Being more familiar with our data than they were, it was fairly easy to brush aside their objections. What I did not understand then was what had motivated them to make it a personal issue for themselves to mount a crusade against studies such as ours. (Maybe they thought our title was too flippant? It announced that the study was "Of Mice and Mites.") They went on to complain that the use of species lists rather than data collected by one's self was invalid. Silly me, I thought that was what Baconian induction was all about: to gather observations until some sense could be made of them.

Is an actual biological census generated by the investigator himself any less fraught with oversights, errors, or just plain natural peculiarities than a tabulation of already published data? True, tabulating the data of others is tabulating their errors, also, but there is not a difference in kind here. The errors are passed through without magnification. If there is bias in the original studies, that bias will come through in the tabulation—or, if the biases of different investigators differ, they might possibly compensate and cancel each other out when the studies are combined.

It is also, in fact, easier to check up on a study based on published records than one with data generated by its author. Unlike the case for studies based on compendiums of other's data, which should be right there in the literature for all to see, as are any errors in their subsequent compilation, the only way that data from experimental or field studies can be checked is by replication of the study. That would require repeating an experiment exactly or redoing censuses, something that is common in other sciences, but is just not done in ecology. Once someone has done a study of a system—the Galápagos finches, for example, or the rocky intertidal zone—it is either gingerly avoided, or is the focus of "pile on" studies that elaborate and extend some aspect of the original system, often in collaboration with the original scientist, but without ever actually retesting the original study. This latter is particularly true for the graduate students of the original scientist. Is there a basic insight here? At least as far as ecology is concerned?

There is a less speculative insight that I have after years away from scientific fields of battle. This is the "robustness" of the theory of island biogeography. Robustness is a term often used to describe the way models appear to apply to so many different situations. This is a bit different from the use of robustness to describe the way a model or theory stands up to challenges. The robustness of a theory in this sense would make it something like the "catholicon" that Don Strong accused competition theory to be. It stands up to all challenges, because it meets none. The species-area relationship allows one to find an equilibrium between immigration and extinction in all manner of ecological communities, from Darwin's finches to insects on oaks to mites on mice. The theory is indiscriminate. One might say, loose.

Yet this is the theory upon which some ecologists have staked the reputation of their science in the eyes of the public. And the stakes are high. The Era of Ecology in which we find ourselves, at least in the United States, means that ecologists now have a say in decisions that is often on a par with that of practitioners of that other unreliable sibling science: economics. During my years as a graduate student, ecologists were admonished to take an economist to lunch as a way of learning their point of view, along with something about the real world. Joan Roughgarden (perhaps the fight was taken out of her in the controversy over null hypotheses) provided a brief etiquette for getting along with them.

"Economists," she wrote, "are not about to cede the moral high ground to ecologists just because humanity is contained in a giant ecosystem."

They were, however, "often on our side," she noted.

Her remarks indicated a sea change in attitude. It was not the attitude of the ecologist toward the economist that had changed, but the other way around. I bet the economists asked her to lunch.

Ecologists now do have power. They also have money. Starting with the funding the Odums received from the old Atomic

Energy Commission right up to the present, big ecology gets big money. Moreover, where Eugene Odum had to politic for inclusion of a course on ecology as a fundamental discipline in biology, college and university campuses—from large to small—now can boast of several ecology courses and often several ecologists on their faculties. The Davis campus of the University of California system topped out last I looked (2004) at 197 ecology or evolutionary biology faculty, for example, with Oregon State and Ohio State Universities each topping 100 and Cornell, Wisconsin, and Penn State rapidly approaching the century mark. Ecology *is* big. The trend in ecology away from papers with single authors to papers with multiple authors from different institutions is a clear sign of it being a "big science." Where prior to Earth Day a single Paul Sears or Aldo Leopold might have pointed out the hazards in man's unecological ways, it now takes well-funded study groups with up to several dozen authors to do so. The lonely naturalist has been (almost) displaced by research teams. Ecology definitely is no longer for little old ladies in tennis shoes. They could never garner the funding it now needs.

Given that the theory of island biogeography applies to all sorts of situations that are not at all islands, there would seem to be no limit to its useful applications. This is especially true when it is in the hands of non-ecologists. Habitat islands in urban and suburban ecosystems? Why not if you are a geographer or an urban planner? A park or a vacant lot is as much an island as a woodrat, more so from a geographic perspective. How about nature reserves? Absolutely and essentially, if you are a conservation scientist or a refuge or park manager. How about the entire earth? Isn't it a true island, but—My God! —one that can only suffer extinction. (Colonization events from outer space might as well have a zero probability.)

To apply island biogeography theory to the whole world is not only tempting, but also urged on by reputable ecologists. So is its application to parks and refuges. No less a personage than Lord May brought the possibility of refuge design based on ecological theory to the general scientific community in a news brief, while bird biogeographer and kidney physiologist Jared Diamond brought the same idea to the attention of conservation scientists, helping to spark the controversies of the previous chapter.

Reasoning from species-area data, Diamond concluded that if a habitat area were reduced to 10% of its former size, only half the species that were initially on it would remain once a new equilibrium was established. Although he did take into consideration that small habitat islands can exchange species with each other and thus maintain species that might otherwise have gone extinct in any one fragment, Diamond used the bird data with which he was most familiar to argue that some "species would be doomed by a system of many small reserves, even if the aggregate area of the system were large." In illustration of his ideas, he used the same birds for which he developed his assembly rules. He then offered some design principles, which, as was his wont, he labeled with letters of the alphabet, even though their discussion took up a single page. They can be summed up in a sentence: a large, circular reserve is better than a small one or an elongated one and fragmented reserves should be avoided, but if necessary, should be close to each other, preferably in an equidistant arrangement with connecting strips. Diamond published his ideas in the journal *Biological Conservation* in 1975.

Diamond had jumped into the conservation arena a bit too fast with his ideas, however. There were issues that should have been dealt with first. In simple words they had to do with testing those applications he proposed before recommending their implementation. One thing that should have led to caution on his part was that Diamond's design recommendations were based almost entirely on the biology of birds in the Bismarck Archipelago near New Guinea.

Who could be better to remind us in ecology that clever ideas should first be subjected to rigorous test than Dan Simberloff? Not only that, but Simberloff had to be the world expert on the theory of island biogeography. Simberloff and his Florida State colleague, Larry Abele, quickly went into print in the journal *Science* to point out that there were neither theoretical grounds nor evidence from data to justify the way that island biogeography theory was being applied to conservation. They showed mathematically that two smaller areas would support more species than a single reserve of equal area based on MacArthur and Wilson's theory. What is more, they had data: Simberloff's mangrove islands. Besides fumigating some islands in order to defaunate them, he had also cut barriers through others and censused them until they reached their new equilibrium numbers of species. Fragmented islands held more species in combination than any of the fragments, an entirely expected result, but for one "archipelago" of mangroves there were four more species in total than before the island got dissected. And this was despite the fact that the total area of the archipelago was less than that of the original island. Simberloff had needed to remove some 5% in order to create suitable water barriers between the fragments. The other archipelago Simberloff had created in this way had fewer species than its original mangrove island, an outcome that Simberloff and Abele attributed to just the effect Diamond had postulated. Those fragments may have been so small as to become "refuges" rather than reserves and had suffered a loss of species due to the inability of some of the species to maintain a critical population size.

Simberloff and Abele could not stem the rush of wildlife conservationists to embrace Diamond's ideas, however. Simberloff's data from his island studies—one of the few true experimental studies of the theory—was not enough to bring caution to conservationists. Not only that, they came gunning for Simberloff and Abele's scalps, doing so also in the pages of *Science*. It led to a very visible airing out of differences that could not have been good for the status of theoretical ecology as a useful scientific tool, and it might have added venom to the null controversies. Simberloff, who does not shy away from his principles nor from a fight, lamented with Abele for having been "cast as the *bêtes noire* of conservation" by the conservation scientists, but they stuck to their principles.

Why the uncharacteristically stubborn adherence by those in the conservation field to a tenuous theory? It is not exactly a mystery. Scientists behave in stubborn ways all the time, but only if they have something personally vested in the object of their stubbornness.

MacArthur and Wilson had the strongest claim to a professional stake in the theory of island biogeography, but MacArthur was dead by this time and Wilson quietly stayed out of the fray.

Jared Diamond, in his dissent of Simberloff and Abele's paper, pleaded from the outset: "As human destruction of remaining natural habitats accelerates, *biologists have felt intuitively that most wildlife refuges are too small* to avert extinction of numerous species. However, because there has been no firm basis for even approximately predicting extinctions in refuges, *biologists have had difficulty convincing government planners* faced with conflicting land-use pressures of the need for large refuges."

The italics are mine. The theory of island biogeography gave scientific credence to an intuitive belief.

Diamond went on to caution that "Because those indifferent to biological conservation may seize on Simberloff and Abele's report as scientific evidence that large refuges are not needed, it is important to understand the flaws in their reasoning."

The flaws, according to Diamond, were that Simberloff and Abele had not taken into consideration that some species have poor dispersal ability while others are especially susceptible to extinction at the low population sizes on small islands.

Diamond offered as a compromise: "one refuge as large as possible plus some smaller ones."

The best of all worlds. Despite the flaws in Simberloff and Abele's reasoning, Diamond did allow that putting your eggs in one basket may not be the best strategy.

John Terborgh, another ecologist specializing in birds, continued along the same vein as Diamond.

"If it is agreed that the primary objective of a rational conservation policy should be to preserve viable populations of as many as possible of the species that inhabited the pristine landscape," he wrote in an article following Diamond's, "then at least some large reserves are a necessity."

Terborgh, from Princeton, already had a goal in mind that superseded any application of science. The theory of island biogeography might have served his goal well.

"Species loss is area dependent," he wrote without qualification, citing his own work with birds. "Preserves should be designed to minimize extinctions."

These latter two statements had no need of a theory of island biogeography as justification. Terborgh's argument was clearly not what island biogeography theory predicts when the reserves represent several islands. Neither was there guidance in the theory for those species that do not disperse well, Diamond's concerns, or that have a fragile population structure. Island biogeography theory is as absolutely silent on the identity of species on islands as is ecosystem theory. If there are species that need special attention, they should get it by way of models fitted to that species' particular ecology. Autecological studies, in other words, as Simberloff and Abele had suggested in their original paper, instead of community theory.

A group of mostly wildlife ecologists employed in positions having to do with the conservation of species added their voices to Diamond's and Terborgh's. They also had a goal they were trying to serve with their dissent: the "preservation of entire ecological communities, with all trophic levels represented," which, they added, "requires large areas." As such, it was a goal obtained from ecosystem ecology, not island biogeography. Of course, ecosystem theorists were still puzzling out the relationship between diversity and stability.

Along with Simberloff's studies of mangrove islands, data soon appeared from British chalk quarry reserves, identical areas of different sizes. That data also could not support a single large reserve as an optimum strategy. The chalk quarries happened to be real nature reserves, the subject of the controversy. It was found that more species were preserved on any two small chalk pavements than on one of an equal area to the two combined.

It did not take long for Princeton University to send its young champions out to try to slay the dragon that was awakening at Florida

State. At stake were the most treasured ideas of ecology's most influential theorist, its James Dean. George Sugihara had already been out in theoretical defense of the biological underpinnings of island biogeography theory, now Blaine Cole, urged on by Bob May and aided by Sugihara, set out to restore its place in conservation biology. Like Sugihara, he found a special ecological situation to use to launch his attack in a paper in *The American Naturalist*. The details are perhaps not important, but Blaine found what were mathematical conditions under which one large reserve would hold more species than two smaller ones of equal area to it. Unlike Sugihara's, Cole's paper was not acclaimed as an instant classic.

Simberloff, ever the scientist, suggested in response that data should be looked at for answers, rather than mathematical models.

He and Abele pointed out that, "As we said in 1976, this is an empirical matter, and must be addressed anew for each taxon and region. What is surprising to us is that, for a majority of cases examined to date, it is the right strategy."

"It" happened to be several small, rather than one large island. Their further study of data in the literature found "not a single case where one large site unequivocally excels several smaller ones, and many where several small ones clearly contain more species than the large one."

They were particularly gratified that among those studies was one by Gilpin and Diamond, since Diamond had dismissed their argument against the use of island biogeography theory for reserve design as "scarcely relevant." (The data had to have been Diamond's.) It was Cole's model that was irrelevant, instead.

Simberloff and Abele had gone to the data in particular, according to them, after having "been admonished that the urgency of conservation to some extent obviates the need for scientific rigor."

There's that word again, rigor. It is so important to science. For some.

Oh, and Diamond's rule about the shape of preserves needing to be circular? British conservationist Margaret Game, using geometric reasoning spiced with some calculus, showed that it was not necessarily so.

Eventually, conservation minded ecologists found rationales other than island biogeography theory for large preserves. It was not too difficult, really. Biodiversity and the species-area effect are never far beneath the surface of their justifications, however.

That there is attractiveness to island biogeography theory for conservationists is unquestionable. It's a great theory. Want to build a road through that forest? Sorry. That would fragment the forest into two islands and result in x number of species lost. The exact number depends on the particular taxon and a number of other considerations, but in the end, numbers can be put through an equation and *voila!*—scientific justification.

One marine reserve planner, for example, argued for the resemblance between the coral reefs of the Great Barrier Reef of Australia and archipelagos "in the hope that island biogeographic theory will become a management tool." He had a rationale for his hope.

"Where resource management is more an art than a science, it is useful to apply biogeographic theory as an aid to the decision-making process," he continued.

It can't be denied that it is always useful to bring science to bear on management problems, particularly if the science can be used to argue for larger reserves and less fragmentation. The goal is a noble one.

Michael E Soulé has written, "Finally, an apology and a caveat. Some biologists will be appalled by the blanket prescriptions for survival suggested here, especially in view of the heterogeneity in population structure, genetic variability and, probably, in genetic load that exists among species—even closely related ones. Indeed, criteria such as those recommended here lack precision. The caveat is that the luxuries of confidence limits and certainty are ones that conservation biologists cannot now afford, given the rate of habitat destruction documented in many chapters of this book. Constructive criticism is welcome, but to embrace the purist's motto of 'insufficient data' is to abandon the bleeding patient on the operating table."

The book to which Soulé referred was his own, *Conservation Biology*, one of the first attempts to bring theoretical advances in ecology and evolutionary biology to bear on issues of conservation. Among the authors of its chapters were Jared Diamond and John Terborgh, who by its 1980 publication had taken firm positions on the controversy over applying island biogeography to conservation.

"That current ecological theory is inadequate for resolving many of the details should not detract from what is obvious and accepted by most ecologists: habitat fragmentation is the most serious threat to biological diversity and is the primary cause of the present extinction crisis," was a conclusion of a 1985 study on fragmentation and conservation strategy in *The American Naturalist*.

It is true. Something so obvious needs no ecological justification. There are also non-science reasons for preserving biodiversity. However, the obvious need to protect a given habitat should not lead a science astray. Besides, the reader might agree with me that, today, the loss of habitat is probably more critical than its fragmentation.

"I think the refuge design literature is still not very instructive and has lots of irrelevant papers in it (for instance, on corridors), and some poor science," Simberloff reflected on it after the dust of controversy had cleared somewhat. "It is often more political than scientific."

Political correctness is not a concern in the science of ecology alone. Nuclear energy use split physicists and engineers as much as any schism in ecology. It did not, however, affect the basic science and its methods. Ecology is somehow different in this regard. Maybe it is because of ecology's ties to evolutionary theory. The prospect of dangerous heresy has always been a concern, but as long as it was a sleepy little science, ecologists were relatively insulated from worrying how their science might be put to use by creationists. Earth Day and the need for ecology to manage the environment removed that insulation, and not just against creationists. Ecology as it abuts environmental causes is as riven by politics and philosophy as Darwinian theory was before the Modern Synthesis.

"I've basically decided not to fight those battles any more," Simberloff continued, "as it's hard to get very far (since the battles are more political than scientific.) If I had the right angry graduate student or postdoc, I could probably change my mind."

Later he added, "It's depressing, and it still happens. Anyone arguing against even the most cockamamie idea, so long as that idea is supposed to benefit conservation, is viewed with suspicion, at best."

That may be an explanation for how David Quammen treated the controversy over SLOSS (single large or several small, as the debate has come to be known), in *The Song of the Dodo*.

"I am particularly disappointed in *The Song of the Dodo*," Simberloff told me. "Quammen is a friend, came out and interviewed me at length, and my student Barb Zimmerman in Toronto. Then, for no obvious reason, the book concludes that island biogeography theory is a great framework for dealing with conservation problems."

I think the reason was obvious.

Do Owls Live on Islands?

Once a species is gone, it is gone forever. There is no recovery from the dead except in *Jurassic Park*. We lament the carrier pigeon, dodo, European aurock, all hunted to extinction. We fear that the fate of the right and blue whales may be the same. We congratulate ourselves in saving our raptors, the eagles, osprey, and falcons, from what we almost did to them through the use of DDT. We rejoice over the wolves and buffalo now in Yellowstone. No call upon the science of ecology is needed for the feelings aroused by these examples.

Given that we do have as a goal the preservation of species, what can ecology tell us about attaining that goal? First, let us assume that we can decide to preserve a species on an individual, case-by-case basis. Let us preserve the spotted owl. Let the snail darter be damned. Eradicate smallpox—if you can. The snail darter, even after being dam(n)ed refuses to go extinct. The spotted owl gives us a different avenue from that of island biogeography theory into how science can be put to use for conservation purposes.

In the United States, the Endangered Species Act specifies legal steps that must be taken to identify species that may be threatened with extinction, then to protect them. Some of these steps are based on what can be described as ecological knowledge, but the decision to set the legal mechanism that is the Endangered Species Act into motion was made by the public on something other than a scientific basis.

In 1973, Senator John Tunney of California said in support of the Endangered Species Act that "to allow the extinction of an animal is ecologically, economically, and ethically unsound. Each species provides a service to the environment; each species is a part of an immensely complicated ecological organization, the stability of which rests on the health of its components."

This is ecology by way of Barry Commoner's Laws. (No doubt, Aldo Leopold's *Almanac*, also.) It is not the ecology of Robert MacArthur.

In the wording of the Act, it was recognized that endangered species were "of aesthetic, ecological, educational, historical, recreational, and scientific value to the Nation and its people."

There is no question that ecology should be put to use in the management of a species that has been identified as endangered, but which ecology? The ecosystem approach, as far as concern for a species' connections to others, is always of utmost importance, but it has been shown that ecosystem theory is not. What is left then that is specifically ecological to be applied to an issue that is from every angle an ecological one? Why, evolutionary ecology theory, as set out by the followers of Hutchinson and MacArthur, of course. Or, simple autecology, as recommended by Simberloff and the Tallahassee Mafia.

Even after assenting that evolutionary ecology might be brought to bear on conservation issues, there is still the question of which level of ecological analysis below the ecosystem should apply. Community theorists, those who study niches and species abundance distributions, justified much of their research by its potential for practical application, but had much of that theory questioned. Population theory, which sometimes can be indistinguishable from community theory, by the early days of the Endangered Species Act and NEPA had built up a remarkable theoretical edifice based on Lotka-Volterra type equations, only to have it knocked down into chaos. Autecology, essentially the ecology of an individual species, would appear to be the choice for conservation purposes, if only by default.

Every species has certain requirements for existence. Specific knowledge about the natural history of the species is more valuable than a sound grounding in even the least esoteric aspects of theoretical ecology. Most often, the threat of extinction comes from the threat of loss of critical habitat. That habitat can easily be identified without resort to the theoretical baggage of the Hutchinsonian hyperniche. Once the old-growth forests go, for example, so goes the spotted owl, along with them. Fairly simple stuff.

Man can also modify a habitat without seeming to destroy it. Inadvertent poisoning, changes in amounts and types of carcasses found, and changes in vegetation, all probably led to the decrease of California condor numbers to that point at which they existed only as part of a captive breeding program in two zoos in California. (They are back now spreading their wings along the coastal mountain ranges of southern California.)

Degradation of a habitat can sometimes be subtle, but the science for assessing habitat loss should be non-controversial. When it is controversial, the suspicion must arise that the science may not be up to the task. Of course, controversies can always be based on disagreements over interpretation of scientific knowledge. Or differences in philosophy. Or all of those reasons that make people disagree in general. Ecology as applied to the extinction problem has had each kind of disagreement.

The spotted owl is a good example. Should it be on the endangered list and does it really need old growth forest habitat? Controversy over these questions was one that on first consideration should have been resolved without the need for theoretical ecology. The bird has certain nesting requirements that are critical; they can be understood and interpreted without an advanced degree. Studies showed that the bird is territorial. Like many birds, it establishes an area that it announces as being its own through songs and aggressive displays against transgressors. That's what all of that singing early in the morning outside your window is all about. The owl nests and forages within that territory it marks off. In general, the higher the quality and the larger the territory, the better are the resources available to the bird and the more likely it will be to raise a successful brood. It takes no leap of logic to conclude that there must be a minimum size for a territory of a given quality below which the owl is unable to breed (and another below which it is unable to survive at all).

That minimum size turned out to be roughly 1,000 acres, a goodly amount of timber, for each breeding pair in the forest. But things are never that simple. As ecologists have shown, no individual exists in isolation. What about population effects? Inbreeding, for example? What about the fragmentation of the forest? Here is a place for theory, after all.

In his version of events in his book, *In a Dark Wood*, Alston Chase uncovered a case of what could only be described—if he was correct—as criminal scientific malfeasance. If so, it would be a clear example—not the only one in ecology—in which philosophical belief trumps scientific rigor. According to Chase, biologist Russell Lande, under the urging of Andy Stahl, then of the National Wildlife Federation, concocted a paper, "Demographic Models of the Northern Spotted Owl (*Strix Occidentalis caurina*)," to show that the spotted owl was indeed threatened with extinction. It was based on data fed to him by Stahl and on a mathematical model by Richard Levins, half of the Marxist team of Levins and Lewontin. (The order of names, as Lewontin has always insisted, is alphabetical.) Chase mistakenly identified Levins' model as being based on island biogeography theory. It was not, but it was definitely one inspired by Robert MacArthur. Chase missed the whole controversy over use of island biogeography theory in conservation, but he would have been off base with that criticism for the case of the spotted owl, anyway. The mathematics of Levins' model and MacArthur's happen to be unavoidably similar. Both involve patches of habitat and both attempt to model colonization and extinction events. If we take MacArthur's differential equation from *The Theory of Island Biogeography* and let λ and μ be colonization and extinction rates for populations, rather than for species, we have the basis for Richard Levins' model.

Levins publication of his model in 1969, like Rachel Carson's only research report, was part of the gray literature. He published it in the *Bulletin of the Entomological Society of America*, which is no longer published. (Chase, chronologically challenged with the entire sequence, gave Levins' paper a 1979 date in his book.)

According to a former editor of the journal, Levins' paper would have received "no review." It was gray literature because, not only was it not refereed, but the journal was also too insignificant to be held by most libraries, even those with agricultural schools having entomology departments. Reports in the gray literature usually have little more impact on their science than memoranda circulated internally within large, for profit, research laboratories. Some of those reports do occasionally find their way into the hands of those external to the original loop of circulation. They then become part of the gray literature, potentially important, but gray from not being exposed to the full light of day. Usually they disappear into the darkness a few years after discovery, their utility being timely, rather than timeless. Levins had properly chosen the *Bulletin* as an outlet for ideas he had previously given at a symposium of the Entomological Society. It got them into the hands of those who needed them in a timely and effective way.

Lande's 1988 publication in *Oecologia* was paralleled by a publication in *Science* and preceded by one in *The American Naturalist*. The bird in the title of the *Oecologia* paper is identified by a full trinomial: genus, species, and subspecies. By this time, the subspecies concept was out of favor, but Lande had a reason to be out of fashion. The fading out of subspecies in taxonomy had been a change made mostly in deference to social ideas about races. Human races, or lack thereof. The subspecies, a geographically identified population synonymous with race, had been disavowed by anthropologists in reaction to Nazi excesses. Ashley Montague made this clear in an address to the American Association of Physical Anthropologists, published during the early stages of World War II. Biologists, finding the subspecies to be a convenient taxonomic unit, but not a useful evolutionary one, soon followed suit.

Chase had not read the literature he cited on the matter very carefully. The article to which he pointed by Paul Ehrlich and Richard W Holm as discrediting the subspecies concept was titled, "A Biological View of Race," and printed in a compendium organized by Montague to examine the concept of human races. Had Chase read even further into the literature, he would have discovered that even the "species" concept had become problematic for evolutionary biologists. In a more technical article than that read by Chase, Ehrlich and Holm had argued for the term "species" to be replaced, other than for pragmatic uses, with the term "evolutionary unit." No doubt Chase might have found a slant on that to discredit the entire idea of preserving any species.

Chase is correct, though, that there was no biological basis for race or for subspecies. He was also correct that the Endangered Species Act did protect subspecies, as well as species (and also populations), by defining species to include "any subspecies of fish or wildlife or plants, and any distinct population segment of any species of vertebrate fish or wildlife which interbreeds when mature." (The vertebrates get better protection under the Act than the majority of species going extinct. This is perhaps as it should be, given that ecology provided only one of the numerous rationales for the Endangered Species Act. The public can choose to protect whatever it wants to, regardless of scientific basis. Given that choice, protection of distinct populations of vertebrates obviates any technical and semantic issues over the categories of subspecies and species. [The subspecies, or geographical race, concept has not entirely disappeared today, even given its various philosophical complications. People today pay to have their genome tested essentially for its racial composition, usually in the same categorical guise by which special endemic populations are protected. Henry Louis Gates, Jr, uses exactly that categorization {albeit without mention} indirectly to "Find Your Roots." Physicians, meanwhile are grappling with how to treat indications of diseases that differ among poorly categorized population groups that once might have been called races.])

I bring all of this up because Chase saw some sort of conspiracy among ecologists that ended up having the Endangered Species Act protecting a concept that the ecologists had already disavowed.

As for that conspiracy, according to Andy Stahl, now of the Forest Service Employees for Environmental Ethics, "Chase sees a conspiracy behind every tree."

Subspecies and races were once pretty much synonymous in the biological literature, but subspecies had a bit more formality to it.

Both terms simply identify particular populations. To a taxonomist, a geographical race of a species is one that is found in a particular area. Often its location was given as part of its common name, but sometimes other features were used, or it was named for its discoverer. (That latter possibility is a surprisingly strong incentive for keeping the subspecies designation. Donors and amateurs are much more likely to find their name appended as a subspecies than to have anything to do with the discovery of a new species.) However, the taxonomist's subspecies had to have distinct differences from specimens of the same species found in the main or other parts of the species' range. A geographically distinct butterfly or fish might then be formally named in the scientific literature as a subspecies with a trinomial, rather than just the binomial, Latinate designation. When are differences enough to require the designation of a subspecies? Or a distinct population? That's the problem. There is no "bright line" that can be drawn to tell when the differences are enough. Policy makers and debaters love to have that bright line difference, however.

The Endangered Species Act requires that decisions be based on "the best scientific" data. The act of publication makes good scientific data better and marks one bright line.

Invariably, a later taxonomist comes along and "lumps" all of a "splitter's" subspecies back into a single species. (This is a permanent tug-of-war in taxonomy caused by something in the human psyche. Some people, the "splitters" tend to find finer and finer distinctions between categories, while the "lumpers" seem to find gross similarities to be much more important than fine differences. [Or, perhaps it is some sort of Hegelian dialectic process in action?] One side just has to clean house in order to straighten out the mess of their predecessors and prepare the ground for their own mess.)

In practice, local populations do have federal protection under the Endangered Species Act through the subspecies designation. Fearing a genetic bottleneck that could cause problems through in-breeding, protecting distinct populations is thought to maintain enough genetic diversity to allow species to survive either the effects of in-breeding itself or changes in their environment. The Act actually anticipated the biodiversity term, at least for vertebrates.

Lande had to limit his study to the Oregon and Washington population of the owl, the northern spotted owl in common parlance, for the purposes of the Endangered Species Act. According to Simberloff at the time, more research was needed to determine "whether the California and Northern Spotted Owls are isolated subspecies." It is not clear at all how the results of the model would have turned out for the species as a whole. Differently, I imagine, but not out of any chicanery.

According to Lande, however, "My paper in *Oecologia* was not based on any consideration of the Endangered Species Act. There are ecological differences between the different subspecies of spotted owls (somewhat different habitat types), and I wanted the data to be from as ecologically homogeneous sources as possible to put together all the necessary pieces for the analysis."

In addition, all of the best data Lande could obtain was from the region occupied by that subspecies. Much of it was the result of funding from the Forest Service for research aimed at obtaining information necessary to manage the owl effectively.

The use of subspecies in conservation biology is an interesting phenomenon in itself with which I can only tease the reader. Subspecies are protected when threatened (the spotted owl is an example) and ignored when expedient. The subspecies that represented the extinct peregrine falcon of the eastern United States, for example, was replaced by a subspecies from Scotland that had never before been found in its new area. It is doing quite well, too.

The case of the spotted owl illuminates how ecological science gets mangled by popular writers. It seems not to happen to the same extent with other sciences. Maybe it's because the main principles of ecology are so accessible to the lay person that a false confidence is developed for those other principles and techniques that are beyond the quick grasp of the non-specialist.

The spotted owl also lets me set some things straight. First, there is that confused chronology. Lande's 1988 paper could not have represented the scientific basis for a 1985 management report on the owl. That report needed, according to Chase's inferences, validation by a published, refereed scientific document. It could not have gotten it from a report published three years later. There had, however, been a preliminary version. Simberloff alluded to it in his 1987 assessment of the ecology behind decisions having to do with the bird.

"Where conservationists have argued beyond the limits of accepted data and theory they have been abetted by overly confident scientists," he wrote. "For example, several conservationists cited as grounds for urgent concern a theoretical treatment commissioned by the National Wildlife Federation predicting quick demise of the Owl on demographic grounds. The structure and assumptions of the treatment were attacked not only by an industry scientists [sic] but by the scientists of the Forest Service, Washington Department of Game, and universities as well."

Given Simberloff's jaundiced view on the use of MacArthur-style theory in conservation, particularly island biogeography theory, it is easy to understand why he slighted Lande's work. In addition, Lande's model had not yet faced the rigor of peer review.

According to Stahl, "NEPA required the Forest Service to acknowledge and respond to opposing scientific views, regardless of whether they were published in a refereed journal."

"However," he added, "it certainly didn't hurt to have Lande's work published."

That bright line.

Chase also had some of his facts wrong. The paper by Levins had nothing to do with pesticide use. The model it presented was for a predator population used in biological control. Neither was it based on MacArthur and Wilson's theory of island biogeography, except in a peripheral way. Finally, Richard Levins was anything but an entomologist, as Chase described him. But someone else can clear up the rest of those details.

The incident represents the kind of mess that can occur when ecological theory becomes entangled with practice—and shows a way out of the mess. The theories can run ahead of practical needs.

Or behind. In the owl case, the outdated theory was not that of Levins and Lande, but a guess made by a geneticist that a species that drops below 500 breeding pairs is at risk of serious loss of variability due to inbreeding. Concern over variability is something like concern over genetic load, except that low variability implies that there is no load—or it's all load, depending on point of view. A population having dangerously low genetic variability is composed almost entirely of an inferior homozygote, in effect. A consequence of that might be extinction. In the absence of anything else to guide them on the matter, others picked up the idea that 500 pairs are a minimum number for the survival of certain species. Michael Soulé was among them. The number was cautiously presented as a guideline in the influential textbook he co-wrote on conservation biology.

That number also got into the original 1984 management plan to preserve the spotted owl's viability. The number was more guesswork than science (even though based on science) and was never meant to be one that might be applied in general to all species. Simple ideas do not always lead to simple, effective solutions.

"The plan cites Dr Michael Soulé for the proposition that 500 owl pairs are sufficient," Stahl began his explanation of his concerns over the plan in an email to me. "Soulé tells me that he said no such thing to the Forest Service and advised that I find a competent population demographer to do a credible analysis of the owl's disposition. I ask my dad (molecular geneticist and National Academy of Sciences member Frank Stahl) for help. He writes to population geneticist Jim Crow to ask if he'd be interested in the task. Crow declines, citing lack of time, and suggests that his former student Lande would be best suited to the task. He was."

Some of the science that had gone into the guesswork that led to the 500 number came from Russell Lande's doctoral thesis. By ignoring demography, Lande felt it had been "a misapplication of my work that provided another motivation for me to become involved in these issues."

It had not been a case of "do-it-yourself science," as Chase had made it out to be. Nor had it been a misapplication of island bio theory. Lande did use all of the theoretical methods that were at his disposal, along with adding some of his own pioneering touches, such as including territoriality in the model, but he had aimed them at the biology of an individual species. He might have been the first to have come up with a model that was real, general, and precise, all at the same time. The outcome had benefited the owl, ecological science, and the environment, at least in the views of all but the logging interests.

Lande added to me that Crow, a giant in the field of genetics, had been his postdoctoral advisor. "I was then on the faculty of the University of Chicago. When I looked at the original plan, I told Andy I needed to be a demographer, not a population geneticist. Because I had also studied demography and had some interest in theoretical and applied ecology, I ended up doing the demography. I studied Levins' work as an undergraduate and during my first year in graduate school at the University of Chicago, where he was a faculty member; then my advisor Richard Lewontin, a population geneticist, and I moved to Harvard. A year later Levins joined the Public Health division of Harvard Medical School and was a member of my thesis committee."

Understanding the limitations of theory, as Lande did, helps to keep it from going astray in its application. Practitioners in all fields often use outdated theory even when warned by scientists of its incorrectness. Intervention therapy for disaster victims, for example, concern over self-esteem, and repressed childhood memories have all been shown false by clinical psychologists, yet still are in use by psychotherapists. Don't even start with me about facilitated communication.

The owl's habitat needed protection; the mathematical model helped provide it. Fortunately in this case, the reciprocal also occurred: the owl happened to advance the theory.

"Levins' original model is not sufficiently realistic for application to territorial species," Lande told me.

Stahl had to get Lande to nudge theory along.

Looking back on the models of what he called the "MacArthur-Levins school in the 1960s and 1970s," he found them to be "rather simplistic and not commensurate with the complexities of the ecological component."

The net result of Lande's reworking of Levins' model, according to Stahl was that "Protection increases from 5% to 86% of remaining old-growth."

Never examined for the owl was the community matrix or its assembly rules—or the owl's Hutchinsonian niche.

Still Counting

Ecology is simple and ideas are many on islands. Sometimes there are just too many simple ideas that do get off those islands. One may be the use of island biogeography theory to get a handle on the issue of global biodiversity, the number of species, and the rate of their extinction.

Stuart Pimm engaged us in his *The World According to Pimm* by telling the story in a format that might have been part of a Monty Python sketch. Robert May was cast as something like the Minister of Silly Numbers. Numbers were thrown about, some of them patently silly.

100 million?

1,392,485 described?

End of skit. ("Not good enough!" It would have ended on the Python show.)

The number represents the known species on the earth, those that have been discovered and named as separate species. The actual total of species on the earth is more difficult to state with precision. Even the mathematically conservative Lord Robert allowed that the 100 million number might not be that far fetched. There are more species that we have not yet found than there are those we already know about. Pimm's chapter on the subject described how estimates concerning the number that were not known were made and how Lord Robert whittled them down to reality.

Terry Erwin, mild mannered coleopterist with the Smithsonian Institution in Washington, DC, is an expert on the Carabidae. These are drab little creatures that inhabit the soil litter of most of the world. Turn over a large rock or other cover and, along with the millipedes, ants, and worms, ground beetles will scamper out. For reasons of ecological research, I had to census the little critters in cornfields of the Midwest. This required not just counting them, but identifying them to species. There are zillions of species, I found. Lord Robert be damned, it had to be in the zillions, and I had to sort them out into species! Now, identifying organisms to species is incredibly easy if you are familiar with them and there are only a handful of them. Wilson, in one of his books, retold Ernst Mayr's story of the Arfak people of New Guinea, who recognized the 137 bird species there with almost the same exactitude as had trained taxonomists.

However, Wilson found no such precision for ants among the Saruwaget people of New Guinea. To them, an ant is an ant. Let them try ground beetles, I say.

Actually, tropical ground beetles are a different story from their litter-dwelling cousins. Erwin showed me some of his specimens, when I had brought over mine. Mine were squat and dark and, except for a few over-sized specimens, uniformly boring. They had been collected as part of a comparison of organic and conventional cornfields done by Barry Commoner's Center for the Biology of Natural Systems. I had published my results in what I thought was a suitable journal, but then decided that the ecological community might also benefit from the data. Lee Miller, then editor of *Ecology* thought so, too, but apparently mistrusted my taxonomic skills with this group. The paper had to do with diversity indices, requiring accurate separation of specimens into species. Lee must have caught wind of my estimate of there being a zillion of them, for he put me in touch with Erwin. Fortunately, before I left St Louis for UCLA, I had deposited every specimen collected for the study with the Museum of Science in St Louis. Doing so is known as leaving "voucher specimens," but usually only some, not the entire collection is deposited with a museum. The St Louis museum was not one of the world's great museums, so the curator there was not aware of such traditions and—God bless him—he took the dozen or so specimen boxes I had given him, then was willing to have them available to me by arrangement on my way from LA to DC. It had all worked out spectacularly conveniently for me. My brother had just relocated professionally to Georgetown University, allowing me to combine an unreimbursed business trip with a stay at his new home in Bethesda. Everything seemed to be working out. I managed to book a convenient flight from LA to Washington with a two-hour layover in St Louis and a return flight through Cincinnati that allowed me to do a site visit on one of UCLA's Environmental

Science and Engineering interns at the Battelle Memorial Institute. That let me recoup the airfare. Perfect!

Except—what does one do with some dozen boxes of specimens on an airplane? It was to have been a short stay, overnight only, at both DC and Cincinnati, so I had stuffed overnight things into my briefcase, knowing that it would be risky to check the bags through, given the arrangements I had made. Now there were those boxes.

I hauled them back to the cab I had left waiting for me at the museum in hope for the best. I could put some under my seat, I thought, and my neighbor's seat, and the seats ...

"What's in the boxes?" an attendant at the check-in counter had asked as I started stacking them up in front of him.

I was taken aback by the question. I had expected the airline's concern to be how to handle all of them, not what was in them.

"Dead insects," I had to blurt out after a pause that was much longer than the question had needed.

"How much do you want to insure them for?" he had asked after some hesitation of his own.

Nothing, I thought to myself. They were both worthless and priceless.

"Never mind," the attendant had said, not waiting for an answer. "We'll take care of 'em."

Flying was different in those days. With a completely matter-of-fact attitude toward my befuddlement, he took the boxes into his custody and let me know that I should go catch my flight. The airline must have done that sort of thing all of the time. When I got to DC, there were all the boxes waiting for me, neatly bundled together.

Erwin rather quickly put appropriate binomial labels on the species for which I had thrown up my hands and simply listed by genus. He spotted a few specimens I had misidentified, and corrected the nomenclature on one or two others. (My reference sources were both out-of-date and wrong, Terry had admonished me.) Terry's corrections hardly changed the conclusions I had made based on my own identifications, but it was nice to have the taxonomic exactitude that Terry gave, so we ended up publishing the paper jointly. The ecological community apparently ignored it. They had more serious issues with diversity than which index was the most appropriate for its measurement.

Quite honestly, my research findings had not been new. They represented something more like piling on in football when a runner was already down. Or, maybe even, flogging a dead horse. They did, however, let me meet Terry Erwin and take a look at his beetles.

Terry had been quick and business-like in going through my beetles. It was when they had been pushed aside to be shipped back to their repository in St Louis and Terry turned to show me some of his beetles that I saw why he was a taxonomist and why he was specializing in this group.

His beetles were colored, often brightly and iridescently so. Even the dark ones were of much more interest visually than their Midwestern cousins. Having specialized appendages and long necks, they were graceful acrobats compared to my plodders. And they did not live in soil litter or under bark and rocks. These species were aerialists. They lived high in the canopies of forest trees in the Amazon. And only in the canopies—one would never find them on the ground.

"It must have taken an awful lot of hazardous climbing to accumulate these," I had wistfully remarked as I gazed at his marvels.

Terry had changed remarkably in the switch from my beetles to his. His somewhat somber countenance as he examined my specimens changed only when he had spotted an error of mine to correct. Then he had taken on the gentle look of a kindly schoolteacher. Now his face turned suddenly playful. These beetles, in row after row of neatly pinned precision, much like a coin-collector's displays, were his hobby. The beetles I had brought him were only part of his job.

Except that they were in that part of the Smithsonian devoted to natural history, Terry's beetles in their drawers had looked like art to me. They could easily have been construed as some sort of mixed-

Still Counting

media modernity. Like Andy Warhol's multiple images of Marilyn, nature had provided the beauty, while Terry had arranged it in an artistically pleasing way.

"Not at all," I remember him answering my remark, then going on to explain his method of collection.

Climbing forest trees with the safety of ropes and assistance still made for excruciatingly slow collection, and it probably required special life insurance coverage. What Erwin did instead was send an insecticide fog up into the canopy. It was a lot like what Dan Simberloff had done with his mangroves, except that the hazard in enclosing the trees was now one of a fall from height, rather than a dunking. (I never asked who climbed up into the trees with whatever was used to "bag" the trees.)

"I don't bag the trees," he later corrected me. He "fogs" them. "There are various ways to fog," most of which he invented himself. By fogging, Erwin meant enshrouding the tree of interest in an insecticide mist.

Once the insecticide did its trick, Erwin could then pick up the bodies as they literally fell to the ground. Actually, it was to specially prepared tarps and funnels that they fell. There were LOTS of bodies. The method allowed for thousands of specimens to be collected without risk to life or limb for the entomologist.

He described for me a recent operation in an equatorial rain forest in the Ecuadorian Amazon. Essentially, a very large funnel was constructed by putting up a 3-meter by 3-meter sheet "with a pint bottle of 75% ETOH suspended beneath." (That's 150-proof grain alcohol, in the vernacular. It doesn't kill the insects, the fog does that, but it preserves and protects them.)

"At 4 AM, I stood at the edge of the sheet and fogged upward," Erwin continued his explanation. "At that time the fog goes straight up. Of course it doesn't just stay in the three by three column, but what comes down on the sheet is my sample, the rest is ant food."

What Terry found amazed him. One tree had 163 species of beetles living on it. This does not sound like a remarkable number except that these are beetle species that live only on that species of tree and on no other kind. They are endemic to that tree species, just as there are birds and butterflies that are endemic to islands in the Hawaiian chain, and just as the tortoises of the Galapágos are endemic to those islands. If all tropical trees had this number of endemic beetles, the 50,000 species of trees in the tropical forest would mean some 8,000,000 species of beetles! The math and reasoning are as simple as estimating the number of species of Galapágos tortoises from the information that they were found on 15 of the islands. (Did you estimate 15? If not, then you probably did not realize that each island had a distinct species resident on it. Or you are confused because there are no longer 15 species? Some went extinct. Whalers ate them.) Factoring in the relative proportions of other species of arthropods (mainly insects and spiders, of which there are also very, very many species) in the tropics, the total number of tropical arthropods might well be 30,000,000.

By that time, Erwin had gone through more than nine million specimens of beetles, by his own estimate, and he was still busy crunching the numbers of species. "I can tell you that there are more than one hundred thousand species per hectare and six trillion individuals in that hectare."

Up Against Limits

Ecologists rose to the surface in that first Earth Day's foam of slogans, facts, and ideas, because people rightly associated the science with concern for the health of ecosystems and the extinction of species. *Silent Spring* and the disappearance of the Peregrine Falcon from America still weighed heavily on the public's minds.

"Rachel Carson's book had been published a few years earlier and it was still the major topic of discussion," David Pimentel of Cornell University once recalled for me about his service on the White House staff in 1969 and 1970 as a consulting ecologist.

Ecologists also found that their voices were listened to on issues besides endangered species. World population, the condition of our air, our waters, and our forests—even the status of our mineral resources—all were areas on which the natural historians and students of the evolution of adaptations were making proclamations warning of potentially dire futures.

The reasons why ecologists rose to this position of leadership are both surprising and intuitively obvious. First, a number of ecologists, Paul Ehrlich perhaps the best example, found that, like Rachel Carson, they had literary voices that could reach people. Strong voices such as those of Garrett Hardin, Barry Commoner, and René Dubos rose to their places in the public consciousness through their insights into the human predicament as seen through their science, biological, if not quite ecological. They were all literary successes. But what of LaMont Cole, George Woodwell, David Pimentel, and Eugene Odum? All were reputable ecologists whom no one would have accused of literary aspirations, I am sure. That they and many others like them rose in influence was because theirs was the only science that could at that time address our concerns. Add to that the natural appeal of the science to audiences of readers who saw ecology as underpinning the popular works of Carson, Joseph Wood Krutch, and Edwin Way Teale, and it is clear why ecology had the public's ear ahead of the other sciences. Those others were the sciences thought to have brought on the predicaments that were the subjects of laments on Earth Day.

Paul Ehrlich was a butterfly ecologist, but he was also a population ecologist, as was LaMont Cole and David Pimentel. Demography is the heart of population biology. Of course ecologists should speak up on issues of population growth. Their science was in possession of expert knowledge on that subject.

Meanwhile, Woodwell and Odum, with their ecosystem focus, were expert on the scientific knowledge behind Barry Commoner's laws of ecology, in particular the laws that "there is no such thing as a free lunch" and "everything is connected to everything else." Had Barry been a bit less literary, he might have simply stated that changes in one property of an ecosystem often result in unexpected and unwanted consequences elsewhere. It was those unexpected consequences that were the scientific concerns of Woodwell and Odum. It was Woodwell whose efforts served to tie DDT use to the decline of eagles. Ecosystems do have interconnections. It was the ecosystem scientists, also, who were concerned with biomass, populations, chemicals, and the air, water, and soil media with which and through which they interact. Of course, ecologists should also speak up on pollution issues.

It was ecologists who best made the connections between human activities, natural resources, and population growth. By that first Earth Day, ecologists had solidly established that exponential growth of populations—from bacteria in test tubes to moose on islands—cannot continue indefinitely in nature. If not cropped by a predator, damped by competition, or subdued by parasites (the most ecologically interesting constraints), populations will eventually cease growth upon the depletion of some abiotic factor: mineral nutrients, water, or simply space for more individuals. More interesting are populations whose halt in growth is caused by something akin to wallowing in their own filth, conditions found only in artificial situations such as microbes in test tubes. Readers can make their own analogies between the bacterial test tube and an industrialized society. The consequences are quite sobering. Ecologists did not shy away from that analogy then—nor do they now.

The S-shaped curve that Gause found for his laboratory populations of paramecium turned out to be a foundation of environmental thinking. It was first described and named by French mathematician Pierre Francois Verlhurst at about the time Darwin was taking his voyage on the Beagle. Readers with a historical bent, along with proficiency with math and languages, may want to track down Verlhurst's work for themselves. None of my bents or proficiencies incline me to do so, but I can direct the reader to Joel Cohen, who cited a 1977 translation into English of Verlhurst's 1838 paper in his own book, *How Many People Can the Earth Support?* GE Hutchinson, in *An Introduction to Population Biology*, gave Verlhurst's second memoir, "*Recherchés mathematiques sur la loi d'accroissement de la population*," published in 1845, as the one that introduced the logistic equation. Hutchinson went on to note that Verlhurst's work "seems to have made no impression on any of his contemporaries."

One-time Rachel Carson mentor, Raymond Pearl, and his colleague, Lowell Reed, rediscovered Verlhurst's equation almost a century later, fitting it to US population growth between 1790 and 1910, and again for the period 1790-1940. They played a bit fast and loose with their data in doing so. Cohen pointed out that the tripling of the US land area in the time interval between 1790 and 1910 invalidated the use of the logistic in its strict sense. Pearl and Reed ignored that little nicety. The equation describe what is called a "closed system" only, one that has neither external inputs nor outputs (no immigration and emigration) nor variation in its r and K parameters. Both r and K are required to be constants. There is no reason why the two parameters should not vary—as many investigators have allowed. However, one then has a different equation than that specified by the logistic.

Application of the logistic equation must have been just as infectious as were the later application of species-area curves. One sociologist identified 219 trends that could be described by the logistic. It gets worse. I don't have the patience to tabulate the number of biological tendencies that could be put on Robert

MacArthur's *r* and *K* continuum, but I venture that it would have to be at least as many as were found for the logistic in sociology. Another example of explosive exponential growth, as if we needed it.

A big reason why Gause found success in using the equation is that bacterial and protozoal populations have exactly the characteristics necessary for logistic growth. They have perfectly overlapping generations and there is almost no delay between changes in population sizes and changes in the rate of growth. Real populations breed seasonally and can overshoot, rather than smoothly approach their carrying capacity. More complex species such as flour and grain beetles only seemed to fit the logistic if the experiment was discontinued when the population leveled off. Grain beetles subjected to Gause-style experiments seemed to overshoot their carrying capacity, then oscillate in abundance. In laboratory growth experiments with more complex species done in the decades following Gause's work, the end result was not always a stable population at its carrying capacity, but was more often a population that fluctuated after the initial overshoot, often with a die-off that could be massive. This was the behavior Robert May told us to expect when time lags and discrete generations were added to the logistic. This is the behavior we find in nature when algae in waters enriched with nutrients, such as phosphates from detergents in untreated wastewater, grow until the algal mat formed is so thick that it blocks out sunlight to algae at the bottom of the mat. That leads to death, decomposition, and oxygen depletion of the waters.

Some algae always survive. And there is always another pond. Is there for humans? That was a question raised by Earth Day ecologists. Their answer was a resounding "No!"

Competition, predation, disease, and fluctuations in weather modify population dynamics in nature so much that it is difficult to find real situations in which the logistic might be applied with assurance. Humans, of course, no longer have competitors or predators reducing their numbers. In addition, the control that medicine exerts over our parasites is what by all agreement has led to the unprecedented growth in the human population. As to vagaries of weather, technology supposedly can shield us from all but the most cataclysmic of events (of which, global warming, seems currently to be of the most concern). Neither do we breed seasonally. Perhaps the logistic really is a good model for human population growth. However, there is one important characteristic of human populations that is not in accord with the logistic. There may be substantial time lags in our responses to changing environmental conditions. It is sort of a population inertia. It will keep our global population growing even after its average birth rate is reduced to replacement levels. The logistic curve, therefore, can be no more than heuristic for human population growth.

In fact, the shape of the human population growth curve to the present is more like the letter "J" than "S." Is that J-curve the first part of an S-curve, as Pearl and Reed saw it? Economists and futurologists associated with Herman Kahn and Julian Simon (and later Bjørn Lomborg) have faith that it is.

Joel Cohen once offered the following forecast that he gleaned from the publications of a think tank: "Our argument concerning population growth is that its rate is approximated by a flattened S-shaped (or logistical) curve, rather than by an exponential one. We do not envisage population growth continuing to a point where it encounters physical barriers, such as lack of living space or resources, and brings about widespread scarcities, famines, and intolerable pollution."

Herman Kahn had been one of the authors of that statement. They had been doing the same sort of curve-fitting to the logistic equation that Pearl had done. Cohen, however, showed in his book that human population growth has been "superexponential," in his words, exponentially increasing exponential growth. Nothing in nature has such continued growth.

That there must be limits is what ecologists had communicated so effectively on that first Earth Day. It cannot be wished away as a doomsday prophecy, nor willed away through some sort of faith in human nature and human technology—or whatever it is that the quote from the think tank suggests may be possible. That populations cannot grow without limit is as fixed a natural law as that an unbalanced force will cause an acceleration or that heat flows from a hotter to a cooler body.

The only question is what that limit might be. Its exact number will determine whether the consequences will be severe or inconsequential. This is where ecologists, especially through their ability to quantify complex natural processes, could most contribute to human welfare.

Some ecologists fear that we are already past those limits. This is a sentiment that especially found voice around that first Earth Day. Its most outspoken proponent had been, of course, Paul Ehrlich, who had declared in *The Population Bomb* that "the battle to feed all of humanity is over." Ehrlich had chosen not to trouble the reader with the mathematical niceties of population growth as Joel Cohen had. Instead Ehrlich had admitted to have "just scratched the surface of the problem of environmental deterioration" caused by overpopulation. He had done so with, if not exactly anecdotal evidence, highly digested data and much jumping to conclusions.

Other ecologists had also been as guilty then of exaggerating the doom that awaited us in the near future. Enthusiasm fostered by the first Earth Day may have led them astray in ways that now haunt the efforts of non-partisan environmentalists. The Internet is preserving and disseminating the more careless Earth Day statements for all the world to ponder.

George Wald, whose inspired teaching at Harvard had turned Dan Simberloff from a career in math to one in ecology, has been quoted as having estimated that, without immediate action, "civilization will end within 15 or 30 years."

We should be out and climbing around in the rubble, then, instead of inside reading by an electric light.

Humanist René Dubos is quoted as having said, "the bomb may one day no longer seem a threat, but a release," for some overcrowded populations.

I don't know how much credence to put into whether the speakers meant what some web sites imply by their words—or whether the words were said at all. Following up on one web site dedicated to capturing extremist quotations, I found at least one that was totally without substantiation. It was attributed to LaMont Cole. Elizabeth Whelan quoted him in her book, *Toxic Terror*, as saying that "to feed a starving child is to exacerbate the world population problem." I have no doubt that the old curmudgeon was every bit capable of such a statement, but out of personal interest, I wondered what the source was for the quote. I remember Robert H Whittaker, an eminent plant ecologist at Cornell, having been attributed with the hope for a short, devastating disaster, rather than a long, drawn out one. That there would be a disaster, he had absolutely no doubt about. The logic on which he had placed his hope was that the more rapid our die-off, the less time we would have to destroy the environment. Whittaker had been much too astute to allow a statement like that get into print, however. Quite honestly, I am not certain whether I had heard it myself in some informal setting, such as a coffee break (highly unlikely based on his personality) or a seminar or lecture (more probable), or whether someone else had relayed it to me and my memory somehow weaved it into my own experience. So it interested me to know how Cole had produced his little gem. I tracked the quote down to a Dixie Lee Ray book in which she called Cole a Yale ecologist. There was no citation or details about the circumstance of the quotation. I suspect Cole may have made a speech at Yale, but more likely, the quote is fabricated. Paul Ehrlich quotes are not, however. They are all right there in his publications.

The most famous doomsday quote from that era was an entire book, the Club of Rome's *Limits to Growth*, produced in 1972 not by ecologists, but by systems scientists at MIT's business school. They had been commissioned to study the consequences of unlimited human population growth by the Club of Rome, an elite, international organization, as part of its Project on the Predicament of Mankind. Their method was to create a computer model, World3, that could simulate the interactions between population growth, resource use, pollution, and economics. It was a standard systems model, all black boxes, with input and output arrows and feedback loops. In spite of its admittedly preliminary nature, it was rushed into print. A number of scenarios differing in assumptions about population growth and technological capabilities were presented and their consequences evaluated. Although according to Donella Meadows, one of its authors, it had never been intended to be a prediction of the future, but rather an estimate of the consequences of choices available, the study became notorious as a prediction of doom. Simply stated, the study concluded that the then-current trends (circa 1970) would reach their limits and result in both a population and economic collapse within 100 years. (Does that seem remarkably coincidental with the 2050 to 2060 tipping point in global climate that is today being warned of by news outlets of various kinds? [You can google it.] Except that now it does not feel so far off into the future as it had in the 1970s?)

"The basic behavior mode of the world system is exponential growth of population and capital, followed by collapse," was their general and easily understood conclusion.

To the public, the study seemed to say that it was already too late. The simple box-and-arrow input-output systems models from which the MIT scientists built up their world system showed not a population rising slowly to its carrying capacity, as in the S-shaped growth curve of the logistic, but one that was overshooting it and collapsing. Not only did the MIT team identify limits, but those limits seemed to be catastrophic ones—and we had already passed them.

The reception of the study, despite its sales success was not what the authors had hoped. Instead of feeling pride at triggering real change in the world, Donella Meadows "was hurt and awe-struck by the wild denial" that was directed at the book. I think that may have instantly made her something like an honorary ecologist. Much of the criticism of the study was along Capitalist-Marxist grounds. Embedded in it was a conflict between two diametrically opposed points of view that have come to be known as the neo-Malthusian and the Cornucopian. Neo-Malthusians believe that resources needed by humans are in limited supply. The cornucopia, in contrast, is the goat's horn of plenty, overflowing with fruits and grains, emblematic of inexhaustible abundance.

Robert Thomas Malthus, an English thinker whose *Essay on Population* helped to create the science of economics and is the "Malthus" within Malthusians, also had a seminal influence on Charles Darwin's theory of natural selection. It was from Malthus's analysis of human growth outstripping food growth that Darwin was given a mechanism for evolutionary change. That historical intellectual transfer also linked ecology to economics right from ecology's birth.

Malthus's analysis led to the conclusion that men must always live on the verge of starvation. Human population growth must always outstrip its food supply. In the Malthusian view, every increase in food supply leads to an increase in the number of people to consume it. Every child added is but another hungry mouth to feed. Codify that human misery in terms of wages and profits and economics is, indeed, the "Dismal Science." Neo-Malthusian environmentalists added the effects of pollution due to population and technology growth to the misery. It is the acceptance of the concept that there are limits to natural resources—either direct or as a result of pollution—that limit population growth that is seen as the neo-Malthusian position.

Environmental scientists, especially ecologists, who are used to thinking in terms of limiting factors to growth, welcomed the publicity that the study brought to the idea that there should be concerns about approaching environmental limits. However, other than this publicity factor, the collective judgment of the establishment of those in the know was that, in policy evaluation "large scale economic-demographic simulation models" (such as World3) "are of little use." Or perhaps it was the results they had little use for.

However, the warning in *Limits to Growth* is the same warning, albeit cloaked in systems science and computer simulation, that Paul Ehrlich put forth in *Population Bomb*, or Robert Malthus in his essay on population growth. There are limits to human population growth. Malthus, then David Ricardo, another turn-of-the-18th-century economist, saw the limits as adjusting population growth in ways that would keep the greatest portion of the population just above subsistence levels, while Ehrlich saw the limits as creating ecological disaster, and Dennis and Donella Meadows and their colleagues, economic disaster.

To ecologists, the announcement that the human population had reached its limits was totally unsurprising. In fact, it seemed overdue.

One group of scientists took data on human numbers up to 1958 and applied a model to it that had positive feedback between population and growth. As the population doubled, so did the growth rate. This is exactly what Cohen called superexponential growth. The model turned up the estimate that the population would go essentially infinite—a mathematical limit in the extreme—on November 13, 2026. The results caused quite a stir when published in *Science* under the title, "Doomsday: Friday 13, November, AD 2026," in 1960. Clearly, super-exponential growth cannot be sustained.

In 1994, two ecologists calculated that the human population fell below the "Doomsday curve" for the first time in that year. No one was reassured by the finding.

Potatoes Made of Oil

Justus Freiherr von Liebig, a 19th century German chemist, discovered that plant growth was always limited by some mineral that was in short supply. Addition of that mineral would increase growth until some other mineral became in short supply. His principle is usually explained by analogy to a barrel made of staves of different lengths—or to a leaky bucket. With a bucket full of holes, the amount of water that can be maintained is determined by the lowest hole. Plug this and you can get more water in the bucket, but only until the next lowest hole allows water to flow out. So, you can water a plant and provide lots of sunshine, but you may find that, eventually, nitrogen has been depleted from its soil and the plant shows all the signs of growth stress. Add nitrogen and the plant will come against its next limit to growth due to phosphate or potassium. Carry the analogy from a single plant, perhaps in a laboratory in which growth and nutrients can be carefully monitored, to a field continuously in corn, and the need for fertilizers to maintain high yields becomes obvious. So does the realization that there will eventually be a limit to how high that yield can get. Perhaps the final limit might be sunshine. Think globally and it is clear that humans will eventually run out of space or water or some nutrient or the energy to provide space, water, and nutrients to grow crops.

Eventually is not very precise, however. To some—given the wonders of technology yet to be invented—eventually might as well be infinity.

Determining what the actual limits to human population growth are is no longer the specific provenance of ecologists. But ecologists can offer their peculiar insight into the question. For example, Joel Cohen had shown that the question of how many people the earth can support is not an appropriate one. It is under-specified, a term mathematicians use to describe a problem that cannot be solved because not enough information has been given. Cohen's argument went something as follows.

I can't tell you how many people the earth can support until you tell me what diet the people will have. And then I can't answer that until you tell me how much land and water you want to use to grow food for that diet and what type of technology you will use. Liebig's simple minimum is not appropriate to the concept of human carrying capacity.

More importantly, Cohen argued, resource quantities are not all there is to the reckoning. There are political and spiritual issues to be considered. He quoted John Stuart Mill, hardly an environmental activist, as having written that there would not be "much satisfaction in contemplating the world with nothing left to the spontaneous activity of nature, with every rood of land brought into cultivation, which is capable of growing food for human beings; every flowery waste or natural pasture plowed up, all quadrupeds or birds which are not domesticated for man's use exterminated as his rivals for food, every hedgerow or superfluous tree rooted out, and scarcely a place left where a wild shrub or flower could grow without being eradicated as a weed in the name of improved agriculture." This is before Deep Ecology, the Sierra Club, or Marsh and Thoreau. Mill called for a stationary population long before necessity compelled it.

On Earth Day 1970, Paul Ehrlich predicted that the biosphere could not sustain the 7 billion people projected for the year 2000.

"There can only be death, pestilence and famine to reduce the number," he was quoted as saying.

Barry Commoner predicted that the earth could sustain between 6 and 8 billion. I don't know whether it should be gratifying or unnerving that we are right at the populations that Ehrlich and Commoner had warned about. Theirs are but two of many estimates.

The history of estimates for the earth's maximum possible population is an extensive one. Cohen assembled and annotated what amounts to four centuries of estimates for the human carrying capacity in his book. I have borrowed liberally from his labors.

The first estimate was by Anton van Leeuwenhoek, the father of microbiology. It was Leeuwenhoek who discovered the teeming life in a drop of pond water when he looked at it through the newly invented microscope. Used to extrapolating to large numbers from small samples by his microbiology work, Leeuwenhoek obtained a maximum human population by estimating the land area of the earth and multiplying it by the density of his native 17th century Holland. Many other estimates are variations on the theme of area times density initiated by him. His estimate, 13,385,000,000 is instructive. It may well be the best estimate of a sustainable, if not absolute, limit to human numbers.

Raymond Pearl estimated between 2 and 2.6 billion based on the logistic equation and then-current population data. This was his estimate of the K for humans. It was not far above contemporaneous population estimates. My *Encyclopedia Britannica* (1958 edition) lists a United Nation's estimate of $1,813 \pm 60$ million for 1920. By the time that Rachel Carson was working in Pearl's lab, but for the depression, the global population should have been pushing at that limit.

Harrison Brown, a Manhattan Project physicist who later turned his attention to concerns over population growth in what was then called the Third World, gave one of the higher estimates of maximum population size, 50 billion. He based it on intensive irrigation, new acreage in the tropics and deserts, advances in crops and technology, plus a diet heavy in algae and yeast. This is probably as good an estimate for a reasonable absolute maximum as any. Scripps Institute oceanographer Roger Revelle, one of the first scientists to recognize the possibility of global warming due to carbon dioxide building up in the atmosphere, working on a committee commissioned by Lyndon Johnson, gave a similar estimate, but based on grain crops, rather than yeast and algae. Revelle's estimate also assumed a continuation of two decades of technological improvements, the "miracle" strains of high-yielding crops, irrigation, and chemical fertilizers, for example, all the sorts of things that make Cornucopians complacent. Estimates that exceed Revelle's 40 billion are universally based on at least one patently unrealistic assumption. Most often it is that only one resource is limiting, and that all of that resource can be put to use for food production. One early estimate, based on planting all of earth's land area, including polar regions, was 132 billion. It was given in a report entitled "The Mathematical Theory of Population, of its Character and Fluctuations, and of the Factors which Influence Them, Being an Examination of the General Scheme of Statistical Representation, with Deductions of Necessary Formulae; the Whole Being Applied to the Data of the Australian Census of 1911, and to the Elucidation of Australian Population Statistics Generally, Appendix A., Volume 1," by a GH Knibbs for the Commonwealth of Australia. Again, we have Joel Cohen to thank for bringing this bit of Edwardiana to our attention. A later estimate, essentially based on a photosynthetic limit, the maximum energy that plants can produce from sunlight, was a trillion people, reduced to 146 billion, given the necessity of having space for cities and recreation.

Lower estimates include 7.7 billion by the *Limits to Growth* computer, World3, 3 billion based on competition for resources between food and renewable energy production (mostly solar or solar based), and "far fewer than" at present. The latter pithy estimate was by Paul Ehrlich, who in 1978 estimated that there were "between three and seven times more people than this planet can possibly maintain," and Ehrlich, Anne H Ehrlich, and Gretchen C Daily, who claimed in 1993 that "Earth's nutritional carrying capacity, will number far fewer than today's 5.5 billion people." With a few exceptions, these lower estimates were made without the number crunching that goes into making reasonable approximations of crop yields and acreage. Many are surprisingly below the current world population of 7.7 billion. (Exactly at the limit predicted by *Limits to Growth*. Could the current global immigration problem be telling us something?) In the repertoire of what might be called the "Globally Concerned Ecologist," this has been an ongoing motif: we are either at a limit or have already surpassed it.

As Cohen pointed out, the question of how many people the earth can support is not one that can be answered simply. Cohen even "learned to question the question." To answer it, one must first ask a series of other questions. For a start, what diet is to be expected? How many calories per person? How much animal versus plant protein? What agricultural technology is assumed? How much energy and water use will that require? One of the low estimates, 3 billion, took into consideration that food production needs land and energy, while renewable energy systems (biomass and solar in particular) also require land. One cannot maximize both food and solar energy production at the same time; that's like having your cake and eating it, too. One can, at best, find an optimum allocation of land to both purposes. But that raises more questions. Is all arable land to be used? Or are people to have land in which to live and play? And what about the wild places of the earth? Will they be preserved? And what level of air and water pollution are we to tolerate—or can be tolerated by the agricultural production system? And what will we assume about the economic systems that are to operate and how wealth is distributed? And today, what about the climate?

One thing that we can conclude along with Cohen is that the question of how large a population the earth can support "cannot be answered using only ecological concepts." Population growth by humans does not follow the logistic growth equation. Even if it did, one has to agree with economists Julian Simon and Herbert Kahn that "'carrying capacity' has by now no useful meaning" because technology seems to keep continually raising it. We may have gone from being slow growing K-selected species for most of our history to r-selected species through advances in medicine. If we were Gause's paramecium, we've just found ourselves in a new culture vessel. Not a comforting prospect.

Let's assume that ecology cannot provide a numerical carrying capacity for humans on earth. Paul Ehrlich's guess that the earth is capable of sustaining 1 billion people "in reasonable comfort for perhaps 1,000 years if resources were carefully husbanded," is just that—a guess. If all we can do is guess, then what should be the role of ecological science in population policy?

There is a role. One thing that ecologists can do is to identify and quantify with reasonable confidence the resources needed to feed people. Sunlight, for example, is an obvious resource, indeed the most primary.

Although the maximum potential photosynthetic efficiency has been estimated to be as high as 34%, possible "only under laboratory conditions with unicellular green algae," according to that ecology textbook I would have read in my undergraduate ecology course with Robert Ricklefs, had he written it by then, most landscapes, including agricultural ones, convert 2% or less of incident sunlight into biomass. Even though 2% may represent a natural limit to photosynthetic efficiency, technology has been able to allocate more of that 2% to producing biomass suitable for human food. Short-stemmed varieties of wheat and rice, for example, the "miracle" varieties, put the energy they would have used to produce a longer stem into producing grain instead. Plant breeding has also created strains that are particularly responsive to fertilizers; genetic engineering is creating strains that can resist pests or withstand herbicides. Ecologists can and should maintain a vigil for unanticipated consequences from such advances, but the advances themselves no longer have much to do with ecology. (The situation of genetically modified organisms [GMOs] is best left to future historians.)

Once the maximum yield per unit of land area is attained through agricultural technology, the land itself can be limiting. Wetlands can be drained and deserts can be irrigated with technology already available. Ecologists can warn of what is lost by converting these lands, but the conversion process is essentially one of engineering, not ecology. Estimating the amount of various land types available should be the provenance of the physical geographer, but ecologists just love to estimate the amount and condition of the earth's land area. They have had a long tradition of doing so.

In addition, there is the question of energy. Howard Odum was perhaps the first to realize the relationship between fossil energy and modern agricultural yields when he wrote in 1971 that we were eating "potatoes partly made of oil." A few years later, the Arab Oil Embargo of 1973 led to concern over the dependence of the US on foreign oil, its agriculture in particular. Concern over technology and resource use, energy and land in particular, came to dominate policy discussions. In that same year, EF Schumacher proclaimed that *Small Is Beautiful* in his book of that title. Meanwhile Amory Lovins, fresh out of university, was capturing the attention of his elders by urging us onto *Soft Energy Paths*, the title of the book he published soon after. Meanwhile, David Pimentel independently followed up on Odum's suggestion and tabulated the exact amount of fuel and other energy used in producing corn and feedlot beef, using teams of highly-trained researchers to find the numbers needed for the task. They then did what was really some low-level number crunching. I know, for I was part of those teams of researchers, wondering why I was doing the kind of mindless engineering calculations I had wished to avoid by studying ecology. Why wasn't I out in the Andes studying the Chilean condor?

The reason that I was doing the number crunching at that time was that it was important and no else seemed to be doing it, except, on occasion, people with axes to grind. We know how that works out from the response against *Silent Spring* launched by agribusiness and the Chemical Manufacturer's Association. Did ecologists have an ax to grind, too?

In the decade following Earth Day, the ax that was ground was one of limits.

It was out of a sense of duty that I went to work on tabulating the energy inputs for beef protein production as a fledgling ecologist, instead of watching a soaring condor in the Andes. The weekly meetings around David Pimentel's desk at which we compared our progress in digging up and crunching numbers became something akin to weekly church attendance. I walked away from them feeling quite as absolved as I was by confession. (I know that I am an agnostic, but confession absolves me of that, too.)

Those studies led by Pimentel that tabulated energy and resource use in agriculture found much company. Spurred perhaps by *Limits to Growth*, the late 1970s and early 1980s were times of confronting what lay ahead for population, food, and society. The confrontation was a quantitative one. Systems analysis—often as ecosystem analysis—provided the method of quantification, and the logistic growth curve, a philosophical or theoretical basis. (Whether it was theoretical or philosophical depended on whether you saw the little equation as realistic or heuristic.) Water and soil resources were of great concern. As might be expected, water supply, water purity, soil erosion, arid-land crop systems, and, of course, the interface between water and land systems were issues that were examined in that era. Many of the studies rightly fell to non-ecologists. Many of them called for multidisciplinary approaches to the problems. And it is that that was and is the strength of an ecological point of view. The systems ecologist in particular was interdisciplinary all by himself (or herself).

David Pimentel continued to almost yearly gather around him ensembles of young scientists to take up one issue after another. He was a turbo-charged compact man of indeterminate middle age when I arrived to work and study under him at Cornell in 1973. He had an easy smile that was genuine and eyes that could penetrate you with their gaze while at the same time taking in everything that was going on around him. They were the eyes of an Air Force pilot, which he had been. He was as fit as a pilot, too. His slight body was that of an athlete who refused to let himself get out of shape. In Pimentel's case that meant daily jogging without exception, even one time in the Philippines during an insurrection. (According to him, he jogged so early in the morning that warnings of hazards due to insurrectionists were unnecessary.)

That was the genuine part of him. There were also those British affectations. I seem to recall "Right-Oh" and "Cheerio" peppering his conversations with me. The Britishisms were holdovers from the time he had been at Oxford University working with Charles Elton, among others. The fellowship that had taken him there had expired some twenty years before my meeting him, but the mannerisms had not.

"I spent most of my time investigating the food requirements of voles in the field," he once summed up his Oxford experience.

Those conversations I am remembering were always brief and to the point; there was no rambling with him. He seemed not to be outgoing by nature. Those mannerisms affected may have been meant to charm. Or he may have been much too busy for that. The man was quicksilver. Without seeming at all in a hurry professionally, he managed to always get there first. It was what made his reputation. He was among the first to have combined genetics with population dynamics in a theory he called "genetic feedback." It was probably what made (and broke) his reputation as an ecologist. Note the hint of cybernetic control in the use of the term *feedback*. Genetic feedback was a theory that had been marvelously situated on the arena of ecological and evolutionary research of its day. The debate over what regulates animal populations was raging then. HSS had been published in 1960. The year that Pimentel published his genetic feedback paper was 1961. It was also right on the heels of Robert MacArthur's Three Influential Papers. It potentially could have given a mechanism for the coevolution of species that became Paul Ehrlich and Peter Raven's most influential idea in ecology and evolution. It spoke to basic issues of distribution and abundance of species, invasive species and biocontrol, the interaction of trophic levels, competition and community theory, and, of course, it actually combined population genetics and population ecology before Robert MacArthur did so with his r and K selection. Had Pimentel a mathematical bent or the patience to devote his energies to a single subject, he would have been at the forefront of the MacArthur revolution. The theory in words was unassailable.

"High herbivore densities create strong selection pressures on their host-plant populations," Pimentel wrote in 1968; "selection alters the genetic makeup of the host population to make the host more resistant to attack; this in turn feeds back negatively to limit the feeding pressures of the herbivore."

Strip it of its modern terms and it is logic that might be found in the *Origin of Species*. Science requires testing, however, and to test an idea requires that it first be modeled. Pimentel's model was disappointing. Like my Salton Sea model, it suffered from an annoying mathematical specificity. (Although, before leaving the subject, it should be noted that efforts to put MacArthur's r and K concepts into mathematical forms acceptable to population geneticists wound up with equations that mathematically looked remarkably like those of Pimentel's genetic feedback model. Once the idea was shifted to MacArthur-type terms, though, genetic feedback sort of just faded away. Partnering up with mathematician Simon Levin failed to help resurrect it. Levin seemed to favor collaborating with the University of Washington's Robert Paine. Perhaps Levin preferred the rocky intertidal environment of Paine's studies to Pimentel's basement laboratory ecosystems. (I would have, too, given the chance.)

Pimentel had also been among the first to weigh in with new observations in the diversity-stability debate. That had also been in 1961. That same year, he had published a field experiment investigating the effects of plant spatial patterns on insect population dynamics. He had also been an early investigator of the effects of differences in space and over time on the dynamics of trophic structure. It had no doubt been a result of a sabbatical leave he had spent at the University of Chicago with Thomas Park. Pimentel had also been one of the first ecologists to serve as a White House staffer. His efforts there in 1969 and 1970 may have pushed that administration to pass the environmental laws from which we benefit today. It did allow him to also write an influential book, *Ecological Effects of Pesticides on Non-Target Species*. And, of course, he was among the first to do bookkeeping studies and economic assessments on various ecological issues.

I arrived at those basement laboratories at Cornell with ability in mathematics and systems analysis that could have been applied to melding population biology and population genetics—and an overriding interest in field research. Basically, I cared not at all what particular ecological item I was to study, as long as it was in the outdoors, preferably someplace with a view. The basement laboratories of Comstock Hall had teams of graduate students studying genetic feedback through artificial systems employing vials of sugar and houseflies. The vials represented plants and the houseflies, herbivores. A plastic container with a lid in which the vials were embedded represented their world. The houseflies, honorary herbivores at best, were the only part of the system that was biological. There was a variety of experiments underway using boxes and boxes of the plastic containers. They filled up several rooms.

In one experiment, a chameleon was added as a third trophic level. He was right there in plain sight when you entered the main room of the laboratories. He (maybe it was a she) sat motionless, absolutely unperturbed by flies that set down on him, even those that did so right on his snout. It was an equanimity toward flies that the lizard might have learned—if lizards can learn such things—from the graduate students sharing its laboratory space. They, too, had to deal with flies (these being escapees) settling on them. Unlike that of the chameleon, the environment of the graduate students included copious amounts of flypaper hanging down seemingly everywhere from the ceilings. (One theory then extant, documented by literature I was told, had it that flies avoided pristine new strips of flypaper in favor of those already spotted with fly carcasses, so it was never clear how often the flypaper was replaced.) There were also copious supplies of truly wonderful wire fly swatters close at hand in all areas.

I suspect that I may have been brought in to be the physical embodiment of the collaboration between Simon Levin and David Pimentel in shoring up the theoretical aspects of genetic feedback. I had first contacted Levin when applying to Cornell, and it was he who suggested the arrangement with Pimentel. Levin at that time was an applied mathematician who had just made a splash, of sorts, in ecology by proving that a community having fewer limiting factors than species could not be stable. Essentially, he was proving MacArthur's stability theorem by disproving its negation. It is a silly little piece of work, in retrospect, although it has the redeeming feature of putting into print in the ecological literature Dr Seuss's peregrinations on the niche.

At any rate, there had been a student present who, successful or not, seemed to be filling the same niche for which I was probably brought in. He was about to receive his degree and leave. Given that background, I knew right away that I was not about to study genetic feedback in the laboratory. Neither did I think—wrongly, in hindsight

—that I would learn any useful new math from the task of elaborating on the mathematical model. I managed to avoid both by studying the basis for a polymorphism in white clover, a rather traditional subject in British ecological genetics. It did take me outdoors, even if it was only around Ithaca, New York, rather than in the Andes. I also did wind up doing some lab work. It was not with flies, but with aphids and clover leaves. The leaves were isolated in vials, although not the same ones that pretended to be plants in other experiments. They pretended to be nothing other than vials that I could use to confine an aphid together with a clover leaf. It all got me a degree in the end. I recount it now because the attitude I have been describing in myself is one that I can absolutely blame—and blame it should be—on Earth Day and the Maine seashore and woods that Rachel Carson so enjoyed.

The graduate students in the lab, or "stable" as it was fashionable to call the facilities of a professor having many students at once, were a mixed bag: ecologists and entomologists. Some of the ecologists were technically entomologists. (A technical distinction. A PhD in ecology had a foreign language requirement; one in entomology did not. The language requirement, which for a native speaker of Russian (me) was hardly onerous, was at the insistence of Robert Whittaker. It drove a number of students to change fields. It also is an indicator of where in the spectrum from engineering to pure scholarship ecology was seen to be at Cornell. In engineering it was not even required that you mastered English for a PhD.)

A number of the entomology students were fine people from countries in the Third World who were in mid-career and already with families. By training them, Pimentel was exporting an ecological approach to pest management. One of them, part of the group that prepared the paper on energy use in corn production, went on to become the Minister of Agriculture, then Governor of Bali in Indonesia. Another of that group was from the foreign country of Brooklyn. (He was actually from that part of Brooklyn that happened to be in Staten Island.) One of the original members of the Peace

Corps, he took his baseball skills and ecological training to South America.

Among the Americans in the lab, there was the expected lot of wildlife-oriented ecologists from the Midwest and West and scions of Ivy League families, along with some who, like me, might have fallen into ecology under the spell of Earth Day. The student I thought I had to replace was solitary and mysteriously theoretical. He had not involved himself in the studies of energy and agriculture. He had isolated himself in a second-floor office and warned me not to keep to myself down in the basement labs. Some flippant remark of the kind that it is my wont to make and the reader may have found here and there in this narrative must have turned him off to me, because he refused any communication with me after leaving Cornell. There also happened to be a devotee of Howard Odum in the stable. Attracted to HT's ideas, but not his methodology, he also located himself in a second-floor office and adopted a similar stance of isolation. Then there was a genuine James Dean, both in appearance, dress, and mood, who proudly proclaimed to have been at Cornell "since the first Johnson administration." He had come to a summer program there while still in high school and may have made it to the Reagan administration before finishing his dissertation, which won, it must be known, a university-wide prize for originality. The prize had money associated with it. Before it could get to its proud recipient, however, it was garnished by the university library in payment of fines due it. A cruel joke took hold that the prize was awarded as the only way to have the library collect on his debts.

There was also always a post-doc at the lab. One tried to shepherd the graduate students through the 1973 study of energy use in agriculture. Another did the same with a 1975 study, only to remove his name from its publication (in *Science*) on philosophical grounds. I do not remember his exact reason, nor can I come up with a conjecture about it now, but for some reason his Marxists beliefs were incompatible with energy accounting in agriculture. This would have surprised many of the Marxists I met who were working with Barry Commoner.

I avoided the genetic feedback studies, but Pimentel had no difficulty in involving me in studies of energy use in agriculture. That, along with the possibility of watching condors soar in the Andes was exactly the sort of thing those Earth Day speakers had inoculated me with a desire to do. It was direct service to the environment. I eventually hoped that I might see condors soaring in the hills above Santa Barbara, but by the time I had gotten out there, all California condors had been locked up for their own good in zoos. From that vantage point, they turned out to be hideously ugly.

The paper on energy input in corn production that came out of the lab appeared in print in November of my first year there. It did not escape my notice that it made quite an impact. Pimentel readily admitted that it enhanced his scientific standing. More importantly to me, "Ecologists liked the study," according to him. Entomologists and agriculturists were less favorably impressed. "They felt that I should stick to entomology," Pimentel later remembered.

The genesis of the paper is "a long story," according to Pimentel. "I was involved in one of the thirteen or more National Academy of Science panels. This one was focused on what would be future research in agriculture." He gave its dates as 1965 to 1968. "Twenty-five years into the future. I was included on it because of my interest and concern about chemical pollution in agriculture. However, as a member of the panel, I could raise any question I desired. I suggested that energy was going to be important to agricultural research in the future."

The idea must have gone over like a lead balloon. "None of the other twelve members of the committee agreed with me. At the same time, I was convinced that I was correct and they were incorrect, but I did not have time to write a minority report."

It is quite typical of successful people that at crucial times they decide to follow their own ideas no matter how unsupported the ideas might be by "those in the know." Gene Likens has recounted a similar episode in his career. It had to do with whether New Hampshire granite bedrock truly is imporous to water, as the assumption went on initiating watershed studies there. Geologists always expect cracks and such to complicate flow, even in bedrock. I wondered if Likens had obtained advice that the geology might not be as simple as originally thought, but had ignored it. That's the only way to get anything done.

"That's absolutely right!" Likens had answered me, laughing heartily. "That's been my view all the way along. Well, it may not work, but let's try it."

But is it also true that for every success story of someone stubbornly sticking to their guns, there is some number who did so in a hopeless cause and faded into obscure failure? It seems like a good topic for some sociologist of science, lest we continue to bring up as an example for young researchers to emulate, one that depends as much on luck and circumstance as it does on insight and hard work.

Anyway, rebuffed by the panel, Pimentel set about to put together the data himself. How to go about it? How to get the manpower together? How to do all this without funding?

Pimentel turned to what he had in great supply: graduate students. The way he got them involved was through a course patterned on those National Academy of Sciences panels. He proposed a graduate level course and enrolled all the future authors in it.

"In 1970," Pimentel remembered, "I started the course and did not give the students a choice [of topic]. I suggested that we investigate energy use in agriculture. It took a year for the study and a year to get the paper published in *Science*."

Part of that time was spent in rewriting the paper "about fifteen times," according to Pimentel. "This is true of all the papers we have published since. Before publication, all the papers are peer reviewed by from 15 to 30 reviewers."

Where did he get the idea for the paper?

"I knew about Tom Odum's book and an earlier publication," Pimentel said, referring to Odum's remark about potatoes made of oil. "However, he did not make a quantitative assessment of energy use in agriculture. Thus, his paper was of no help to us—other than providing a philosophical perspective."

Pimentel's repertoire soon grew from studies having to do strictly with energy and food production to biomass energy, agricultural pollution, water and soil resources, and—more recently—global health. A long list of studies by Pimentel's teams filled many of the pages *Science* and *BioScience* during the decade following Earth Day. This was a busy and exciting time for the practical-minded ecology graduate students associated with Pimentel.

What was common to all of the studies was a quantitative finding, such as the amount of oil energy required to produce corn in that first study. If that finding could be a simple number, such as that 1 kcal of energy invested in corn production returned 2.82 kcal of corn, on which policy or further research could be based, so much the better. The concern then, for those readers not having lived through those times, had been that the amount of energy used to grow corn, petroleum-base energy, in particular, was so surprisingly high. This was during times of energy shortages leading to gas rationing. Every little bit of energy use was important. The energy required to grow crops no longer seems to trouble us as much, even though there is presumably less of a stock of petroleum to draw from.

The tangled web of ecology, then, has brought me back to where I was on Earth Day, 1970. There was general concern in the crowd around me over the fate of the planet and the fate of individuals. Polluted air and water threatened almost everyone's health directly. Ecologists had the public's ear then because of their expertise on issues having to do with population growth and ecosystem connections—and because of their concern.

When Pimentel set out with his team to document energy use in agriculture, settling on corn production as a proxy, he did so in part because energy accounting was in the mainstream of ecology at that time. It was smack-dab in the ecosystem analysis tradition of the Odums. However, ecologists had tended to shy away from studying managed ecosystems such as cornfields, preferring to study untrammeled nature. One explanation given for that preference was that agricultural ecosystems had complexities that natural ecosystems lacked, such as many man-manipulated processes linked to economic forces. Why the subtle, cryptic forces that sometimes underlay natural ecosystem processes should be easier to study than the overt, highly visible, and well-documented manipulations that man imposes was left unexplained. By now the reader should be able to supply his or her own explanation.

Post-Earth Day, systems ecologists began to make contributions on energy resources and economics, both not exactly in the normal curriculum of ecology. One group brought their unique perspective to bear on the US economy, calling for modifications in standard economic models to account for biophysical limits. (They called them "constraints.") It was the ecologists' version of the Club of Rome's study. The ecologists predicted that future economic growth would "depend largely on net energy yield of alternative fuel sources." Being second across the finish line and less notorious, those studies received less attention from the public than did *Limits to Growth*, although they received substantial attention from the scientific establishment. Out of these energy studies, an energy theory of economic evaluation sprung forth in the 1970s, seemingly from the minds of systems ecologists, the first mind being that of Howard Odum. As Pimentel did with Odum's idea of potatoes and oil, Robert Costanza did with his idea of an energy-based economy.

It has not, however, replaced monetary theory. Costanza himself no longer champions it.

The questions raised by ecologist-generated studies such as those of the energy costs of food production were then picked up by scientists in specialties more narrowly related to the subject. Being in influential journals, they were also picked up by environmental advocacy groups, as suited their goals. Or, those groups adopted the methods to generate their own numbers. Lester Brown's Worldwatch Institute is one example of the latter. Ecologists were there first with analysis because it was knowledge that was needed and that no one else was generating. The tradition of energy accounting in ecology may have started them looking in those needed directions, but a sense of duty probably pushed them on.

Systems ecologists, especially those working in the national laboratories, more and more in this era began to take on the concerns that *Silent Spring* had made part of the political mainstream. Rachel Carson's detractors could attack her credentials, competence, and motivation, but when highly credentialed and respected government scientists such as George Woodwell of the Brookhaven National Laboratory brought their quantitative expertise to bear, usually finding support for points raised in Carson's book, industry found itself apologetically on the defensive.

In many ways, DDT had been the pesticide industry's Maginot Line; once it fell, all other pesticides became equally vulnerable. Woodwell's "Toxic Substances and Ecological Cycles" in *Scientific American* in 1967 and articles in *Science* in that same year and in 1970 brought rigorous ecosystem science to bear on the question of persistent pesticide residues in the environment. A decade later, James W Grier's "Ban of DDT and Subsequent Recovery of Reproduction in Bald Eagles" in *Science* settled the issue of DDT's effect on that non-target species once and for all.

One battle was won, but others are always on the horizon. Bjørn Lomborg, passed over lightly on the issue of species extinction rates in a previous chapter, now comes back into the story. In his 2001 book, *The Skeptical Environmentalist*, Lomborg jumped on some of the excesses of the environmental advocacy groups that had taken up the causes that once had been the subjects of ecological research. According to him, the environment was not only not as bad as some made it out to be, but was actually improving in a number of significant ways. It was a refrain of the views aired by Julian Simon. For Simon, recall, the first Earth Day was an occasion for fisticuffs. Simon had not tempered his Cornucopian views by Earth Day 1980. Although he had written a number of books presenting his views, a paper in *Science* that year was perhaps the most appropriate of his statements to take up for our narrative. *Science*, after all, was the forum for much of the ecology-environment intersects.

Simon's title, "Resources, Population, Environment: An Oversupply of False Bad News," perhaps tells it all. Simon was responding to news stories in sources such as *Newsweek* and the *New York Times* that he saw as perpetuating inaccuracies originating from biased sources, such as Paul Ehrlich in *Population Bomb* and Worldwatch Institute's Erik Eckholm's *Losing Ground*. Things were not nearly so bad as they made them out to be in Simon's point of view. He managed to marshal some selected statistics that supported his contention against the selected statistics of the other side.

Simon's conclusions bear strongly on the magnitude of the response he—and later Lomborg—evoked among ecologists. I have done some careful editing of his words for narrative reasons, but their meaning and their impact is unchanged.

We are not running out of energy, natural resources, or raw materials, Simon concluded. "The only meaningful measure of scarcity in peacetime is the cost of the good in question," he also concluded. And that cost for "almost every natural resource" has "been downward over the course of recorded history."

Furthermore, in direct opposition to *Limits to Growth*, natural resources are not finite, according to him. "The term 'finite' is not only inappropriate but is downright misleading in the context of natural resources," Simon wrote. "The future quantities of a natural resource such as copper cannot be calculated even in principle," he added, "because copper can be made from other metals."

Given the possibility of that little bit of alchemy, "even the total weight of the earth is not a theoretical limit to the amount of copper that might be available in the future. Only the total weight of the universe," Simon ended his flight of optimism, "would be such a theoretical limit."

With that. Simon echoed American economist Robert Solow, who once claimed that "the world can, in effect, get along without natural resources."

"Why do false statements of bad news dominate public discussion of these topics?" Simon asked in his *Science* paper. Then he answered his own question. "There is a funding incentive for scholars and institutions to produce bad news about population, resources and the environment," he speculated. Also, "bad news sells books, newspapers, and magazines."

(An astute observation in reality. What would Simon have said about today's "fake" news?)

"There is a host of possible psychological explanations for this phenomenon about which I am reluctant to speculate," he added, but speculate he did. One of those might have struck a target of which Simon was not even aware.

"The cumulative nature of exponential growth models has the power to seduce and bewitch," Simon speculated. Had he realized that those exponential growth models for ecologists were actually logistic growth models, he might have been closer to the mark. Exponential growth models can predict nothing but disaster, starting with Malthus's geometric growth model (a degenerate exponential, if I may say so without offense.)

This is all the same ground that was later covered by Lomborg. It did more than just raise the hackles of ecologists. I have never seen such vituperation from ranking ecologists directed against anyone before. Maybe I just have not been reading the right literature. Or, maybe there is some other reason why.

Revolutions must come to an end, but a milestone date to mark that end, is not so easy to find, although it must have been some time in the mid-1980s for this one. Earth Day 1970, although bringing together many disparate parts of our society, was in many ways an angry event. It is not difficult to pick it out as the culmination of a movement that was sweeping change through society. Earth Day 1980 found the environmental movement with much still to accomplish, but also much in which to take pride. It was more tame, more positive. By Earth Day 2000, in contrast, the event had become a quiet day perhaps best typified by grade school children picking up roadside trash. "Chemistry Goes Green," *Science* proclaimed in its version of a headline. Rachel Carson's enemy, the chemical industry was now making biodegradable polymers from corn, rather than the plastic anathema to environmentalists that they produced from petroleum. The industry went hog-wild green on solvents by developing solvent-free processes and reducing and recycling solvent and water use. What better sign that "The Green Revolution" has successfully transformed society? The only targets that seem to be left for Earth Day environmentalists seem to be GMOs and plastic grocery bags.

Even pop culture reflected the change. The protest songs of the "hippie" Sixties morphed into the protest songs of the "punk" Eighties and then into the serious protests of Rap. Instead of the environmentally friendly peasant fabrics and the simple love beads of the hippie era, punk rockers turned to metal and leather and self-mutilation. Today's rockers are no longer rebels, they are gangsters and their molls.

Joni Mitchell's pop protest anthem of the 1960s, "Big Yellow Taxi," transposed into David Byrne's "Nothing But Flowers." While Mitchell lamented the paving of paradise to put up a parking lot, Byrne bemoaned the difficulty of adjusting to a life style in which parking places and convenience stores were being replaced by fields of flowers. I can only wonder what sort of confused reception Byrne's musical piece would have received a generation earlier—or whether it would have managed to get any air time at all. The popularity of Byrne's satire could be read as a sign that by 1990 the environmental movement—with ecology at its center—was firmly ensconced in the American mainstream, both politically and economically. The time had come when pop culture could afford to satirize what had once been deadly serious.

A revolution had ended. Was another to begin? Who would lead it? What would it be like? There were and are still cornucopians hiding in the bushes around politically-correct campuses (and occupying the White House at present). Would it be an academic? Someone perhaps from one of the more environmentally-conscious countries, such as those of Scandinavia?

Although I am now past the historical period that is the subject of this volume and this episode seems from the current vantage as having been a trivial one, it is worthy of inclusion here, for it reflects back on the ecologists of the 1970s.

Bjørn Lomborg, from Sweden, actually, admitted to having been inspired by Julian Simon's seeing things getting better for "most people, in most countries." It was a point of view that, according to him, Lomborg first tried to contradict, then accepted wholeheartedly. Turncoats often turn out to have the fiercest loyalty to their new causes. His book was, indeed, a repetition of points previously raised by Simon. The only difference is that Lomborg seemingly amassed more data.

As did Simon, Lomborg also turned to *ad hominem* attacks. These might be brushed aside, as one reviewer did, finding it not to be news "that Paul Ehrlich and Lester Brown have perennially exaggerated the problems of food supply," but Lomborg did not stop with those two highly visible targets.

"Most basic environmental research is sound and unbiased, producing numbers and trends as inputs to evaluations such as Worldwatch Institute's *State of the World*," Lomborg admitted at the start.

"However," Lomborg continued, "there is a significant segment of papers even in peer-reviewed journals trying to make assessments of broader areas, where the belief in the Litany sometimes takes over and causes alarmist and even amazingly shoddy work."

"The Litany," according to Lomborg consisted of prophecies of doom. More specifically, he wrote, it consisted of the "usual conception of the collapse of ecosystems."

Besides the print media, Simon targeted popular figures such as Worldwatch's Erik Eckholm and ecologist Paul Ehrlich for criticism. Lomborg's longer work targeted, among others, Worldwatch's Lester Brown and ecologists Paul Ehrlich (of course), EO Wilson, and—David Pimentel. The latter target was totally unanticipated. Pimentel had not been a public person in the way the others had.

Lomborg claimed in his book that he did not want to show "just a single example or pick out a lone error, but to show you the breadth and depth of the shoddiness" he uncovered in environmental studies.

The next sentence started with "Professor David Pimentel of Cornell University is a frequently cited and well-known environmentalist" responsible, Lomborg told us, for too large an estimate of erosion and too small an estimate for the ideal US population.

It was not the first criticism Pimentel had ever had. Any scientist dealing in areas as important and complex as those Pimentel tackled will find himself being questioned as to methods and conclusions. The information available is never so clear-cut that its analysis will pass without having people with axes to grind coming out of the woodwork to question it. That's why research on it was needed in the first place. Eventually, the necessary information is filled in, a consensus is reached, policy is made, and researchers go on to something new.

A few pages before pointing out Pimentel, Lomborg made it clear that he intended to attack a single individual. "Take notice," he wrote, "this is not due to primary research in the environmental field; this generally appears to be professionally competent and well balanced.[70]"

His footnote 70 then cited Pimentel's study of global health as a counterexample. It was equivalent to formally denouncing someone in the Soviet Russia of the previous century. Was it not also the strategy of a Stalin-esque bully to pick an individual out of a crowd in order to destroy it one person at a time? Why battle a whole army if you can more easily take out its leaders by individual combat?

Yet to put a needed perspective on it, remember that this criticism was from someone who early on in the book confessed that "I am not myself an expert as regards environmental problems." That made it even more like a denouncement at some meeting of a Soviet commissariat.

This is not the place to go over Lomborg's criticism point by point for, unlike the argument over species extinctions in the previous chapter, little of it had to do with ecology. According to

Columbia University's Stuart Pimm, "very serious environmental researchers have gone through chapters and found that he practically doesn't get a single point right." At the least, however, it should be noted that, again as in the argument over extinction rates, statistician Lomborg found data at the other extreme from that of ecologist Pimentel. While ecologist Pimentel may have been out of his field of expertise on global health issues, one of the studies for which Lomborg took him to task, statistician Lomborg was even farther out of his field on the subject of soil erosion.

David Pimentel had long been fascinated by soil erosion. Concern for soil loss and desertification had been something almost innate in environmentally conscious ecologists ever since Paul Sears, an ecologist active on environmental issues from the Dust Bowl through Earth Day, published *Deserts on the March*, a critique of the land-use practices that contributed to the Dust Bowl.

An estimate that Pimentel uncovered for the loss of topsoil from a field planted in corn apparently staggered him.

"30 tons," I recall him telling us in answer to his own question. It had been in his basement office in Comstock Hall at one of the many meetings I attended there. He enjoyed gathering all of his graduate students and post-docs around the huge table that filled his office. Sometimes there were more than a dozen of us. The table was actually comprised of a number of normal-sized wooden tables pushed together, but it gave the impression of the single solid substance one might expect in a boardroom at General Motors. It was clear of all clutter, except for a small, neat pile near to where he sat at the far end of the room, by the windows. Most of his documents were either neatly arranged on his desk close by or filed away by a succession of very competent office assistants.

Not having a very good feel for what a ton of soil is nor a hectare of land, for that matter, I gave the issue of 30 tons no more thought than I did Tennessee Ernie Ford's "16 Tons." Less, in fact, for I had been subjected to the latter weight endlessly as a child.

(Actually, the soil loss was in some forgotten number of tons per acre. I have reconstructed the number as 30 in hindsight. It turned out to be only a few inches of soil, not bad for deep soils, but of concern for thin soils being continuously cropped.)

Pimentel apparently chanced upon better data than what I remembered. Lomborg took him to task over it, nonetheless.

Pimentel's current global estimate of 17 tons per hectare, obtained "via a string of articles, each slightly inaccurately referring to its predecessor," according to Lomborg, came from "a single study of a 0.11 hectare plot of sloping Belgian farmland."

Having been involved in studies such as this, which attempt to measure the quantities that could, on a global view, be used to determine the actual limits to human growth, I know how easy it is to get lost in a sea of mushy unscientifically gathered data. Being a "hard" scientist like Pimentel, an experimentalist rather than a theorist, I can understand the appeal of a 0.11-hectare plot over the kind of fuzzy estimates that are arrived at by committee. In science, a well-controlled experiment takes precedence over tomes of approximations based on little but reasoning.

Getting data on soil losses from cropping over the course of a year is not a simple matter. How would *you* do it? Measure depth before and after? At how many places and with what precision? Collect runoff and dry out the sediments? How and from how large a plot? The 0.11-hectare plot may have been unique in its characteristics in the same way that Hubbard Brook was. Each allowed an unmanageable quantity to be measured. There probably was no equivalent study of soil run-off anywhere. If there had been, Pimentel would have found and used it. The "string of articles" Lomborg complains of was nothing but testimony to the study's utility.

"Indeed," Pimentel conceded with a hint of frustration and sarcasm, "as Lomborg states, instead of losing an average of 17 metric tons per hectare per year on cropland, the U.S. cropland is now losing an average of 13 tons per hectare per year."

In Lomborg's own words, he is not an environmental scientist. Although Stuart Pimm exaggerated in that statement quoted above that Lomborg got not a single point right in his book, the fact that Lomborg is not an environmental scientist, particularly, not an ecologist, is telling. Neither is Lomborg an economist. He is a statistician. He in no way has taken the place of Julian Simon. There was no bet between Lomborg and Pimentel. In this day of fake news and social media, it should be kept in mind that, with the "Science Wars" over, science gets things more right than do, for example, global warming deniers such as Lomborg.

(There is a curious parallel here that I cannot resist sharing between Lomborg's career and those of the discoverers of a "cold fusion" device. After their "discovery" was soundly discredited, both of the "discoverers" settled into a well-supported laboratory in France despite their research having become something of a "pariah." Lomborg now heads his own privately-funded think tank. There is a moral there someplace.)

What Are Nature's Services Worth?

What if we got away from that S-curve? What if we stopped concerning ourselves with limits? Sure, those limits are there; we all know that. But what if we are not as close to them as we thought? After all, we have not exactly all gone to hell in a handbasket between Earth Day 1970 and Earth Day at almost 2020. What can the science of ecology do to improve the human situation without letting that Malthusian equation color its views?

George Perkins Marsh, no ecologist—the science had not been invented yet—was probably the first to note that natural environments can be economically beneficial to man. In his *Address Delivered Before the Agricultural Society of Rutland County*, Marsh argued in 1847 that man derives "supplies for his various wants" from nature. The Library of Congress finds this document to be of sufficient historical interest to have a facsimile edition on the web. It is not accessible directly through an http address, but it can be reached through their "Evolution of the Conservation Movement, 1850-1920" web page with just a little navigating. (Click on 1847-1871.)

Returning cleared land to forest, Marsh wrote, "though comparatively worthless to us, would be of great value to them," referring to future generations. Marsh found the functions of forests to be "various," including soil formation, water purification, and humidity equalization, that would otherwise cause a farmer "to drive miles for water" and pasture. These views were elaborated on in his book, *Man and Nature*, first published in 1864.

Ecologist Walter Westman posed the question of what are the values of nature more quantitatively a century later. At the time, Westman was a plant ecologist with the Department of Geography at UCLA. Active on social issues, handsome, athletic, interested in people and solicitous of their welfare, Walt gave his time freely to me on my arriving at UCLA. I would look forward to my walks across campus from the rather somber engineering quad in which I was housed to the more artsy part of campus in which the Geography Department was found. I was trying to determine what was what in my new and confusing situation, and Walt tried to help. We shared picnic lunches on lawns decorated with sculptures. He would listen to my academic and research plans encouragingly no matter how impractical he may have known them to be. Hindsight has left me with no doubts as to their impracticability. Walt kept his doubts quiet as he tried to gently get me to make the right decisions myself. I am sure he went to bat for me at least once in my travails at UCLA without my ever knowing about it.

The words, "The objective of this Act is to maintain the chemical, physical, and biological integrity of the Nation's waters," in the Clean Water Act of 1972 were put there by Westman while he was a Congressional Fellow under the sponsorship of the Ecological Society of America. He had been a surprise choice for it, according to Gene Likens, because neither he nor his adviser, Robert Whittaker, seemed to be anything but academic-style plant ecologists at the time. (Former ESA President Arthur Cooper credited a letter to Senator Edmund Muskie from George Woodwell with the words quoted above in it for their inclusion in the legislation.)

Only a few years older than me, Westman had a career as ecologist and environmental scientist ahead of him that I could only dream of for myself. It ended tragically and prematurely of an unexpected scourge that struck down too many young men in the 1980s and 90s.

Westman, reasoning along the same lines as Marsh, but with a century's more data, examined ways to put dollar values on those "various wants" identified by Marsh that we now consider to be basic functions of ecosystems. Previously, Westman and an Australian collaborator had toyed with the idea of something that they called a "natural resource unit" that could be allocated and traded as a way of taking the environment into account in economic decisions without placing a monetary value on the resources. Although published in *Science*, it turned out not to be a particularly practical idea. Westman therefore tried the next step. Even given the century that had elapsed since Marsh's identification of natural services, economists had not taken them into account in any way but the old way that Marsh argued against. That old way was that they were free. In more modern times they have been called "externalities," because their costs were not borne by commercial enterprises. That made them "external" to traditional economic accounting. Having apparently taken the advice given to ecologists to "take an economist to lunch," in this case his UCLA colleague, David Conn, Westman became more willing to consider putting dollar values on nature's services.

As in so many things environmental, Westman and Conn found themselves on ground that had already been gone over by the Odums, Howard and Eugene. Both had already demonstrated ways in which dollar values could be estimated to cover the replacement of the services provided by wetlands. In addition, Westman and Conn benefited from the pioneering work of economist Lester Lave, who put dollar values on the human misery caused by pollution. Lave's methods of estimating the economic losses due to pollution-caused illness provided an added benefit to ecosystem services. To the cost of replacing the air purification services of nature, for example, was added the health costs of the pollution that was no longer being naturally mitigated.

Nature's free services, Westman wrote in a 1977 paper in *Science*, "include the absorption and breakdown of pollutants, the cycling of nutrients, the binding of soil, the degradation of organic waste, the maintenance of a balance of gases in the air, the regulation of radiation balance and climate, and the fixation of solar energy—the functions, in short, that maintain clean air, pure water, a green earth, and a balance of creatures; the functions that enable humans to obtain the food, fiber, energy, and other materials needed for survival."

Some were apparently expensive. Westman gave an estimate of $205,000 per hectare of wetland in order to duplicate only the tertiary wastewater treatment and fisheries facilities provided by that hectare. The figure had been obtained by researchers at Louisiana State University working with Eugene Odum. Primary wastewater treatment, for those who want to know, removes those things that float or sink, such as oils, grit, and dead bodies. Secondary treatment is accomplished by the natural process that converts sewage into carbon dioxide and bacterial sludge. Essentially, microorganisms found in nature and used in the treatment plants feed on the sewage as they do in natural waters, except that everything is more intense and concentrated in a treatment plant. Tertiary treatment is usually a chemical process that removes important pollutants such as phosphates from a wastewater stream. These processes are not cheap when part of a modern wastewater treatment facility. Think of your local fish-hatchery-plus-sewage-treatment-plus-water-treamtent plant (usually all separate) to envision the kinds of facilities needed to replace the waste treatment aspects of a wetland. Another study quoted by Westman, this time from a business school, gave a value of $1.8 million for the annual services of a 930-hectare river-swamp-forest ecosystem in Georgia. (Think of a hectare as about the minimum-sized lot allowed in a suburban subdivision of the type whose residents would never allow a waste-water treatment plant to be built anywhere near it.)

Westman gave examples of ways in which ecosystem functions could be quantified and evaluated in dollar amounts in that *Science* paper. Like the Odums, he was pointing the way for others. And unlike theoretical ecology, this part of ecology, ecosystem ecology, had progressed to the point at which its knowledge could be put to practical uses. It was economics, rather than ecology that was holding back that application.

A caveat put by Westman on his ideas that "At present, our understanding of ecosystem functioning is limited" can best be interpreted in the same way that Terry Erwin might say that our knowledge of tropical beetles is limited. There was still a lot about ecosystem dynamics that was not known, but there had been quite a bit of information collected—some without consideration either for ecosystem science or theoretical ecology—from which a tabulation of benefits could have been begun. In terms of the analogy to the discovery of beetle species, a lot of beetles had already dropped through the funnels by the time Westman suggested a dollar accounting of ecosystem benefits. There were no surprises to ecosystem functioning in the intervening years that would have affected the analysis.

Traditional economics, however, with its reliance on discount rates, external costs (also called social costs), and marginal evaluation, seriously undervalues natural resources. If there is no market for natural resources, such as clean air or water, they cannot be given dollar values through standard economic means. Getting around this was one of the triumphs of Lester Lave having related pollution amounts to health costs. Putting a price on pollution was not a simple matter, but, thanks to Lave and others, methods for it did exist.

A popular text from 1975, William Murdoch's *Environment*, reviewed the methods of environmental economics, as imperfect as they were, that were available at that time. In theory, environmental costs, once determined, can be passed on through legislation and regulation to the polluters or the consumers of products that pollute. Isn't everything, ultimately, passed on to consumers? A government could simply tax polluting products based, for example, on the damages they produced. It is not a popular option, though. Marketable pollution credits, as allowed for sulfur dioxide emissions in the Clean Air Act Amendments of 1990 represent a more palatable option. They require oversight by government to work. In essence, polluters are allowed to either continue to pollute at some level or transfer their right to pollute to some other company, typically through sale. The pollution credits are something like Westman's proposed "natural resource units," but easier are to allocate.

The method only works if new polluters are not allowed to pollute or are very greatly restricted in the amount of pollution they can produce. Therefore, any transfer of pollution credits is required to result in an over-all reduction in pollution. In theory, a company might find it economically viable to purchase and shut down an old polluting plant in order to be able to build its own newer, less polluting facility. Everyone wins.

In practice, it may be easier to pass on most pollution costs through tort law. This latter method has been popularized in the book and movie, *A Civil Action*, and in the movie, *Erin Brockovich*. Citizens whose lives or well being were damaged took the offending company to court. The monetary award obtained by the litigants should make that company and others look at their polluting ways in a new light, a costly one. The cost of malpractice insurance is illustrative here. A physician who is sued too often will have his or her insurance rates raised or coverage dropped. A company that is sued too often might be forced into bankruptcy. There have been a number of these cases; that of Johns-Manville and asbestos is the best known. On the basis of equity and efficiency, recovery of social costs of pollution through litigation seems not to be the best way, however. Do all pollution sufferers recover monetary damages and are legal expenses on both sides a necessity? In the malpractice example, it has led to the practice of obstetrics becoming excessively expensive, with a consequence that new physicians are shying away from that field, the consequences of that being obvious. Resorting to civil action in court, however, may be the currently preferred way to right environmental wrongs in the US.

These are matters that should be tractable, in principle, through legislation, court action, whatever. Less tractable in environmental economics is one of the fundamental tenets of economics: the discount rate used for return on investment.

The discount rate approach to investment leads to economic decisions that are too short term for most issues of environmental health. Marsh saw this insight while Darwin was puzzling over how best to explain what he had observed on his voyage on the Beagle.

"The increasing value of timber and fuel ought to teach us, that trees are no longer what they were in our father's times, an incumbrance," Marsh reported to the farmers of Rutland. "We have undoubtedly already a larger proportion of cleared land in Vermont than would be required, with proper culture, for the support of a much greater population than we now possess, and every additional acre both lessens our means for thorough husbandry, by disproportionately extending its area, and deprives succeeding generations of what, though comparatively worthless to us, would be of great value to them."

There is not a discussion of the shortcomings of interest rates in the passage—Marsh's concern was over things more basic that we now take for granted—but there is prescience over limits to what nature might supply that economics was not taking into account. Trees being worthless encumbrances reflects an attitude toward wilderness that our society has been fortunate to outgrow. Their value then was in terms of timber and firewood, which Marsh saw as being in such abundant supply that the planting of trees would scarcely be a worthwhile investment. And yet—what if all those trees were removed?

The discount rate is a simple method for determining how much money in hand now would be worth in the future. It is also an obstacle to managing resources for future generations that needs to be considered seriously and critically. Investing in the planting of trees that cannot be harvested for some 40 years, for example, is never—even at present low discount rates—economically feasible. Savings bonds provide a better return with less risk.

Consider a stand of Ponderosa pine on the eastern slopes of the Rocky Mountains in Colorado or Arizona. Given knowledge of the costs of felling, removing, transporting, and marketing the timber for whatever product it can be used, it is not difficult to come up with a suitable price for the standing timber, just as it is not too difficult, with a little knowledge of the local housing market to put a price on the same timber-bearing land for a housing subdivision. On the basis of future value, however, these two market decisions diverge considerably. Although land values for housing can go down, the investor will at least have the land and make the decision to purchase and hold it or not, based on what the future return may be expected to be and the current rate of interest. (The investor might get cold feet and choose to invest in a 3% certificate of deposit, instead, for example.) The trees, however, are not as likely to be still there, even the next year, particularly in areas plagued by fire. What should someone invest for timber in the future in that case? Land values for more typical real estate may go down precipitously, but they may also go up. At best, timber only accrues at its physiological rate. The investment in 3% certificates of deposit will double your money in some twenty years. A forest may need forty or fifty years (often more) to regenerate to the state it was in before a fire. What happened to the investment? Even if the trees are worth ten times their present value in forty years (an economically sound decision based on the 3% alternative, which would only quadruple your value in forty years), except through current extreme youth or future extreme longevity, you will never live to see that return on your investment, as good an investment as it might have been.

Most corporations work on a six-month to one-year investment period and pay their executives obscene salaries to do such short term planning. Not even governments plan for forty years.

One economic analysis not too long after the first Earth Day demonstrated that the most profitable strategy for harvesting whales is to convert them all into money in the bank as quickly as possible. That means harvest them all. Don't leave a single whale alive. Colin Clark, then a professor of mathematics at the University of British Columbia, took Garrett Hardin's reasoning about common resources a step farther to come up with that little bit of economic magic. He showed that it all made economic sense under then-current economic principles. Extinction, Clark wrote, "will tend to result not only from common-property exploitation, but also from the maximization of present value, whenever a sufficiently high rate of discount is used." By "common-property exploitation" he meant Hardin's "Tragedy of the Commons." Clark saw that that was not the only way to ruin a resource. Free market economics had another built-in way to do so—the discount rate.

Hardin's metaphor of the commons was published in the journal *Science* in 1968. It evokes a peaceful colonial New England landscape and the centerpiece of many of today's tourist-attracting villages: the village green. Many settlements in New England clustered houses together around a common green for protection, often herding their animals within that protected area. The one in my village happened to have buildings of the Vermont Marble Company around it. (Were they clustered in defense of striking workers?) Although Hardin's idea of a common pasture was borrowed from a 19th century Englishman, imagine a Vermont village green available to any who might want to let goats, as an example, graze on it. Clearly there would be a limit to the number of goats that could graze the pasture without nibbling the grass down to its roots. Each herdsman, however, would be tempted to graze more goats on the pasture, even though it meant that the pasture would be overgrazed. The individual psychology of each herdsman presses him to maximize the number of goats he can put on free pasture. The result of the individual actions is that the pasture becomes incapable of supporting any goats. Everyone eventually loses. Hardin added the word "tragedy" to the idea, and some simple math. The math was an accounting to show that there is no incentive for the individual herdsman to restrict the number of cattle that he grazes. Doing so put him at a disadvantage with respect to others, who gathered to themselves the use of the pasture that he essentially gave up to them. By going along with the rest and increasing his stock to numbers that result in overgrazing, however, the individual herdsman is no worse off than others doing so—and better off than those who eschew increasing their herd. The eventual result is the ruin of the common resource, the pasture, to the detriment of all. The tragedy is that no other result seems possible. Hardin extended it to fisheries, national parks, and to land, air, and water receiving pollution. Fisheries, before the imposition of catch limits, are a particularly apt example of unregulated use threatening a natural resource. Scientists at Dartmouth soon after produced a highly enjoyable board game, Fish Banks, to teach Hardin's principle in application to North Atlantic fisheries. (It had also been meant to teach principles of something called systems science, but that somehow always got lost in the lessons I taught with it.)

In his analysis, Clark estimated a maximum reproductive rate of 10% per year for blue whales (conceding also that it could be half that). "If in their calculations of profit and loss, the owners of the whaling fleets were to utilize an annual discount rate of 20 percent or greater, they would therefore opt for complete extermination of the whales—at least as long as whaling was profitable," he concluded. "This would occur whether they were competing, or cooperating, in the slaughter."

The 20% discount rate was, by Clark's reckoning, "by no means exceptional in resource development industries." By around Earth Day 1980, short-term certificates of deposit were yielding close to 20% (as were certificates of deposit; both are much less now). It is surprising that any harvestable natural resource in private hands was left standing.

A debacle having to do with the Pacific Lumber Company, also known as Palco, in northern California may be the best example of the absolute inappropriateness of free-market capitalism when applied to timber resources. Alston Chase probably gives the best account of what happened in his book, *In a Dark Wood*. When I first became acquainted with Palco in 1979, I found it hard to believe what I was seeing. I was then leading a study that may have helped (I hope so) to preserve large portions of California's North Coast rivers when they were under threat of development to supply water for southern California. It is detailed in a technical report that probably can no longer be obtained from UCLA and in book chapters in volumes with titles that are sure not to catch the eye of most librarians. If, as I believe, the reports did help those rivers escape the threats they were under, they escaped only in protected portions, those that were placed under the protection of National Wild and Scenic Rivers legislation. The other parts of those rivers will always be there to tempt California water resource planners. In California, most fresh water is consumed in the southern part of the state, but falls on and flows to the sea through some truly spectacular country in the northern part.

Having expertise in both wilderness and water quality and supply issues, UCLA's Environmental Science and Engineering Program garnered a generous grant from the Ford Foundation to make some sense out of the various issues involved with that part of

California's water that falls on its North Coast. The program's head, Richard Perrine, made it my first assignment when he hired me.

The North Coast is a spectacular area of inaccessible mountain wilderness through which rivers funnel the water that is squeezed out of coastal storms as they rise over mountains, then deposit that water back into the Pacific Ocean. In so doing they meander through valleys whose gentle slopes hold the world's redwoods. They are the steelhead trout and salmon streams of California. Their names are magic to fisherman: the Eel, the Van Duzen, the Trinity, the Klamath, the Smith. There are also tributaries of note, such as the Scott and Salmon Rivers. It was almost as good for me to be there as being in the Andes in search of what was more and more becoming for me a rather mythical condor.

Some of the rivers, such as the Russian, furthest south, and the Mad and Trinity had already been dammed at their headwaters. Claire Engle Lake, which stores the "overflow" water of the Trinity, in the quaint term used by hydrologists, was already sending water across the mountains to the Sacramento Valley. Isn't technology marvelous?

No policy analysis on the North Coast rivers can be complete without a consideration of logging. Palco was a small, family-owned company at the time, headquartered in Scotia, California, where their mill and housing for their workers was situated. It was a classic company town, with the exception that no one seemed to complain about it. The company-owned worker housing consisted of small, neat single-family homes of the kind I wished I could have been living in. There was a waiting list for them, though. And you had to be a Palco employee to get on it. The $100 a month rent for a house in the redwoods compared quite favorably to the $475 I was paying for an apartment in the San Fernando Valley at that time. Guess what else? The mill obtained its power from burning the bark and scraps of the trees it harvested. Everything was recycled. It all seemed too good to be true.

Palco had some agricultural interests—cornfields in Iowa, for example—among its assets, but mainly it was a company that produced redwood lumber in the highest of grades. Old growth redwoods are the subjects of much environmental drama and history. I recommend Raymond Dasmann's *The Destruction of California* and the Sierra Club's *The Last Redwoods and the Parkland of Redwood Creek* to those unfamiliar with it. It was the tree that could be said to have helped to launch the environmental movement. It also produced the most valuable lumber. Palco had been reseeding and replanting and determined that they had a rotation that could produce 400+-year-old timber and 70-year-old timber, essentially in perpetuity. It did require logging redwoods that were well over 400 years of age, but it would not at all be the kind of carnage of clear-cuts that Georgia-Pacific and the Miller Logging Company had inflicted on their old redwood groves. Palco's 400 year olds would not be harvested individually from a stand that would then be left relatively undisturbed, however. A 200-foot tall tree needs to land somewhere and it needs to land in a way that will not splinter its long trunk into kindling wood. Although the strength and denseness of tropical mahogany may make it an exception to that, taking down a giant singly within a grove endangers both it and the trees around it. Palco's intention, however, had been to keep the acreage cut at one time as small as practical. There would be the usual sight of timber slash, but less of it.

It **was** too good to be true. Palco had held over a billion dollars in assets and had almost no debt, according to Pacific Lumber's proxy statement of January 22, 1986, as cited by Alton Chase. The assets were standing timber, largely old growth. Corporate raider Charles E Hurwitz, financed by Michael Milkvin junk bonds, found in Palco a temptation too great to resist. The financial shenanigans of Ivan Boessky, Milkvin, Solomon Brothers, and the like, and their relationship to Reaganomics and natural resources (recall Reagan's "killer trees" remark) is a story that needs to be told in full. Chase just skims its surface. What happened was that, using the standing timber assets of Palco to leverage a buyout of the company, Hurwitz showed what the invisible hand of economics can do in action when totally unrestrained. It was pretty much as

Colin Clark had warned. The high interest needed to pay off the junk bonds required a high discount rate. That made the trees absolutely unprofitable to keep. It wasn't even a case of alternative investments; there were debts to be paid off. The rate of tree felling quickly doubled in order to pay off that debt and, one can rightly assume, line the pockets of Hurwitz and Milkvin. The ultimate outcome, was tree sitting, ecotage, arrests, and a possible attempted murder. And felled trees. Read all about it *In a Dark Wood*.

Cost-benefit analysis is the term for another tool for economic decision making. The discount rate plays an important, but hidden function in it. Much in use for public works projects, the technique essentially totes up the dollar benefits of an action and makes a ratio of that against the cost of the action. A ratio more than 1.0 means that more benefits accrue from a project than the money spent on it. A ratio less than 1.0 would cause a prudent investor or administrator to shy away from the project; it would cost more than it was worth. Of course, the costs and benefits have to be given in dollar values. Keep in mind that, right up to that first Earth Day, environmental costs were still considered to be externalities in standard cost-benefit analysis. Walt Westman, in arguing for nature's "free" services to be considered in economic decisions, shied away from using cost-benefit analysis, believing that it "will always skew estimates of nature's value" in favor of development over preservation "because of the limited state of our knowledge of ecosystem function and the difficulties in expressing these values in monetary units." The economist he took to lunch must have pulled the wool over his eyes. It was not something lacking in ecosystem science, but a basic flaw of economics that had Westman seeing his then-current attempts at evaluation as having mainly "heuristic value."

There is that word again. It was much too popular in ecology at that time. Looking back over that era, it is a blot on ecology every bit as much as was the reputation of ecologists being butterfly chasers.

No doubt Westman feared that nature's services would never be found to be worthy enough. He need not have feared. A general accounting of nature's services twenty years later found that they were very valuable, indeed, worth many trillions of dollars. You can go to the Ecological Society of America's web site and click through to *Issues in Ecology Number 2* and check it out yourself. There, the ESA has posted a professionally written synopsis of a larger work edited by Gretchen Daily of Stanford University. The report has been put on the web in order to present consensus opinion on environmental issues. The consensus of course is those of experts, ecologists. The experts in this case were Susan Alexander, Gretchen C Daily, Paul R Ehrlich, Larry Goulder, Jane Lubchenco, Pamela A Matson, Harold A Mooney, Sardra Postel, Steven H Schneider, David Tilman, and George M Woodwell, in alphabetical order. It is a *Who's Who of Ecology, Past, Present, and Future*. Its major conclusion? "Biodiversity is a direct source of ecosystem goods." Defensible or not, it is another argument from ecology for the preservation of diversity. This time without the necessity of resorting to the limits of the logistic curve.

The cast from *Who's Who* go over much the same ground in terms of what nature's services are that was covered by both Marsh and Westman. Although they conclude, "ecosystem services supplied annually are worth many trillions of dollars," they also concede that evaluating those services is a complex and highly uncertain task. Unlike Marsh, but like Westman, they do not shy away from discussing intangible aesthetic and spiritual values provided by ecosystems, although they do shy away from putting dollar values on them.

As difficult as it was to put a value on some ecosystem services, it was a task that was soon accomplished by a group of mostly American scientists led by Robert Costanza. Then with the University of Maryland, Costanza published his account ledger in *Nature* almost simultaneously with the Ecological Society's study. The authors of the *Nature* study were mainly systems ecologists, or people who would have been called that back when the term had not yet lost its luster. It was a valuable task to which they devoted their talents. Their study actually put a value on all of the services provided by the ecosystems of the world. It turned out to be, at a minimum, $33 trillion, roughly twice the global gross national product then and many times the number of beetles that Terry Erwin had collected in the tropics. Their method required a lot of looking up of data to estimate the marginal value of services per unit area of biome and a lot of number crunching in order to sum them up. Computer spreadsheets, no doubt, not only must have made the number crunching more bearable, but they probably allowed the endeavor to be encompassed within the realm of the possible, rather than just the desirable. Too large for the magazine, their spreadsheet was displayed on *Nature's* web site, http://ww.nature.com. (You probably will need to purchase a subscription to *Nature* to see it now. Even though the electronic format is supposed to provide much greater access to research results, there are still out-dated economic issues that need to be dealt with. But try a university library. It may be able to provide access to it for free. Or subscribe to a service called JSTOR. Or go to Google Scholar. If it is important, you will find it for free in full somewhere on the Internet.) The study identified 17 services of ecosystem functions relating to the flow of materials and energy, distributed over 16 distinct ecosystems, encompassing everything from open oceans to wetlands to deserts to urban regions. Some of the ecosystem services, such as climate control, were right out of Marsh's *Address*. Others, having to do with dimethylsulfoxide and ozone, clearly represented new knowledge.

  The methods of the study can and were criticized by economists. "Willingness-to-pay," for example, in which a monetary value is established based on what people would be willing to pay to preserve (or to give up) some service, is not a market value, but serves as at least one estimate of value for goods for which there is currently no market. Costanza and colleagues admitted that "many of the evaluation techniques used" were based on this method.

  Traditional economists are suspicious of economic values that are not set by a market, especially if used, as in this study, to determine a "marginal value," for an ecosystem service. (Marginal value basically represents what you would pay for your next bag of potato chips, in a rather crude example of the concept, given current supply and demand. Of course, you would pay whatever the store was charging—or not buy. While jumps in price seem to come in the dime and quarter amounts by which you make your decision to buy, calculus can determine what the price to the fractional penny would be "at the margin," given real data.) Their conclusion that "as natural capital and ecosystem services become more stressed and more 'scarce' in the future, we can only expect their value to increase," invites the same sort of response from Cornucopians that Ehrlich and his associates received over a coming scarcity of metals decades before in a famous bet with Julian Simon. Simon won it. It had to do with the price of those metals that were to become scarce in the next decade. Maybe this time, however, the advice, "Don't bet along with Ehrlich!" might no longer be good advice.

Traditional economists, it should also be noted, are becoming more and more willing to let go of some of their traditions. Economists at Harvard and Columbia, Martin L Weitzman and Giulio Pontecorvo, respectively, suggested throwing out the old discount rate in environmental benefit/cost analyses. A new "intergenerational discounting method" proposed that the value of natural resources for future generations be taken into account. It is exactly what environmentalists have been asking for in economics since the first Earth Day. It should make expenditures that were previously uneconomical due to their payoff being so far in the future (the prevention of global warming, for example) now be economically feasible. It also means that ecologists have managed to turn around an absolutely fundamental principle in accounting.

Interestingly, one critic of Costanza's study was David Pimentel, who pronounced that the researchers "were giving some things much too high a value," having himself estimated global ecosystem benefits at only $3 trillion. So much for Pimentel being prone to exaggeration.

Ecological Engineers and Environmental Doctors

Lawrence Slobodkin had been doing ecology and watching it being done and evaluating it for more than 50 years, starting from even before MacArthur stepped into Hutchinson's lab. He saw the practice of medicine, not physics, as an appropriate model for ecology.

"While we need theoretical ecology to focus attention on certain problems, we cannot expect ecology to become a theoretically tractable discipline in the sense of physics," he wrote in a thoughtful essay in 1988.

By tractable, he meant having questions that can be answered.

"Tractable sciences, such as physics and mathematics, focus on a few central feasible problems at a time," he explained.

Ecology, meanwhile, is called upon to solve a variety of problems, some of which are not at all tractable. Unfortunately, ecology cannot discard problems that are not elegantly solvable.

"Most of modern medicine is intractable for similar reasons. Neither ecology nor medicine can reject problems within their subject matter on grounds of intellectual nonconformity to theory," Slobodkin concluded.

By "intellectual nonconformity to theory" he meant those problems based on "the sigmoid population growth equation," our mathematical friend, the logistic equation or S-shaped curve. That little equation, of course, was the underpinning for the MacArthur revolution. He accepted it like a physicist accepts Newton's Second Law. Maybe he had physics envy. Had not the Marlboro environment turned him toward ecology, much in the way that Woods Hole turned Rachel Carson's interest to the sea, Robert MacArthur might have wound up collaborating with his brother John and Robert May on esoteric papers in their chosen field of physics, rather than on the papers with which they tried to force that point of view on ecology.

Medicine being somewhat of an art, engineering is what commonly comes to mind for applied science. The view that engineers take knowledge already available and use it somewhat mechanically to solve a problem is common, but not, however, accurate. Principles must sometimes be tweaked before they can be applied and sometimes data needs to be collected. Experimentation may be needed. The engineer's experimentation is often of the "trial and error" or "seat-of-the-pants" sort. He is looking for a solution to a pressing, practical problem. The basic researcher in his experiments, if he does them, is looking for elegant theory through a haze of practical details that are seemingly obscuring his vision.

Robert MacArthur was firmly behind the testing of hypotheses. He did not, however, call for experimental data. He was a mathematician and bird watcher, not an experimentalist. What MacArthur's scientific investigations needed to be complete were field glasses and a mathematical equation. Time and evolution provided the experiments. MacArthur had given ecology a valuable, but incomplete instrument in his hypothetico-deductive reasoning. Missing from MacArthur's instrument was the part that allowed application. In that, it failed ecology.

British ecologist TRE Southwood, who dedicated his career to promoting the use of rigorous methods in his science, already in 1965 lamented: "It is frequently pointed out that ecological theories have outstripped facts about animal populations and I trust that it is not too presumptuous of me to hope that this collection of methods may encourage more precise studies and more critical analysis of the assembled data so that ... we may have 'more light and less heat,' in our discussions."

Southwood had the following to say about some of MacArthur's successors: "Here and there in some of the earlier chapters with their apparently limited knowledge of the literature, as evidenced by the restricted list of references, and their arguments based on a single restricted taxon of animals, I felt a vague unease." He was reviewing ecology's version of Fort Sumpter.

MacArthur was notorious for laziness in delving into the literature of science, right from his undergraduate dissertation. Granted, there are differences between math and ecology, and an undergraduate dissertation should not be held to the same standards as a refereed publication, but one might have expected MacArthur to have mastered whatever Modern Language Association rules for citation were in effect at the time. He didn't. His ten references were all in the following format: "3. Jackson: "Fourier Series and Orthogonal Polynomials." I assume that all of his sources were to books, which might be found in a library by title and author's last name, alone. If they were journal sources, it would have taken the bibliographic skills of the national archivist to locate them. Considering that Marlboro College had been built around farmsteads, instead of a library, they no doubt were books, probably passed on to him by his brother. There is no literature review or discussion of previous work in the document. MacArthur pretty much just jumped into, then out of, the math. His admirers adopted the bad habit of ignoring the literature on a subject much as admirers of James Dean took up smoking cigarettes.

Then there is the simplicity of MacArthur's mathematics. It led to his closest collaborator, EO Wilson, and his mentor, GE Hutchinson, concluding that he "was not a mathematician of the first class." Simple math had benefits, of course. Simple math is easy math to understand. Consider what it is that makes Newton's and Einstein's famous equations so beautiful. It is their simplicity. Einstein's theory, in particular, such a remarkable conceptual leap that the joke went around that only three people in the world could understand it, one of whom was Einstein, is summed up by an equation that can be understood by a schoolboy. MacArthur's theories had that same simple beauty to ecologists.

In their simplicity, though, MacArthur's theories and those of his followers missed something major. Had they been experimentalists, they could not have neglected it. Organisms deal with their environments in not at all a simple way, because their environments are not the uniform creations assumed by the simplest mathematical models. MacArthur did recognize this. To be fair, he also recognized that there was a camp of ecology that, "primarily in making a science of ecology, arranges ecological data as examples testing the proposed theories and spends most of its time patching up the theories to account for as many of the data as possible." He was not critical of it, however.

"It is only by attempts such as this," MacArthur wrote about some theorizing by Slobodkin, "that ecology can hope to become a respectable branch of science."

It was because ecology was so unsophisticated mathematically that MacArthur's theories for it seemed so trivially simple. Simple things for simple folks.

When the complications of the real world intrude, the math becomes more difficult. As they flocked to MacArthur's way toward scientific respectability, ecologists (and evolutionary biologists) chose to ignore what they now call the heterogeneity of the environment, those patches of habitat that Lande had to deal with before he could provide advice on the management of the spotted owl.

In a heterogeneous environment things are not the same from place to place, nor from time to time. Gause found that his predator protozoa and paramecium prey could both survive only if some of the paramecia found a refuge in the sediment at the bottom of their culture vessel. Meanwhile, he had intended that the experiment be done in an environment similar from glass boundary to glass boundary. Believing this assumption to hold yielded the impossible result for him that the predator could survive without prey. The reality was that the prey could reinvade the main part of the culture from its refuge after all of the paramecium there had been eaten. Even the simplest of laboratory results suggested that the simplest of models were only trivially true.

Thomas Park, performing similar experiments with flour beetles, found that, although only one species could exist in any of his culture vessels, different species won the competition under different conditions, with some situations always favoring one species or the other, while other situations had somewhat random outcomes. One beetle species would win more of the time than the other, but on any given trial, which beetle it would be could not be predicted with assurance. His results, published in a number of outlets, the latest being a 1962 issue of *Science*, were well known among ecologists by the 1950s. Park's various vessels were patches of habitat. By simply multiplying the number of competitions possible (each patch has a separate outcome), surprising results were uncovered, much as later Robert May uncovered chaos through mathematical analysis that made models more realistic, but less simple to analyze. Park's results pointed away from the use of deterministic equations. The random winner situation is what scientists call stochastic behavior. It necessitates statistical analysis.

Charles Huffaker did similar experiments with predatory mites and their prey, a mite that feeds on oranges. He found that if he used larger numbers of oranges and spread them out more, he could get coexistence of predator and prey, as opposed to extinction of predator or both.

Clearly, the world must be more like a universe of dispersed oranges than a bottle of yeast or bacterial culture. The single bottle of culture was what had been modeled by the mathematically inclined up to then. The laboratory work of Gause, Park, Huffaker, and others called for models to have subdivided environments. It was a realization that was slower to dawn on most of ecology. Perhaps ecology had to first give up on the promises of the MacArthur school. Once those theories began to reach dead ends, once their lack of practical utility was exposed, other ways of seeing nature became appealing.

The mathematics of patches involved things like diffusion equations, however. These had been worked out and applied to a number of situations in physics, chemistry, and hydrology. They worked. To ecologists, they provided a statistical way to describe how organisms spread. But they did not excite many ecologists. The equations turned out to have no biology to them. The idea that wolves could be nothing more than a statistical abstraction, like nitrogen atoms in a bell jar, will never excite many biologists.

In addition, the mathematics involved in applying diffusion theory to most biological situations is well beyond that provided in the training of PhD ecologists. Differential equations, Fourier series, Legendre integrals and Bessel functions are not in the ecological curriculum. Most biologists would be driven from the room in which such was taught. I happen to know this math. Like Alexander Solzhenitsyn before he saw the light and turned to literature, I, too, have solved a Bessel function. It was not that hard, actually. All I had to do was look up a root in a handbook and plug it into a formula. The rest was multiplication and division, essentially. Still, now that I am no longer in engineering, that math would drive me from the room just as fast as it would any other ecologist. In order to use diffusion theory, ecologists needed to hire personal mathematicians. It all just does not seem to be worth it. Better leave it to the mathematicians.

If there is one topic on which ecologists have functioned as engineers, it is the greenhouse issue, now known as climate change or global warming. It is also an area that is as far from MacArthur-style theorizing as one can go and still be in ecology. But yet …

Concern that increasing carbon dioxide concentrations resulting from fossil fuel use might effect the earth's radiation balance was raised by atmospheric scientists as far back as 1938. Ecologists were pretty much asleep on it then. There was no ecosystem concept, no mathematics of the niche and competition, no productivity measurements. No science at all, in the view of some, as we have seen. Limnologist Hutchinson saw the ocean as essentially an infinite sink for carbon and voiced "grave objections" in 1948 to early reports that industrial CO_2 was accumulating in the atmosphere.

Yet there was a science to ecology then. It was vegetation analysis, and it had been a large part of ecology since it emerged as a distinctive science in the US. It was data on plant communities that found its way into greenhouse issues.

Oceanographer Roger Revelle and Hans Suess, colleagues at the Scripps Institute of Oceanography in La Jolla, California, warned with understatement in 1957 that "Human beings are now carrying out a large-scale geophysical experiment of a kind that could not have happened in the past nor be repeated in the future. Within a few centuries we are returning to the atmosphere and oceans the concentrated organic carbon stored in the sedimentary rocks over hundreds of millions of years. This experiment, if adequately documented, may yield a far-reaching insight into processes determining weather and climate."

Did the science of ecology have a place in assessing the experiment? It did. The rates at which CO_2 enters and leaves the atmosphere were an issue from the earliest modeling attempts. Generally, modelers divided carbon flow between the atmosphere, oceans, and the biosphere. Three boxes, with arrows interconnecting. This is where the ecologists came in. Measuring carbon uptake in plants as part of ecosystem productivity studies is standard ecological procedure. Estimating totals for the entire globe is difficult, as we have already seen, but not impossible. There were estimates made in the 19th century (one by Liebig, he of the minimum) for the amount of carbon tied up in biota. Strangely enough, those early estimates, based as they were on guesswork, at best, are not remarkably different from current estimates.

The biota, when it burns or decomposes, produces carbon dioxide. Growth, on the other hand, takes carbon dioxide from the atmosphere and puts it into living tissues. The productivity measurements made by ecologists on various ecosystems therefore are a parameter in those box-and-arrow models of global carbon dynamics. Simply stated, if the standing biomass of the globe's ecosystems is increasing, corresponding to positive net productivity, the world's biota is then a sink for excess carbon dioxide. If the converse is true, negative productivity, then the biota is itself a source of excess carbon dioxide in the atmosphere. Recalling that plant biomass greatly exceeds animal biomass, the question that modelers need ecologists to answer then becomes: Do plants, forest trees, in particular, release carbon dioxide to the atmosphere or remove it? Is the biota a net source or a net sink for carbon into the atmosphere? (The role of the ocean's biota in all this has been ticklishly difficult to unravel. The biomass in the oceans is less than 1% of the earth's total, but the productivity of the oceans is substantial, approximately a third of the total. Where it all goes is unresolved. In the past, some of it went into producing the petroleum deposits that now help to aggravate the carbon situation.)

Dan Botkin reviewed the possibility of the biota being a net sink for carbon in the 1970s in an article he published in *BioScience*. Much of his information was the result of having his JABOWA model simulate the effects of increased atmospheric carbon dioxide. Carbon dioxide is a growth requirement of plants. It is well known from laboratory and field experiments that enriching a plant's environment with carbon dioxide gas can have great positive effects on its growth. This is the CO_2 fertilizer effect. All other things being equal, increased carbon dioxide concentration should result in increased plant productivity. Botkin estimated that a 3.6% increase in the world's rate of photosynthesis "could absorb the total amount of carbon added by burning fossil fuels and kilning of limestone." After a digression on how r- and K-selected species (fashionable ecology at the time) might respond to carbon dioxide enrichment, Botkin concluded that, although primary productivity could represent a "short-term carbon sink," man's activities at reducing biomass were too rapid to be compensated for by carbon dioxide fertilization. More important was Botkin's concern that "interactions of species in natural ecosystems may have important effects on the ability of terrestrial ecosystems to act as sinks for anthropogenically produced CO_2." This was the point of the r- and K-selection discussion.

There are themes intersecting here that have been developing throughout the Earth Day era. There is the use of the systems approach, the need for data over theoretical insight, and questions over what kind of ecology is best for practical applications. Is it MacArthur-style ecology? The early theory it spawned has not held up well—but what about the r and K idea? Is it Odum-style ecosystem analysis? His ecosystem was a concept fraught with difficulties—but do all of those problems go away when the earth is considered one large ecosystem? And how does theory unsubstantiated by experiment serve to help in any way? And what about the diversity-stability issue? That last brings us to a more modern perspective that is aimed right at global warming: can productivity depend on diversity?

So the diversity-stability argument, with connotations seen in *Silent Spring* and Deep Ecology, morphed first into a diversity-resilience argument and now into a diversity-productivity argument. What baggage came down to us with it in those transformations? For one, ecologists seem loath to accept the biota as a sink for carbon. Hutchinson argued as early as 1954 that the biota must be a net source of atmospheric carbon. This has been a consistent mantra from biologists in debate about the global carbon budget ever since.

Data, of course, is crucial to resolving any scientific disagreement. Data suitable for inclusion in the accounting of global carbon required big funding. The International Biological Program, the paragon for "Big Science" in ecology, helped to provide much of the early data. The goal of that multiyear program had been essentially to measure and describe the biota of the earth. Given the strong influence of ecosystem science on it, measurement of productivity was a large part of the IBP's effort. In the United States and other sites this was accomplished by breaking the globe up into units characterized mainly by their vegetation, such as desert, tropical forest, etc, biomes, in other words. Primary productivity, being one of the baseline parameters for ecosystem studies, was high on the priority lists at all sites. Results found their way into a symposium at the Second International Biological Congress of the American Institute of Biological Sciences in Miami, Florida, October 24, 1971, another at Brookhaven National Laboratory, published as *Carbon and the Biosphere* in 1973, and a book-length technical publication, *Primary Productivity of the Biosphere*, that came out in 1975. Influential chapters by Robert Whittaker and Gene Likens, both then at Cornell, "Carbon and the Biota," and "The Biosphere and Man," respectively, in the two latter volumes are almost identical and seemed to have been used interchangeably in the literature of the day. It was important data that found its way into many studies of the time. Authors would cite whichever of the two papers must have been handy to them.

This being the era of the Cold War, the productivity of great expanses of forests in communist countries had to be estimated based on the publications of Russian scientists. Although Russian data was used by the American scientists, it was thought by them to be erroneously high. There was a later echo of this at Kyoto, Japan, as Russia sought large pollution credits for its supposed carbon-accreting forests.

Citing data from Whittaker and Likens as "the best estimate available," which it was, Bert Bolin used it and data from his native Swedish forests to argue that deforestation is causing the biota to be a source of carbon to the atmosphere. Carbon modelers had thought the biota to be a net sink for carbon; vegetation at the time was thought to take up more carbon by them than it released. Bolin's report came out almost simultaneously with an overview of the carbon cycle prepared by a team of scientists at Oak Ridge National Laboratory. Its conclusion was "that man can have significant influence on fluxes between the land and the atmosphere," but the Oak Ridge scientists threw their hands up on what the exact effect was at that time. The whole issue needed—of course!—more research.

Into this breach stepped a team of highly productive ecosystems scientists, led by George Woodwell. Woodwell, another byproduct of our atomic testing program, as many applied ecologists were, of which the Odums were foremost, made his reputation with his study of the effects of radiation on ecosystem "behavior," as he called it. It was supported by the Atomic Energy Commission and carried out at one of their National Laboratories, this one at Brookhaven, on New York's Long Island. He had gone on there to trace the fate of DDT in the environment, confirming much of what had been labeled by critics as speculative when Rachel Carson had presented it in *Silent Spring*. The carbon cycle as it affected greenhouse warming was a natural step for Woodwell, given his interests and expertise.

He did not stay at Brookhaven. Stubbornness in holding to his scientific principles got him fired. It also got him fired from his next situation as head of the Ecosystems Center at the Marine Biological Laboratories at Woods Hole. It was a position that had essentially been created for him. He didn't have to move very far, though, creating for himself the Woods Hole Research Center, at which he finished out his career.

Others in the team included Dan Botkin, Gene Likens, and Robert Whittaker. At the time, Whittaker was a giant in ecology. He was not, however, a leader. There was no Whittaker "school," no Whittakerites. He had, however, manage to put the final nail in the coffin of the Clementsian superorganism.

Whittaker's career had gotten off to an even less auspicious start than had GE Hutchinson's. It is an inspiration to press on, regardless of the adversity. Before settling into his final prestigious position at Cornell, Whittaker's resume had been spotty, to say the least. He had received his undergraduate degree from Washburn Municipal College in Topeka, Kansas. It was a school deemed so inadequate by the University of Illinois, that its Department of Botany had rejected his application for graduate studies. He had to enter his chosen field of plant ecology through the back door—the Zoology Department at Illinois did take him.

Starting with that rejection, Whittaker set off on a series of professional frustrations and failures that would have driven a lesser man away from research. His first position was with Washington State College at Pullman, Washington. Walt Westman, one of his doctoral students, once confided to me that Whittaker's leaving Pullman after three years had not been entirely voluntary. He had been "fired," as Westman had put it. The Aquatic Biology Unit at the Hanford Laboratories (a nuclear weapons related facility) in Richland, Washington, where high-level radioactive wastes are today still stored in wait for more permanent storage facilities, picked him up.

When Whittaker moved on from Richland in 1954, he was convinced that his work there with radioactive phosphorus would kill him. Whether it did or not, Likens reported him as saying on his deathbed: "It finally got to me."

Brooklyn College, where Whittaker turned up after his experiences with radioactive tracers, was not exactly a bastion of research then or now, but he stayed there for a decade, all the while doing and publishing research and being very active in the affairs of the Ecological Society of America. That professional service must have finally paid off, because he left Brooklyn College in 1966, first for Brookhaven National Laboratory, with the assistance of George Woodwell, then for the University of California at Irvine, then a new part of the University of California system with big designs for research and scholarship. Hating it there, Whittaker eventually settled at Cornell University two years later.

Although his many collaborations were perhaps even more productive than those of Robert MacArthur, Whittaker did not have the personality of MacArthur or Hutchinson. It was easy for me to see that he was the one who made the decisions in ecology at Cornell, but it was not easy to see him doing it. He was the power behind the throne of whoever happened to be chairman. He had chosen it to be that way.

"He knew that he didn't have the social skills to be departmental chair," Gene Likens once said to me about Whittaker. "Even when he was elected to the National Academy of Sciences, I don't know if he ever went to a meeting."

Whittaker was rather severe in his dealings with people. It was probably what had sent him packing from Pullman. He was not someone who would put you quickly at ease.

Severe might not be quite the right description. Socially inept leaped to my mind, but I have little memory of Whittaker in social situations except for around-the-coffee-maker or at-the-copying-machine types. I always thought he just had a low tolerance or trust for students, which is what I was one of at the time.

Likens was surprised when Whittaker's widow told him that Whittaker considered him to be his best friend. He had never demonstrated anything of the sort in Likens' recollection.

"Whittaker was a good friend of mine," Likens remembered, "but he was a strange person in many ways. Very difficult to approach. He scared secretaries silly. That's no lie. He'd be in his office and they would come knock on his door, which was always closed. 'What do you want?' he'd say, just like that—" Likens had affected a very believable growl in imitation "—and scare them half to death."

Whittaker came to life, though in his formal lectures and presentations. Otherwise, he was someone who left the impression that he felt ill at ease and out-of-place no matter where he might be. And he infected you with the same feelings. It was when he was in front of an audience—and probably face-to-face with a manuscript he was preparing—that his knowledge and ability came to the fore. You knew there was a special intellect before you at the blackboard or lecture podium. He was not a hero to worship, though. And like many other ecologists of his time who were not exactly unaffected by the revolutions going on in ecology, he went on doing his work despite the trends and fads that were carrying his discipline this way and that.

Based on the same data used by Bolin, Woodwell's team surprised carbon modelers with their conclusion that the biota may be a source of carbon dioxide "as large or larger than the fossil fuel source." They dismissed carbon dioxide fertilization as having too small an effect, even if the laboratory experiments were correct for the real world.

This came as a shock to those involved in modeling the global carbon budget, for they had been operating under the assumption that a small proportion of carbon produced by fossil fuel production, about 10-20%, went into greater biomass production. Where does all of the extra carbon go then? One critic argued that the ecologists must have found a source "that was not there to begin with" by assuming that the annual rate of forest removal was 1%, an overestimate. The actual rate of deforestation is an issue that is still with us at present. It is also tied to the rate of extinction of species.

According to scientists involved in modeling the global carbon cycle, the newly found releases of carbon dioxide from the biota could not have gone into the oceans, the only other possibility except outer space. Although not measured, the escape of carbon dioxide into space was unlikely on basic principles of physics. (It is too "heavy" to escape the pull of gravity at the speeds at which its molecules must be moving up there in the atmosphere.)

The geoscientists responded that estimates of ocean uptake were too "firm" to allow very much more carbon dioxide to dissolve in the seas. Their studies were more rigorous, they seemed to say. In addition, the "significance of regrowth has not been adequately treated," according to the carbon modelers. This was unarguably true. Much research and many millions of dollars have been devoted to the carbon dioxide fertilization effect since then. William H Schlesinger of Duke University, for whom Robert Whittaker had been a mentor at Cornell, made a career out of experiments in which portions of forests were actually fertilized with excess carbon dioxide. One of them, accomplished as might be expected by a large team of scientists, put an upper bound on the amount of carbon that would be sequestered in the biota as a result of carbon fertilization. It was not enough to solve the global warming problem. To date, the results hark back to Dan Botkin's warning from his JABOWA results that the interaction of species affected by the nutrient enrichment needed consideration. A large change between then and now is that simulation is being replaced by experimentation. The search for "natural experiments" in ecology, such as those on islands, has been replaced by the search for funding for large-scale manipulations of ecosystems.

Nature's services are worth something, it is now clear. And they are worth the effort to do a full accounting on. And all this without concern for the consequences implied by the logistic equation and community theory. Or so it might seem.

The carbon studies are applied ecology on a global scale, a matter suitable for the attention of leaders in the field. Was there also an ecological engineering to be used by those lower in the disciple coming out of ecology? Not if the theorists could have their way. In 1968, Richard Levins had rejected one effort at a systems-based ecological engineering. It's failings were that it focused on energy flux in ecosystems and paid inadequate attention to the mysteries of the logistic equation. Levins had also bemoaned the funding and university positions being won by the systems ecologists of the time. What he recommended had been that applied problems "be approached in a fundamental way." As in theoretical physics, one could conclude was what he meant in reading between the lines, or in the ecology of the MacArthur school. How?

To take my own circuitous route to an answer, in 1978 I risked the adventure of moving to Los Angeles and becoming involved in UCLA's nascent Environmental Science and Engineering Program. The risk had not been with the move. Los Angeles had never been and still is not one of my favorite places, but California was and still is an ecological and recreational Mecca. I lived there as long as those benefits were outweighed by the well-known drawbacks. The risk had been instead in launching an academic career as an untenured assistant professor with no home department. Departments award tenure. Being interdisciplinary had its drawbacks. It has fewer now, I hope.

The UCLA program was the result of an idea by Willard F Libby, who thought what was needed was an "Environmental Doctor." The efforts of Richard Perrine helped to make Libby's idea into a graduate program. It was launched soon after the first Earth Day. It took trained scientists and engineers (a Master's Degree was a requirement for admission) and provided them the breadth of knowledge needed to tackle environmental problems. That was accomplished through a variety of required course work. Interdisciplinary, team problem-solving put the course work into practice. The degree conferred was a Doctor of Environmental Science and Engineering (DEnv). It had been intended to be analogous to an MD or JD, rather than a PhD. There had been no pretense when the program was conceived that it would make scholarly contributions. The Environmental Doctor was to either fix an environmental ill based on current knowledge or refer it to a specialist for management, the idea went. Training in ecology (mostly in the form of a single course) was a required part of the program. Interdisciplinary environmental research was not. A "doctor" could, the reasoning went, use well-established research results or, with slightly greater care, new discoveries by specialists. It was neither science nor engineering. Care-giving comes to mind, although it was not always clear to me whether it was the environment or the firm employing the doctor that was getting the care.

Willard Libby, who discovered carbon dating (for which he was given a Noble Prize) and the gas diffusion method for enriching uranium (for which, according to his widow, he received nothing, not even patent royalties—it was highly classified research done for the Manhattan Project), must have been thinking of Eugene Odum's superorganism. That was just the sort of thing that could be treated by an environmental doctor. It was also an idea that was in the air, as the expression goes, around that first Earth Day. Speaker after speaker had urged the need to treat a sick environment.

Some ecologists, just as enamored with it as he was, took the analogy a quantum leap beyond what even Libby had envisioned for it. In a 1978 symposium that included such champions of the homeostatic ecosystem as Eugene Odum and Ramón Margaleff, along with a sprinkling of Oak Ridge-IBP types like Stanley Auerbach, one of the papers bore the title, "Diagnosis, Prognosis, and Treatment of Ecosystems under Stress." The title's inspiration came from the idea of an adaptive stress syndrome that physiologist Hans Selye described in mammals, calling it "the general adaptation syndrome" and "eustress," rather than "distress." (*Stress Without Distress* was the title that Selye chose for his 1974 book on the subject.) The ecologists inspired by him, Canadians mostly, called for diagnosis, prognosis and treatment—including a little preventive medicine—of ecosystems. The analogy to ways a physician approaches a human patient could not be more direct. They were still promoting the idea as of 1985, when I let this loose thread drop from my attention.

Most of those doctors turned out by the UCLA program had little more than a vague concept of what an ecosystem was—it was not exactly holistic medicine that they practiced. I can attest to that. The environmental doctor concept is still alive and well at UCLA, something over which I have mixed feelings. In its current form, the program has incorporated many of my curriculum suggestions, something that should make me feel proud. On the other hand, then and now, it is a profession without a point. With the death of the patient, the superorganism, the doctor became a bureaucrat or technician, albeit one with doctoral credentials and more than ordinary skills. I note more and more that some of its graduates list a PhD on their résumés, rather than the DEnv that they actually earned.

Hired early in the program as its only ecologist, I thought my mission was to nudge the program toward a more biological (if not exactly ecological) point of view. There had been a growing number of biologically trained students joining the program. Some of them being proper Earth-Day-inspired environmentalists, they began to chafe at the engineering orientation of the program and its applied research. I heeded their call, abandoned research in theoretical or evolutionary ecology, let some of my thesis work stay unpublished, and set about finding that body of ecological work that an environmental doctor could apply, if not to a superorganism as a patient, then to environmental decision-making on a specific project. Applying theory to problem solving had been my original orientation; it was only the particular theories and problems in engineering at the time that had caused me to abandon that subject. A handful of quantitative results and relationships trump all manner of philosophical tomes for me, as long as the problem at issue is an inspiring one.

How did I do? I managed to get those future environmental doctors out along streams, into the woods, mountains, and out in deserts. Some with backgrounds in the engineering and health professions had to be dragged kicking and screaming into the environments they were purportedly doctoring. One had trouble stowing her amply equipped make-up kit into her pack before a wilderness hike. Another badgered me with the need for 190° F temperature wash water for my camp skillet. She had been trained in microbiology. Still another insisted on driving his Porsche to meetings, rather than car-pooling in a Volvo. My life was a difficult one then.

What did I learn about my chosen science from this experience? I found out that most environmental decisions had no ecological foundations and could not have. Ecology was either irrelevant or lacking in any applicable laws but those in *The Closing Circle*. No input from ecology at all was needed a decade after the first Earth Day for decisions regarding most pollutants. They were either toxic or not, persistent or not, due mostly to characteristics of their chemical structures. Toxicologists could work that out. If they were dangerous chemicals, then they legally had to be reduced, but deciding on the amount of reduction was a legal and political process, not an ecological one.

Why not ecotoxicology instead of toxicology?

I scoured the literature. I learned the history of my science. Early on, it was agreed that a loss of diversity was an indication of a degraded system. Ruth Patrick told a compelling story of polluted streams losing species that were quite common in similar, unpolluted streams. She published the work in 1949 in the *Proceedings of the Academy of Natural Sciences of Philadelphia* under the title, "A Proposed Biological Measure of Stream Conditions Based on a Survey of Conestoga Basin, Lancaster County, Pa." A child could make a value judgment based on her proposed measure. Unpolluted streams had fish, diatoms, and crustaceans. Polluted streams primarily had algae and worms.

Initially, diversity was used in its simplest and present sense, numbers of species. Patrick, however, did some elegant work that showed that in stressed streams, not only were the number of species and individuals reduced, but their relative abundances were also changed. Her paper on this has been so variously cited that I know I should have looked up the original sources, but I have another excuse: *Tempus fugit*. Ruth Patrick, who did much research in support of MacArthur's ideas, was what I would like to call the "Queen of the Atavistic Publication Outlet." Although she did publish in more traditional journals, much of her work was published in what, although not exactly the gray literature, was almost as inaccessible. She placed important results in outlets such as *Transactions of the New York Academy of Sciences* (also cited as *Annals of the New York Academy of Sciences*, although they may have been separate publications), *Proceedings of the Pennsylvania Academy of Sciences*, *Technical Publications of the American Society for Testing and Materials*, and a truly remarkable atavism, *Acad Nat Sci Phila Not Nat*. I am certain in my conviction that *Not Nat* does not stand for "not natural." (Actually, it is *Notulae Naturae*, naturalist's notes or notes on nature, although others who should know have cited it as *Notulae Natural, Notulae Natur* and *Notulae Naturalis*.) These atavisms are the way that science was published in the 19th century and earlier. Back in those days, learned societies would publish work by their members as pamphlets distributed to libraries and those having subscriptions or requesting a particular serial. I imagine the request was filled after a payment, but, perhaps as in the more recent past, reprints were sent *gratis* to interested scientists.

The sources I have actually used are a lecture by her that I attended and a book she edited, *Diversity*.

At any rate, Patrick showed that Preston's canonical distribution was shifted away from rare species toward a few dominant ones. Patrick's observations, along with others, led to the introduction of indices of diversity to identify pollution stress.

Excellent, I thought. Let me get started.

I had not taken Robert Whittaker's class on community ecology in which all of this diversity stuff was covered. I had passed it up for what I thought to be a more useful course, although I no longer remember what that might have been.

Fortunately, Whittaker had shouldered his way to intellectual leadership in community ecology, at least as far as plant communities were concerned, and had left plenty of written material on diversity. I turned to his 1972 review of species diversity. It had a confusingly large number of ways of defining and measuring diversity. Several indices were available, one of which, the Shannon-Weaver Index, was the most popular, mostly in high school and undergraduate ecology courses, perhaps.

It should have been called the Shannon-Wiener Index. I learned that by tracing its history while working on the paper that had brought me to Terry Erwin's beetles. Ecologists call it the Shannon-Weaver index because they usually cite its source as Claude Shannon's book with Warren Weaver, *The Mathematical Theory of Communication*. Claude Shannon, along with Norbert Wiener and Ludvig Von Bertalanffy, was a founder of information theory. The index was developed by Shannon from work done by Wiener and probably published somewhere in Wiener's book, *Cybernetics*. The genesis of its name is as confused as a bibliography of Ruth Patrick publications, but its use in ecology started with one of those Three Influential Papers by MacArthur. The attraction was (and is) that the index gives in a single number an indication of both the number of species present in a community and the evenness of the distribution of individuals among the species present (or lack thereof). The ease with which the index could be interpreted based on Preston's work was somewhat pleasing, a marriage of the practical and the theoretical.

"We may note first the extraordinary development of the study of diversity during the last fifteen years," Whittaker noted in the paper I read. "Articles by Hutchinson and MacArthur offered promise of a new area of a different kind of ecology, one of an orderly, formal system of mathematical relationships by which diversities and the importance-value of species should become predictable."

Elsewhere in the paper, almost as an aside, Whittaker also noted that "Diversity is generally reduced by chronic environmental stress on the community—overgrazing, air pollution, and gamma irradiation." End of topic. Ecology had still very much been a theoretical subject, even after that first Earth Day.

Unfortunately, as we have seen, the whole area of community organization which Whittaker trumpeted, even as he cautioned against its mathematics "obscuring rather than clarifying relationships," turned out to be irrelevant to the study of nature. Several years before the null hypothesis controversy broke out into war, Whittaker warned that interpretation of community interactions based on species-abundance curves was "a tactic of weak inference" that had given less insight than had been hoped.

To return to practical issues, however, in the end, the Shannon-Wiener index gave misleading results about community diversity. I found with my own data that simple numbers of species and numbers of individuals were more sensitive to environmental differences related to chemical pesticide use than the ever-popular diversity index. This was especially true in communities with subtle differences in their environments.

Given large stresses on a community, such as those stream communities studied by Ruth Patrick, differences are obvious. Flames on the Cuyahoga River and dead fish washing up on lake shores do not require community analysis by ecologists, however. There is no need at all to turn to ecological theory in such cases. What was hoped could be developed through community theory was some kind of early warning indicator of environmental degradation. In reality, of course, we need no such indicator to give us information about habitat changes currently under way. The evidence is there before the eyes of any traveler. Forests are being logged, rivers are being dammed, industry and urbanization encroaches on farmland which then is displaced to more marginal areas resulting in more insult to the environment. Since diversity and all of its associated paraphernalia from theoretical community and ecosystem ecology is unnecessary even were it to be reliable, I think that ecologists were forced to take the species diversity argument to a higher plane. This is the concern over global diversity and global productivity. It is all that is left once the rug had been pulled out from under considerations of stability and diversity.

It was difficult for ecologists to give up on some sort of community structure properties for use as indicators, however. John Cairns, Jr, a former colleague of Ruth Patrick at the Philadelphia Academy of Natural Sciences, went on to great prestige as an applied ecologist at what is now Virginia Technical University. Tracking the evolving changes in his views in the professional publications of this time is instructive. As the director of its Center for Environmental Studies he first urged personnel involved with water quality monitoring to use diversity indices, including those "derived from information theory," none other than the Shannon-Wiener index.

By 1974, Cairns was arguing for some sort of structural ecosystem characteristic as an index of pollution over individual indicator species in a paper meant for an ecologically-naïve but not non-technical audience—it was published in the *Water Resources Bulletin*, a journal for engineers and water quality technicians. Changes in the abundance or the presence or absence of particular organisms in an ecosystem had been the rule rather than the exception. Those changes, however, could depend on so many factors unrelated to pollution that as a pollution indicator, the signal-to-noise ratio in the numbers for almost all organisms is highly unsatisfactory. That is what diversity indices had been held out to offer, though, clear, but still more sensitive signals.

Cairns spent considerable effort trying to argue away results from a 1968 paper in *Ecology* that tested the relationship between stability and diversity. It concluded based on laboratory data that the diversity of microbial communities (those most often used for pollution indices) is not simply related to their stability. Its authors warned against "a widely held opinion" that had "almost reached the status of an axiom" based on MacArthur's 1955 paper. It could not be supported by their experiment. Increasing the number of different paramecium predators, for example, destabilized the system, while MacArthur's conclusion from his theorem, remember, had been that maximum stability could be expected when there is one species on each trophic level "eating all species below," a condition of maximum diversity.

What Cairn's reluctance to give up on diversity indices had shown was that the hope for some biological meaning to community structure was and is embedded deeply into the thinking of ecologists. He then turned to something called cluster analysis. Cluster analysis is a rather unwieldy hi-tech statistical technique, used by plant community ecologists and numerical taxonomists, but it was available as a statistical program for computer data analysis, making it less onerous. Also, it was recognized as a technique used by ecologists. The clusters (or dendograms) generated are not easily interpreted by any except those trained in the technique.

No one technique for assessing the ecological effects of pollution, except for the much-maligned indices of diversity, managed to gain dominance. The same is true for techniques that might act as early warning indicators. Every team of ecologists seemed to have its own preferred methodology, no doubt fine-tuned to their own particular systems.

By 1977, Cairns was still championing the use of measures of structural integrity such as diversity measures, but not without attention also being given to functional measurements of ecosystems. "As a screening method for locating trouble spots in most flowing systems, it is superb!" he gushed in favor of diversity measurements. However, there was something else now that needed assessment. It was called biological integrity. Cairns defined it as "the maintenance of the community structure and function characteristics." Methods for assessing functional integrity, all of those ecosystem properties identified by Odum, were still badly needed a decade later, however. They are probably still being anticipated by some today.

Measuring and then tracking things like the productivity of an ecosystem or the cycling of nutrients through it is a prolonged and intensive process. The chemical industry, particularly pesticide manufacturers such as Dow and Monsanto, however, had a need to simply and quickly identify chemicals of environmental concern and be able to easily monitor their effects. Testing protocols from that era that were suggested by ecologists could be fairly elaborate, including not just characterization of the pollutant, but also the receiving system—the ecosystem being polluted—followed by laboratory and field toxicity testing. This is research on a scale that is totally impractical for most ecosystems.

The documents that described the protocols used by manufacturers, however, looked and read very much like the most standard parts of engineering handbooks. Assessment could be done through their use by technicians with but rudimentary training in theoretical ecology.

All of this time, as could be expected, there were other ecologists who continued to emphasize the use of ecosystem models. Two University of California scientists, one being Robert Rudd, who shared much of his information on pesticide use and misuse with Rachel Carson, developed an ecologically sophisticated, but practical model based on DDT residues in Clear Lake. It was not what might be called theoretically elegant. The ecosystem model benefited from over two dozen studies performed on Clear Lake, "a site of intensive investigation for 30 years," they pointed out. Few other sites can be expected to have as much data available on them. The model, successful though it was in predicting the fate of DDT in the Clear Lake ecosystem, emphasized how important reliable data is. With good data, more than a simple, linear model and a computer to crunch the numbers is unnecessary.

The Clear Lake modelers made no attempt to put an economic valuation on 30 years of data collection. The attempt would be futile; no decision-making process can survive a gestation time of thirty years. Yet industry operates based on a bottom line that has dollars on it. I made a time and manpower estimate using information put out by John Cairns for a complete ecological assessment. It amounted to several years of investigations and a minimum of 2500 person-days for a complete evaluation of a chemical. Given that back then a consulting firm would bill the time of their technical people at about $500 a day, ecological testing was serious money in any company's budget.

Naturally, companies such as Dow were receptive to less extensive investigations. They pushed away from ecosystem analysis. Company scientist Eugene Kenaga argued for toxicity testing of but a few species, suggesting that acute toxicity testing of a rat, one species of fish, and one species of aquatic arthropod would be sufficient for most purposes. In fact, according to him, the fish and crustacean could also serve as surrogates for aquatic plants. Fancy that! Even less need for data.

Indicator species were found in abundance in this era. Some have justifiably come down to the present. Others may have been championed more from their representing a particular scientist's expertise, than from any intrinsic qualities that made them good indicators. Ground beetles have been used extensively to detect and measure the extent of pollution. One ingenious team, jumping off from reports on the effects of pollutants on honeybees, turned the problem into a tool. Honeybees could be effective biological monitors for a variety of pollutants, according to them. Bees act as sampling devices over large areas as they fly from flower to flower and are exposed to and presumably concentrate gas, liquid, and particulate pollution. Beekeepers could then send samples from their hives to test for a variety of pollutants.

Lichen species, having a variety of tolerances to different kinds of disturbances, have been and still are used as pollution monitors, in particular for sulfate pollution. Google lichens and air pollution on your computer to get the latest on them. Given the information on the web, schoolchildren can go about setting up their own air pollution monitoring surveys using lichens.

No one would suggest that an endangered species be used as a biological monitor, but there they are. They can be, as in the case of the gray bat and chlorinated insecticides, along with the many bird species that succumbed to DDT, exquisitely sensitive to pollution. From the standpoint of an environmental activist, an endangered species serves not just as an indicator of environmental degradation, but as an effective legal instrument for pollution control and abatement through its protection by the Endangered Species Act. It is not only a canary in the cage going down with the men in the mine whose silence will warn of toxic air, but also a canary in court that can sing to the judge about evil deeds. Recently frogs have also served the role of canaries quite well. More recently, their global declines led to more confusion than than utility.

Somehow, though, the indicators are never quite good enough. The search for newer and better ones continues to this day. Ecological health is a difficult thing to monitor, perhaps, in the absence of knowledge of the patient's identity.

Willard Libby's idea of the environmental doctor died before he did. It died when the ecosystem as a superorganism concept died. Ecology did not give his doctors any dependable and simple early-warning indicators for the physiological state of an ecosystem. They couldn't just march into a problem with a few instruments, take a few measurements and present a diagnosis and treatment plan. The spotted owl, for example, needed three papers that also served to advance ecological theory in order to be protected. Even though Lovelock's Gaia keeps the superorganism on life support as an idea, it can no longer benefit from an environmental doctor. Still, the idea that there is something besides individual species and obvious deterioration for ecologists to use is tempting and will not die. The hope is that those indicators of ecosystem health lie somewhere in the functional characteristics of ecosystems.

A step up from using individual species for toxicity testing is the use of *microcosms*. Model ecosystems, those worlds in a jar used by Gause, were proposed soon after the first Earth Day for use in assessing pesticide biodegradability and biomagnification and the effects of power plant effluents on aquatic ecosystems. Here is something for an engineer rather than doctor. Analysis of bench top systems is standard in the engineering sciences. A theoretically feasible idea for a chemical synthesis is first tried out in test tubes and beakers, then a contraption simulating the chemical plant that eventually results is built literally on a laboratory bench top and tested. There is usually an intermediate scale, called a pilot plant between the bench top and the final facility. Not all processes, though, scale up as anticipated from the bench top to the multi-acre plastic and metal behemoth of valves, pipes, and vessels that is a chemical manufacturing plant.

We did have something like a pilot plant for the global ecosystem in operation for a while: Biosphere 2. What we seem to have learned from it is how difficult it is to create and maintain a closed ecosystem at this scale. At least two ecologists, however, could see through all of the technical malfunctions to find that the Biospherians "had to make enormous, often heroic, personal efforts to maintain ecosystem services that most people take for granted in natural ecosystems." The implication, of course, was that those services should not be taken for granted, a goal of the research of at least one of the ecologists.

On Biosphere 2, nitrous oxide attained concentrations at which there was the potential for brain impairment or damage. Loss of pollinators, nutrient loading of waters, and the ascendancy of ants and cockroaches in Biosphere 2 are all items to ponder.

A Brook Runs Through It

The role of vegetation is central to pollutants other than carbon dioxide. Think back to former President Ronald Reagan's "killer trees" remark during his 1980 election campaign. Reagan's trees actually died for his environmental sins. Dead and dying conifers absorb oxidant molecules from southern California's pea soup of smog through their leaves, thereby harming themselves while detoxifying oxidant through their action on plant tissues. True, trees do release chemicals through their leaves that themselves contribute to photochemical smog. Establishing the amount of terpenes released from plants into the air should have been a task for ecologists that was accomplished instead by chemists. As Earth Days accumulated, air pollution studies, like pesticide studies, more and more fell under the attention of scientists in disciplines other than ecology. Less and less could theoretical ecology be made relevant to practical environmental problems.

Acid rain, the public perception goes, was discovered at a small brook in New Hampshire. Ecologists studying the forest around Hubbard Brook uncovered the phenomenon in the US and led to its understanding as a worldwide problem. Gene E Likens, F Herbert Bormann, and Noye M Johnson are the names usually associated with the discovery, given in the order in which they appeared in their 1972 paper in the journal *Environment*. Only the first two are ecologists, and they were unfazed by the struggle in ecology between the Odum-style ecosystem approach and the MacArthur-style community theory approach that was raging.

"It was just a continuum," Likens summarizes his view of ecology then. Along with Robert Whittaker, Likens had fashioned what were called "core courses" then required of ecology graduate students at Cornell. The series of four courses covered all aspects of ecology from autecology through population and community ecology to ecosystem ecology. There was none of the MacArthur-Odum dichotomy in those courses.

That dichotomy had little effect on the science being done at Hubbard Brook. That work, in essence, combined Howard Odum's "macroscope" of systems science with a tradition of experimentation.

Theirs was systems ecology at its most unglamorous. Take your black box of a lake (at Hubbard Brook it was part of a watershed) and determine its inputs and outputs, then measure them again when the system is perturbed in some known way, such as clear-cutting. What they got by doing so was an ecological and environmental classic. In combination with the null model fracas, they set the standard for the ecological research of today.

The Hubbard Brook study grew out of ideas that came up when Bormann and Likens put their heads together at Dartmouth College. "Bormann was at Dartmouth and he had brought up a class or two here on a field trip. He thought that this would be a really good place to start up a project," Likens recalled for me. Both were to move on from Dartmouth, Bormann to Yale and Likens to Cornell, before the study bore most of its fruit. Neither really ever left that watershed.

Their idea, unprecedented at the time, was to use what they called a "small watershed technique" of controlled experiments that would allow them to construct a nutrient budget for a forest ecosystem. Luck presented them with opportunities for success. Or maybe you could say they made their own luck.

"I believe in what I call serendipity," Likens will tell you. "Good things happen when you keep your eyes and ears and mind open."

Things happened at Hubbard Brook.

I caught up with Likens there, where he was still active, even though most of his time then was spent directing the Institute of Ecosystem Studies that he created at Millbrook, New York, for the New York Botanical Garden.

Likens, although admitting a continuum of philosophical approaches to ecology, came out of a tradition of experimentation that he continued in the forests of New Hampshire.

"My major professor's major professor, Chancey Juday," Likens recounted for me, "had fertilized lakes trying to increase the fish production. Before people were worried about eutrophication, they had added fertilizer to lakes trying to increase fish production.

Hasler, my major professor, did a lot of experiments. Many of them, I was involved with. They focused on these small, acid, brown water bog lakes. We would manipulate them in various ways like adding lime to them."

Likens laughed before adding, "This is long before acid rain."

Interesting.

"Oh yeah. Very interesting," Likens continued. "No, it's a big kind of hysteresis. We'd add lime to them to clear the water. Lime will precipitate the humic colloids in the water column and clear it so that more sunlight can penetrate. Hasler did these really elegant early studies in which he found a lake that looked sort of like an hourglass. This was on the property of Notre Dame up in northwestern Michigan. He got permission to bulldoze this narrow part of the spectacle and make two lakes, which essentially were very, very similar to one another."

Replicates?

"Not quite, but very similar," he answered. "I fuss about the use of the word control at the ecosystem level. I fuss about using only reference, because I think you can't ever, really, fully replicate it. And so he then added lime to one—these lakes were called Peter and Paul. He added lime to one and not to the other. It's a classic experiment. Again, it caught the fancy a lot of people. We did a lot of different things. We added different kinds of lime and one of the students even added hydrogen peroxide. Which worked very well, but—"

Hydrogen peroxide, well known as a bleaching agent, is a very unstable molecule, so unstable that "explosive" is often used as an adjective for it.

"He was doing just this little pond," Likens continued. "Actually, the pond was probably about the size of this building in circumference. It wasn't very big at all. He had, oh I don't know, a 50- or 100-gallon drum of hydrogen peroxide on the back of his rowboat. He was rowing around the lake, letting this run into the lake, and it fell off the boat and went to the bottom of the lake, the drum. So he went home and he came back the next morning. The lake was not there. The hydrogen peroxide had leaked out overnight and produced enough gas that it just floated the bottom right to the surface. So it was covered over. That experiment never got published."

Did that student's professor keep him on at least?

"His name was MC Sparr," Likens answered. "I don't know whatever happened to him.* We added—again there was great concern about increasing the productivity of these lakes after we'd cleared them, so we added anhydrous ammonia once to a lake. Turned it green as can be."

By training, Likens was geared up for large-scale experiments.

"Absolutely," he agreed. "I was greatly taken—enamored probably is the right word—with the idea of being able to experimentally manipulate big, big systems, just like other scientists do on their bench tops. I really thought that was a powerful approach and I'm convinced it is. We had started here in '63 so about in '64—it didn't take me long—I was pushing my colleagues very hard. We need to do an experiment. We need to do an experiment."

The Hubbard Brook site is entirely unspectacular-looking on my scale of condor-viewing sites. It is located in the middle of the White Mountains of New Hampshire just south of the Presidential Range. The Presidential Range, which includes Mt Washington and is spectacular, is barely within view of its study areas. The forests are similar to those throughout northern New England, where the lumbering and farming pursuits that had once dominated the economy of the region had faded away in the early 20th century, hopefully to now be replaced by tourism. The red spruce that once dominated there has given way to sugar maple, yellow birch, and beech. According to Likens, the similarity of the site, officially called the Hubbard Brook Experimental Forest to those throughout New England, was a major factor in its selection.

---

*Publication of an earlier version of this book unearthed a daughter who was tickled to learn of the incident Likens had described and let us know that he had had a full life working professionally overseas. Sometimes editing out extraneous details will simply bar the discovery of unexpected fruit.

Amey Bailey, a forestry technician assigned to the Northeast Forest Experiment Station there showed me around. She took me first to two of the monitored watersheds, then to a lovely stretch of Hubbard Brook. Amey was particularly interested in having me see and understand the working of the weirs and she urged me to meet with "Herb," the other half of the Likens-Bormann team. He was clearly her favorite. The weirs had concrete basins behind them that were attached to gauging stations through floats. They recorded continuously, using strip-chart equipment that was already old hat when most were put in in 1963. As with any stream flow, the more volume, the more depth. Weirs are concrete dams with v-shaped notches through which water flows. With a bit of calibration and geometry, the height of water on the sides of the "v" can easily be related to stream flow.

Both of the watersheds Amey showed me, watershed 2, the now-famous clear-cut site, and watershed 1, a "reference" site in Likens' terminology, seemed remarkably small to me. Likens later corrected me on the matter of reference sites. Watershed 6 is their undisturbed reference site. My notes, however, are clear on my having visited watershed 1. It does not matter that I missed the reference site. Many years spent in various forests in New England have provided me with an internal reference for what is typical. The forests around Hubbard Brook held no surprises for me.

What did surprise me, though, were the sizes of the watersheds. I had expected each site to cover a portion of a valley from ridge-top to ridge-top. A watershed covers the entire area in which a drop of water, should it land anywhere in it, would find itself united with all other drops of water falling on the watershed when it eventually reached some larger water body. I was thinking of whole valleys. Instead, each watershed at Hubbard Brook was on one hillside of the valley through which Hubbard Brook flowed. As I stood in the middle of a depression through which water was flowing mostly underground and unnoticed in late July when I was there, I found it difficult to discern any watershed boundaries, even though I should have had a clear sight line to them through the tree trunks. I should not have been surprised, though. My vision of a landscape-sized watershed was impossibly large for experimentation. Instead of something resembling a trout brook, what I saw was, at most, an ephemeral stream.

Amey had been almost indignant when I remarked about the small size of the watersheds.

"Oh no," she had countered. "They're quite large."

It's all a matter of perspective, I guess. Watershed 2 was listed as being 16 hectares and watershed 1, 12 hectares in size. To put things in a different perspective, had the Forest Service decided to sell the land for someone to put up a hillside house of the type being built all through privately-owned New England forests by refugees from the city, both sites would have been deemed too small by such a prospective purchaser. A profit minded developer, however, may have found ways to "squeeze" one or two houses onto each site.

Although, originally, the entire effort was pushed forward by three Dartmouth scientists, it soon encompassed over "50 senior scientists and scores of graduate students," from various institutions. Their success—even without regard to the acid rain discoveries—was remarkable. It overshadowed the concurrent and much bigger budgeted International Biological Program. In 1973 alone, IBP-related projects provided ecologists with almost $11 million in support. Total expenditures were reported as $57 million. Meanwhile, overall support from the National Science Foundation for the Hubbard Brook studies between 1963 and 1976 was under $2 million. The outcomes? Critics were not kind toward the IBP. In a study by a private think tank commissioned by the NSF, the funding source for both, the output of the IBP was compared unfavorably to that of Hubbard Brook. It ruffled some feathers, especially at Oak Ridge. It was embarrassing to the Hubbard Brook people.

There have been a number of histories and studies of the IBP published. It is a very important part of that tangled web that is ecology that I warned the reader about at the outset of this book. I can only give it an aside here, but I suggest that the reader might benefit from reading Joel Hagen's *An Entangled Bank* and Frank

Golley's *A History of the Ecosystem Concept in Ecology*. The IBP came at a time that, in reaction to his proselytizing for ecosystem ecology, evolutionary ecologists began to gather in revolt in camps around Robert MacArthur. According to Golley, an acolyte of Gene Odum's, one reason for the division into camps was that "ecosystem concepts were not presented as hypotheses to be tested or questions to be asked."

Here, Golley was touching on personalities. One was that of Gene Odum, the sermonizing son of a Southern preacher man when it came to his ecosystem concept; the other was that of Robert MacArthur who, in emulation of GE Hutchinson, perhaps, made others feel more brilliant in his company.

"Rather," Golley went on about ecosystem principles, "they were derived from authority figures who frequently were the professors of the key investigators. These ideas were often presented in authoritative language and, most important, as principles in the textbooks used to train the next generation."

Those textbooks were, of course, Odum's. It was his principles and systems science that led systems ecologists to their dubious achievements in the IBP.

Meanwhile, the Hubbard Brook studies, also ecosystem studies, brought prizes and accolades to its two originators. By 1977, over 200 publications resulted from the work, one of which became one of the most cited publications in its field. Likens and Bormann attributed that success to the "individual research freedom" that the administrative structure at Hubbard Brook afforded investigators.

In fact, the Hubbard Brook researchers had consciously opted out of the IBP.

Clearly identified as an ecosystem study, it has to be placed in the Odum camp, even though Likens insists that he saw no such camps. Still, Bormann and Likens had the following to say about their studies right at the outset of their book, *Pattern and Process in a Forested Ecosystem*: "We have no grand computerized model where all the animate and inanimate components and processes of the dynamic ecosystem are elegantly linked and where details of the interactions can be spilled forth by a conversation with the computer." Neither of course did IBP researchers, even though, unlike the Hubbard Brook studies, that had been their goal.

That model that they did use was not mathematical. There were no computers to describe its system dynamics.

"I am very proud of it," Likens told me. "I strongly believe that science progresses in that way, that you start with a conceptual model and then you go to a more mathematical rendition of that, if you like, and then a computer simulation model or whatever. We've done that."

The JABOWA model came out of Hubbard Brook.

"And many other models," Likens added. "That conceptual model really has served us very well. We've actually used it from day one."

His answer on their modeling approach had been straight out of *Pattern and Process in a Forested Ecosystem*, the book on Hubbard Brook ecology. There they had written, "Our general philosophy was that, at this stage in the evolution of ecosystem science, a carefully defined and documented case history would provide the best vehicle for generating principles regarding structure and functions of an ecosystem. Some principles, rigorously based, could then be tested by ecologists and extrapolated or modified for other ecosystems."

It is a systems science approach without the systems analysis.

"It is," Likens had agreed. "You can look at it and say: well, I need to measure this or I need to measure that or if I measure that then that means—you know you can actually use it to guide the research, and we have over the years. There's a new conceptual model that's just been prepared and sent to *BioScience*. We'll see if it gets published. I'm a part of it but I'm not the lead in it. But the old model really served us very well and I am proud of that fact. It's a way of thinking about complex ecosystems. And these are very complex, so its also a very transferable way of thinking. I can show that model and describe that model and have it be understood very quickly.

"Let me give you an example. I'll never forget this one. When I moved to Cornell—again these were very, very interesting times. The environmental movement and Vietnam and all the stuff was going on—so I moved there in '69 and I think it was about 1970, I was invited to give a lecture in the engineering school. I think you had said a little earlier, it's just an engineering-type model."

It certainly was. It was the exact approach I had been trying to escape in abandoning engineering and becoming an ecologist.

"I was invited to give a lecture there," Likens continued. "Clearly the reason was, here's this new ecologist. Let's see what they do."

Likens was describing a common event at research universities. New appointees are "checked out" by the research community for future reference, in a sense. I can remember my surprise on coming to UCLA and being asked to give a sample lecture to what I thought was going to be a small group of biologists. As it turned out, the audience unexpectedly filled a 200-seat lecture hall. They wanted to know who the hell I was.

"I gave a talk," he continued, "but it started and focused on the model. I called it my black box approach to understanding this complicated system. Then I hung real data on it and made a story out of that. I had engineer after engineer come up to me right after the lecture or later saying, 'I always wondered what ecologists did. Now I know. Well, that's cool!'

"They didn't use cool then," Likens had gone on to admit. "'That's neat stuff!' And I think that story for me captures—the what? The transferability? I can tell you about that model and you can understand it. Then I can throw up a picture of a forest or a lake and I can say, 'Well, that's so complicated I can't get my mind around it, but this conceptual model helps me do that and now I can do quantitative ecology.' That was another thing at that time. A lot of engineers thought that ecologists just chased butterflies with butterfly nets and that's fine. I'm not dumping on them at all. But the fact that you could put quantitative numbers and then you develop gadgets to measure the hydrology or the chemistry—or the whatever, so that you could put quantitative numbers on that conceptual model and then go with it where you want."

Likens, I realized, was doing what the engineers were doing.

"Sure!" he had agreed to that. "But I think whether you were an engineer or a mathematician or an evolutionary biologist, it was something that you could relate to and that's what I mean by the transferability. You could say 'Oh yeah. That's what I do,' or whatever or however you might want to relate to it. So I think a lot about this. I think one of our successes probably was being able to put real numbers on that very simple model and make management applications."

It also reminded ecologists that they were experimental scientists. That experimental approach had come from his graduate training at Wisconsin, obviously, but the inspiration for his conceptual model was out of Gene Odum's *Fundamentals of Ecology*, the second edition to be exact.

Likens had himself gotten to Wisconsin by a circuitous route, but one that was fortunate for him, starting with his choice of undergraduate college. He chose a small college close to home over a baseball scholarship to Michigan State.

"I chose Manchester College,' he reminisced, "which is one of the better decision I ever made, because I'd have been totally lost at the big school. Totally lost. Then I went to Wisconsin from there. Madison. I wasn't going to go on, I was going to be a basketball coach and an elementary school teacher.

"So I went to Wisconsin and I had a wonderful time there. I had a very good education. I was extremely naïve when I went to Wisconsin.

"I still am naïve," he continued, laughing, "my major professor getting me to speak English. I grew up in northern Indiana where the language really was Kentuckian, from the hills of Kentucky. There'd been a big influx of Kentuckians in that area when I was growing up. I spoke that way. That's the way the newspaper wrote."

Likens at the time was only an older version of himself at the beginning of the Hubbard Brook studies. His small frame, which I took to be that of a third baseman, was just as fit as it must have been the two years that he pitched and played outfield in professional baseball in a class somewhere between A and B, in his own estimate. His brown hair seemed a bit thinner and darker to me, but was still combed in the same way. Where I grew up outside of New York City that hair was a mark of someone who had moved up from the South. All that Likens was missing was the Elvis-style sideburns. It's a haircut that I've also seen on the Odums, parted on the left and combed slightly upward on the right, except Gene Odum's tended to curl back down. There is a picture from the 1980s of Gene Likens with both Odums and Gene Odum's son, William, at some sort of award ceremony, I am sure. William Odum, with a 70s-style hairdo, is the one who looks out of place in it, not Gene Likens.

Although there had been lots of interesting things, as Likens put it, going on in the early 1970s when I got to Cornell, I couldn't imagine him being involved in them. He had looked like a Midwestern insurance salesman. My engineering professors at Penn looked more hip than he did. Now I know that it was Kentucky coal miner that he should have reminded me of.

"'He don't know no better!'" Likens laughed in imitation of someone from his past, then fetched up an even deeper accent to imitate. "'Ain't ya had no fetchin's up?'"

That small college helped him to flower intellectually. "I had a professor at Manchester who insisted that I apply to go to graduate school," Likens had continued. "And literally to get him off my back, I did. I applied to Cornell and I applied to Indiana University and the University of Wisconsin, because there were professors there who were from Indiana and Wisconsin. Cornell turned me down. I took great pleasure to remind them of that when I was a full professor and endowed a chair."

Speaking about the Cornell rejection, Likens said, "They turned me down because I was from a small school. They told me so.

They couldn't be convinced that my undergraduate education was equivalent to some other place."

Strangely, eastern schools (such as Marlboro?) seem not to be too small for Ivy League faculty when considering applicants. Nonetheless, Likens made up for it with a little help from Gene Odum.

"When I was in graduate school, Odum's second edition," he told me, "*Fundamentals of Ecology*, was just published and was focused on ecosystems. One of my professors used it as a course in general ecology. I was just blown away by it. I thought it was really neat. You know you could do terrestrial and aquatic, and not have to do just one or the other. At that time I majored in zoology and minored in botany at Wisconsin. I never could decide which one I liked better. At that time I liked them both, so I couldn't decide. Doing ecosystems was a wonderful way of doing them both. That's what I did."

He was never an Odum student or a postdoc or a collaborator or any of those things.

"No," he said. "No. It was being introduced through that second edition. And it was the second edition. I never liked his third edition. The second edition is about that thick." He held his thumb and forefinger less than an inch apart, then tripled the distance. "And the third edition went to about that thick. And that's significant, because this one was manageable." His fingers came closer together.

"I think that was a lesson I learned from the second edition of Odum. It was the ideas in that book. It was the fact that you could deal with nature out there as one piece and I found that extraordinarily compelling."

By collecting, measuring and analyzing water percolating through one of their small watersheds, the Hubbard Brook scientists determined both the hydrological and chemical budget for a forest ecosystem. Successful early ecosystem studies of aquatic systems, Linsley Pond, Eniweetok Atoll, Silver Spring, Lake Washington, to name a few that are classics in ecological literature, were not chosen accidentally. The water-air and water-land-interfaces provide well-demarcated boundaries for an ecosystem. The genius of Likens and Bormann was their linking of a forest to a stream or pond in terms of nutrient outputs.

Serendipity received a little bit of the credit.

"The concept of the small watershed approach," Likens had continued, "that's been applied widely around the world, so that was nice. But it was rediscoveries that were very controversial and led to just a continuation of the research not only here but at other places. The first of those was that we did an experiment cutting down all the trees in one of the watersheds. We did that and found that it totally changed the nitrogen cycle of the system and the nitrate came running out in the stream water."

The watershed was the sampling device for the forest and the pond or stream into which it flowed was the collecting point of all the samples. That was the "out" arrow from their system's black box. Inputs could be obtained from precipitation data. That was the "in" arrow. For the terrestrial part of their ecosystem (the trees in the forest) the Hubbard Brook investigators were on less uncharted ground, studies of successional changes in terrestrial communities being as old as the discipline of plant community ecology itself. In particular, they acknowledged British research from a generation before, summarized by Alex S Watt in a paper entitled "Pattern and Process in the Plant Community," that took into account the patchy nature of forest ecosystems. That research was in agreement with parts of their model. The pattern of ecosystem succession presented so authoritatively by Eugene Odum received less support from their data. The problem, according to Bormann and Likens was that forest ecosystems never really attain an equilibrium or steady state in their properties. They are a mosaic of patches resulting from various small disturbances. Eventually, the JABOWA model successfully simulated the processes that led to the forest formations seen today.

The Hubbard Brook study location in the White Mountains may have been favorable due to its proximity to Dartmouth, but it also allowed for replication by having a number of small watersheds in close proximity to each other. One could have more than one data point, where an entire ecosystem is a data point. The Hubbard Brook system had a number of small watersheds for which precipitation could be measured as input, while outflow could be measured through weirs as output. Chemical inputs and outputs could be measured, of course, by collecting and analyzing samples. Except that the focus at Hubbard brook was on nutrients, rather than energy, it is in principle the same approach as the Odums used on their atoll or Howard Odum for his influential study of a Florida spring. Likens readily acknowledged that other watershed studies such as the Coweeta Hydrologic Laboratory in the mountains of Western North Carolina contributed to his ideas.

There was an important difference at Hubbard Brook, though, he had stressed. The Hubbard Brook researchers manipulated their ecosystems and measured changes in structure and function. That took their study from the realm of the descriptive to that of the experimental. The most obvious manipulation of a forest is removal of timber. They had clear-cut one watershed, strip-cut a second, and allowed a commercial forestry operation to cut a third.

What they found was that forests held on tightly to their calcium, potassium, and nitrogen nutrients, but released them through erosion and leaching on disturbance. This had consequences both to downstream ecosystems and the regeneration abilities of the forest soils. After 10 to 20 years, the forest begins to recover and no longer loses nutrients. In my walking about on watershed 2, I was surprised to see no evidence remaining of the clear-cut. It was a particular surprise because none of the timber had been removed at that location. It had been left lying on the ground. Here and there I saw a little hummock that might have been the remains of a stump, but nothing indicating that tree trunks had covered the area forty years earlier. It was a different experience than I had at the Andrews Experimental Forest in Oregon. There the reference sites were in old growth forest that you could not see through for the huge trees. Nor could you even walk through it because of the downed trees in your path. You could hardly get down to a river through all that inconvenient vegetation to fish. Neither would one have difficulty distinguishing logged from old growth sites at Andrews. The differences were quite obvious.

Unavoidably, in addition to the effects of clear cutting, the Hubbard Brook researchers found that they had an experimental system to study acid rain.

"The very first sample of rain we collected we knew was acid," Likens told me, "but we didn't know why. We didn't know what its extent was. Then we published the first paper in 1972, and that really has been very controversial."

Controversial and very well covered in the popular press. Too much so for a digression by me on it now. The clear-cut also had a surprise. Likens refers to it and the acid rain discovery as "serendipity."

"We expected all kinds of things like the water would increase and whatever," he told me, "but we didn't expect that the microbes would come jumping in. They were much more important, and so, with decomposition and then nitrification—that's where the nitrate would come from. We didn't expect that at all. And that was very controversial.

"It was at a time when there was a great deal of relevance about environmental issues. The Viet Nam War was very controversial at that time. So the results here even got extrapolated to Viet Nam. I was sitting in a lecture once at Cornell in—whatever did they call it? Remember that program? The 'Biology and Society' program?"

I had been there for it but did not take part in it. I thought I was doing my part for society then just by studying ecology. Likens had participated in it in a major way.

"I don't remember who was speaking," Likens continued, "but he was saying that Agent Orange had sort of destroyed the forests, that the forests were gonna become sterile like they had at Hubbard Brook. I almost fell out of my chair. But those results got involved in Senate testimony and congressional testimony, because clear-cutting in the West was a very big and controversial issue at that time. They were extrapolating our results and not only did it look bad folks, but there's all this nutrient loss out of the system and look how terrible that's gonna to be. All kinds of Congressional comments were made about Hubbard Brook, to the extent that it caused the Forest Service—this is Forest Service property, a Forest Service facility—a great deal of concern about these professors that were there doing this work, because clear-cutting is a stated policy of harvesting timber by the Forest Service and somebody was saying that, well, it may not be such a good policy after all. That was very controversial. It spawned huge amounts of research all over."

In an unanticipated parallel to *Silent Spring*, the discovery of the extent of nutrient leaching after a clear-cut faced the same kind of industry attacks as did Rachel Carson's criticism of DDT. In eastern forests small clear-cuts are now the recommended practice.

"It's true," Likens said. "We've actually, I think, promoted that idea. We published a paper in 1978 in *Science* in which we talked about the conditions of clear cutting and how you could do it and minimize impacts. We proposed a 75-year return interval. If you don't cut too often. If you don't cut too large. If you don't cut the blocks too close to each other. And a whole series of management proposals in that paper in *Science*, which, looking back, is kind of interesting, too. I don't think that they'd publish an article like that today. But yeah, I think that's true, and particularly the timing. The Forest Service policy now is a cutting interval of about a hundred years. That drew heavily on our proposal. You need to have a long period of time to recover naturally, unless you did something like spraying fertilizer or planting or whatever. But, naturally, it takes about that time for it to recover and, again, like you said before, it's just plain common sense. It's like your garden. If you just keep doing this to your garden, you're gonna run it downhill."

The results? Data. Also an ecosystem model acceptable to the field during an era when ecology was trying to define itself, with some trying to define the ecosystem concept right out, while others were defining it with sets of differential equations. No doubt because their model was relatively unsophisticated, it was also relatively unintimidating. Some of it was presented with actual figures of trees and soil representing compartments, rather than boxes and arrows. There were no partial differential equations, no matrices, no Bessel functions to solve.

And it provided guidance for management decisions.

Thus, the Hubbard Brook studies, more so than the IBP, brought about the realization of the hope, first planted in the public mind largely by *Silent Spring*, that pollution problems are ecological problems and have ecological solutions. The extent of acid rain problems in the United States was discovered by an ecosystem study in the woods of New Hampshire. Ways to harvest timber sustainably in the northeastern United States came out of the same study. The discoveries of the MacArthur school, which served more to confuse than enlighten controversies over broad environmental issues, now had an alternative besides the cybernetic science of Eugene Odum.

This was at the same time that the need for the application of ecology to the practical purposes of protecting the environment under NEPA was seen to be outpacing the ability of the science to provide it. While the Hubbard Brook scientists were preparing their first compendium volumes of results for publication, Canadian ecologist David Schindler wrote an editorial in *Science* that called the environmental impact statement a "boondoggle."

Impact studies, according to Schindler "are often done by scientists who cannot successfully compete for funding from traditional scientific sources. In general, their methods are ancient, descriptive 'textbook' techniques, which do not reflect either the many scientific advances of the past decade or the problems unique to the study undertaken." The end result, in Schindler's view was that "Enormous sums are therefore spent with little or no scientific return," while placing in jeopardy scientific integrity and credibility.

"Bunk," replied a team of scientists involved in impact assessment at Oak Ridge, led by Stanley Auerbach. What else could they have been expected to say, given the circumstances? Yet the situation that Schindler described could only be expected, given that much of the science of ecology was following the MacArthur-school theorists, the mathematically inclined, rather than the Odum-style environmentalists, those who did experiments.

As with his friend, Robert Whittaker, there was no Likens camp in ecology. There were no Likenites. There was just science.

MacArthur's Ghost

The Serengeti is that part of Africa that the great white hunters would go to on safari in order to shoot various docile grazing mammals and be frightened by a few lions and rhinos. It is now a National Park in Tanzania and the location of numerous tours and nature documentaries. The natural condition of the grass of the Serengeti Plains is one of being grazed on by the 1.4 million wildebeests, 600,000 gazelles, 200,000 zebras, 52,000 buffalo, 60,000 topis, and some other grazers, according to estimates by SJ McNaughton of Syracuse University. (The buffalo, of course, more closely resemble cattle than our American bison.) Drought is always a problem for the living things of the Serengeti. For the grazing animals, it leads to those long lines at the watering holes and drought-related mortality. The grasses have to deal with ever-hungry grazers and little water with which to sustain growth. What McNaughton found on the permanent study sites he began to observe in the 1970s was that the grazers would eat less of the sites with more diverse vegetation in some, but not all cases. Also, sites with diverse vegetation recovered more rapidly to control levels with the onset of rains. There was much in the way of complications having to do with the identity and behavior of particular grazers, but McNaughton saw ways of analysis past the peculiarity of his critters. His control plots helped.

One conclusion McNaughton made was that "greater community diversity tended to stabilize community functional properties, reducing the relative magnitudes of such fluctuations." He found this to be "due to the greater diversity of growth patterns within more diverse species pools." It was a new way of looking at diversity, through functional diversity, rather than species diversity, per se. From this perspective, the number of species does matter, for a greater number of species being present is more likely to insure that particular types of species, necessary to ecosystem function, will be present. Functional diversity means exactly what McNaughton said above, different growth forms, different metabolisms, all of those not-so-subtle important differences that can be seen in plants. It might not matter from this perspective that a stand have four highly similar species of grasses (which, of course, competition theory would claim could not happen), but that it instead have a grass and a clover, in a simple example. The clover will fix nitrogen, thus stimulating the growth of the grass.

McNaughton's paper summarizing his work came out in *Ecological Monographs* in 1985 and set off studies around the globe. These have taken the ecosystem idea (if not the original conception of it) to heart, as they have also the example of successful ecosystem-level experiments at Hubbard Brook and the Tallahassee Mafia's arguments for the use of null models in interpreting them. Experimental designs have been impressive. David Tilman, starting in 1982, has been studying and manipulating over 200 plots in four grassland fields in Minnesota. Another study was comprised of 480 plots distributed over eight sites in seven different countries. The report of its results in *Science* required the listing of the names of 34 investigators.

"The current global extinction rate, which is 100 to 1000 times greater than prehuman levels, and the loss of local diversity due to management practices have the potential to affect ecosystem processes strongly on both a local and global scale" is the justification given the research.

Interpreting a study known as BIODEPTH, Tilman concluded, "Their results suggest a rule of thumb—that each halving of diversity leads to a 10 to 20% reduction in productivity."

This new way of looking at diversity was not without controversy. Debate seemed to "explode," in the description of its participants, with the publication of a dissenting opinion in the *Bulletin of the Ecological Society of America* in 2000. An official publication of the Ecological Society of America, part of a series produced by the society and available as *Issues in Ecology*, brought on the dissent. An article in the series concluded that there was experimental support for a cause-and-effect relationship between diversity and ecosystem functioning without mentioning some serious questions that had been brought up about the methodology behind the experiments. Its critics, an international group, raised the oversight to a breach of the ESA's *Code of Ethics*, and called the paper "little more than a propaganda document." The controversy became public, reaching even the pages of the *Chronicle of Higher Education*.

It led to Harvard's Fakhri Bazzaz having to call on the disputants to "cool" the debate "because of the great importance of the relationship between diversity and productivity to conservation policy." His tone is that of a schoolteacher admonishing unruly schoolboys.

"Heated arguments among leading ecologists about the relationship between biological diversity, the total number of plant species in the world, and ecosystem productivity have the potential of harming ecological science," he wrote in the *Bulletin of the Ecological Society of America*. "To call ecological experiments 'irrelevant' and 'politically manipulated' is not helpful. Ecologists should refrain from using such language."

In tone, the argument was a return to the "barroom brawl" of the null models. The advice given echoes that given by Diamond to Simberloff over refuge design. There is a difference, however. The ghosts of Popper and Wittgenstein no longer haunt the ecological literature. Ecologists have perhaps decided that if Descartes was good enough for Galileo and Newton, he is good enough for them. After all, look at the revolution those two accomplished.

These later arguments, as spelled out on the pages of *Science*, had to do with the interpretation of data, especially in light of difficulties in controlling factors other than diversity. The experiments, both Tilman's Cedar Creek studies and those of BIODEPTH, consisted of starting plots with different composition of plants or selectively removing plants from more mature plots. The rules for choosing how to include or remove particular species of plants and arguments over them echo similar concerns over null models more than a decade before. Needless to say, the field plots being experiments, statistical analysis incorporating null hypotheses was standard practice. That part of the debate differed.

The new ways of looking at the diversity-stability relationship no longer saw diversity to be the cause of ecosystem stability. Rather, diversity was a "passive recipient" of important mechanisms inherent in ecosystems. There are things that contribute to the stability of ecosystems that are reflected in the diversity of the ecosystem, in other words. Those "things," as I have rather loosely labeled them, have to do with how ecosystems function to capture energy and cycle nutrients, for example. Maximize one (diversity) and you maximize the other (productivity and nutrient cycling) is a conclusion that might be drawn, if a bit hastily.

Realizing that the complex ecosystem interactions in nature are non-linear and rarely at equilibrium—those two assumptions so crucial to the Hutchinson-MacArthur-May revolution—one can also take a new look at stability. Stability of populations came to be seen not so much as constancy, but as lower variability. Populations are expected to fluctuate, but stable populations are those that avoid extremes. Using concepts such as resilience and resistance, ecologists began to attempt defining stability as an ecosystem's ability to defy change. This was not so much new as rediscovered.

Of course, the theoretical ecologists (mathematical modelers) hardly went away while these long-term and widely spread experiments were being performed. Nor did they change the justification for their research. So why are there so many species? Stephen Hubbell of the University of Georgia and the Smithsonian's Tropical Research Institute tried to work out a mathematical theory that he thought would unify biodiversity and biogeography theories. If biodiversity is an important issue in the environment, and Hubbell firmly believed it was, working out what controlled it was a high priority.

"In view of the genuine possibility of a global collapse of biodiversity in the near future," Hubbell has written, "it is unconscionable that we still have no serviceable theory of biodiversity."

What Hubbell did was find a way to add speciation to MacArthur and Wilson's theory of island biogeography. Falling out of the effort was a dimensionless number, much like dimensionless numbers in engineering, such as the Reynold's number in fluid flow. It was Hubbell's hope, no doubt, that his fundamental biodiversity number would be as useful as the Reynold's number. That quantity can be used to predict the drag that a flowing fluid will exert or to scale up airplane and ship designs from bench top to the real world. Hubbell, however, was not yet ready to explain how his number could be applied to improving the human predicament. Of course, the preservation of global diversity is always a justification that can be wrapped around the most esoteric of theoretical ecology.

What else was going on in ecology on the subject of diversity as the 20th century came to a close?

I have a friend who was finding much satisfaction in studying the relationship between local species richness and regional biodiversity. How did what happened in a local area depend on what went on in the surrounding region, or globally, was the question he asked. Again, although there was no immediate application, it was basic research on the dynamics of biodiversity. It let him do research that was at the forefront of ecology, while getting him to places such as Samoa and the Great Barrier Reef of Australia. As with many other ecologists working on questions of diversity, his approach was right out of the MacArthur tradition. Hubbell's neutral theory seemed to have an answer to some of his questions, but my friend has yet to determine how to test it. I find myself changing the subject with him when in conversation on it; it all seems so far removed from the concerns of that first Earth Day that had led me to ecology.

One surprising new finding from investigating food webs mathematically was that weak interactions may be more important than strong interactions in stabilizing ecosystems. A strong interaction would be one between two closely related species eating the same food, as with some of Darwin's finches in times of drought. A predator and its favorite prey, wolves and elk, for example, might represent another strong interaction. A weak interaction is well described by Darwin's famous relationship between cats and flowers. A number of studies have been cited as showing that food web interactions in nature are, indeed, weak. Darwin was right. (Isn't he always?) This is an interesting and potentially explosive finding! It is exactly the type of evidence, if confirmed, that would support the radical ecology argument that every single creature, no matter how lowly and uncommon, is absolutely necessary for global functioning. In fact, the weak-interactions hypothesis means that the lowly and uncommon may be more important than the common.

A contrary view, however, is that ecosystem processes are governed in nature by the particular characteristics of the organisms making them up, rather than the number of different types of organisms. This view echoes that of Simberloff in trying to turn ecologists away from using community theories for conservation purposes. It, too, has new evidence that can be marshaled in its support. The criticisms that set off the new diversity-stability debate tend to turn on just that point. It is the identity of the species rather than their numbers that matter. It is not easy to change the number of species in an ecosystem, according to the critics, without altering a host of other properties that were inherent in the species chosen for inclusion or exclusion. While two species of grass might be substitutable for each other, clover, with its ability to fix nitrogen, is in no way a substitute for crabgrass. Why not then just study the roles of crabgrass and clover in ecosystems and forget about diversity?

The more things change, the more they stay the same. This time, however, differences appear to have been mostly reconciled. Ecologists have learned from their past. Those who had conducted the "full-blown war" have agreed in print that human domination is increasingly causing the depauperization of ecosystems and that the consequences of biodiversity loss are of considerable interest because ecosystems collectively regulate the biogeochemical processes of the earth.

They also agree that, "In a few cases, the opposite relationships emerge—increasing biodiversity can actually decrease primary production and other ecological processes."

Surprisingly, the explanation for this seeming contradiction is found in the work of—Guess who?—Robert MacArthur! Right there in the pages of MacArthur and Wilson's *Theory of Island Biogeography* is the discussion based on the "r" and "K" of the discredited Lotka-Volterra type equations with which Stephen Pacala and David Tilman crafted a suitable explanation for their findings.

"There is consensus that at least some minimum number of species is essential for ecosystem functioning under constant conditions and that a larger number of species is probably essential for maintaining the stability of ecosystem processes in changing environments," both sides, when assembled by the peacekeepers, agreed. "Determining which species have a significant impact on which processes in which ecosystems, however, remains an open empirical question."

In addition, there was agreement that "From a strictly functional point of view, species matter so far as their individual traits and interactions contribute to maintain the functioning and stability of ecosystems and biogeochemical cycles. Although species richness is easier to measure, a more predictive science might be achieved if appropriate functional classifications were devised."

These are sober, carefully considered words. The strongest statements are qualified with words such as "might" and "probably." This is quite unlike the following, from another group of ecologists with similarly sterling credentials.

"Human alteration of the global environment has triggered the sixth major extinction event in the history of life and caused widespread changes in the global distribution of organisms," this group wrote in *Nature*. "These changes in biodiversity alter ecosystem processes and change the resilience of ecosystems to environmental change. This has profound consequences for services that humans derive from ecosystems. The large ecological and societal consequences of changing biodiversity should be minimized to preserve options for future solutions to global environmental problems."

Note that we have come back to ecosystem ecology. Note also some of the questions the new work is hoping to answer: "How do species coexist? Why do we see so many together, and why don't we see more?" And: "What awakens MacArthur's ghost?"

Meanwhile, Gene Likens was expanding the time scale for long-term research to encompass our entire new century. When I arrived at Hubbard Brook two undergraduates from Syracuse University were setting up what might then have been the mother of all Global Positioning Systems (GPS) on the lawn outside the Forest Service Building. It was day-glow bright yellow in color—no doubt so that it would not be lost—could be packed away into briefcase-size containers for easy portability, and was precise in identifying its location to within 2 centimeters. That's a far cry from the precision you can expect from an $80 discount store special. The price is also a far cry from that in the discount store. It was $300,000 according to the students working with it, even more based on hints Likens gave me. Now it has probably been superseded by a microchip with an even more discount-store price.

Calculus and Commoner

Two letters in my Pulitzer Prize winning local paper (not at all about to disappear, I hope) caught my eye as I sipped my morning coffee before turning to how to complete my history of ecology's transition in the latter half of the 20th century. Almost side by side, they were occasioned by Vermont's "Champion Lands" controversy, as it was then called. The Champion lands are well timbered, but now recovering acreage in the northeastern corner of the state. The thinly populated region, none of its forests more than 100 years old, is called the Northeast Kingdom area of Vermont. In it are spectacularly clean lakes and vast tracts of sparsely populated forests. Alas the streams are mostly too warm in summer for the state's native brook trout to thrive, but there are lots of big fish in the lakes and in the Connecticut River that borders the area. Perhaps the years of intensive and continuous logging have deprived the streams of the cover that might have kept them colder through the hot days of August. Or perhaps it is because the mountain elevations are relatively unimpressive, even for the East, and they have never supported the kind of "brookies" with the careless feeding habits I prefer. Or maybe I just haven't learned how to fish for them yet.

When a timber company decided to abandon its forests by selling them off, the state of Vermont, in conjunction with private foundations, managed to gain control over a huge swath of land at a bargain basement price. The enabling legislation that appropriated the money for the state to take advantage of the situation promised to maintain the land in a sustainable manner for its current uses. Those current uses were recreational and commercial: camping, hiking, hunting, fishing, snowmobiling, and, of course, timbering. Champion International, whose land comprised a large part of the forest in Vermont, had harvested trees in the area to provide paper pulp and wood for doors. Somehow, the Vermont Agency of Natural Resources and the Nature Conservancy, that marvelous spin-off of the ESA, decided it would be appropriate to set aside a small portion (less than 10%) of the land as wilderness, protecting it from such depredations as logging, motorized vehicles, and—initially—bait fishing. This was what had set off the controversy that was the subject of the letters in my morning paper.

At the time, there were letters almost weekly in the paper coming down firmly on one side or the other. The specific points of contention need not concern us here, but the tenor of those two particular letters may be instructive.

One stated that we took it for granted that access to Vermont's wooded areas for hiking, hunting, etc, would always be there, but we may not have noticed how more and more houses and "No Trespassing" signs were filling up our back roads. "Our desire for living along a secluded road, and our need for timber, will only increase, never decrease in the foreseeable future. In short, the demands for what is best for Vermont can cause the death of Vermont," its author argued.

"We need areas where the trees will grow as they will," he went on. "We need wilderness ... We need areas that are open to anyone regardless of their wealth and ability to purchase their own Shangri-La ..."

The other letter argued that large blocks of roadless, unmanaged land were "necessary to protect and enhance the diversity of the interdependent web of all existence of which we are part."

We need wilderness, he wrote, "because our native ecosystems evolved in old forests without management ..." Only in public land, he concluded, "can we establish wilderness areas where mature forest can evolve ..."

I recognized one of the letter writers as a highly respected ecologist who was at the University of Vermont at the time. The other letter was from a layman, I presume, from the bucolic village of Chester. Which letter writer was which? Granted, this exercise may be somewhat unfair, because I know the ecologist to be an excellent writer, some of whose books have graced best seller lists. Still, shouldn't the answer to my question be obvious?

The first writer was the University of Vermont's Bernd Heinrich. Are you surprised? Is there something amiss here? It is not that the ecologist should be capable of writing clear, unpompous prose. These things happen, and God Bless him for it. Why is it, though, that the layman's argument should be so instantly steeped in ecological jargon, when the issue did not warrant a plunge into ecosystem theory?

Does ecology still deserve to be the fundamental science of the environment? Is it through ecology that we want to integrate all the other diverse sciences, not to mention the law and social sciences, that we bring to bear on the environment?

I am beginning to think not. I am beginning to think that the geosciences, earth sciences in the vernacular, geology, atmospheric science, oceanography, and hydrology, are the fields we must look to for integration of our efforts at understanding assaults on our *environmental* systems. Perhaps a new environmental systems science can develop from the contentless approaches of cybernetics and systems science. Or, more likely, perhaps, an engineering science of the environment has already developed, but not yet been recognized. The ensemble voices of ecologists crying for more resources, more attention, more protection, more and more, might not be letting other voices to be heard.

Gene Likens's *Biogeochemistry of a Forested Ecosystem*, in which with his Hubbard Brook collaborators he summarized the first fifteen years of their study of nutrient cycling, has a curious epilogue appended to it. It is too long to quote here. Likens and his colleagues cite pages 104-8 of the 1949 version of Aldo Leopold's *A Sand County Almanac* as its source. You will find it easier by turning to an essay entitled "Odyssey" that follows the actual almanac, whatever pages it may turn out to be. It describes the imaginary path of an atom of calcium as it weathers from its parent limestone deposit to be passed from organism to organism until—after stops like in an eagle feather in an Indian's headdress—it eventually finds its way down the Mississippi and into the sea, where its fate is to eventually become limestone once more. A rather poetic flight of fancy, it is the type of writing for which Leopold has become justly popular.

It was Likens who added that passage to the text. "Since I learned about Aldo Leopold at Wisconsin," Likens once said to me, "I've been a great fan of his way of thinking."

He was particularly fond of that passage detailing the fate of Atom X. "I think that really captures very nicely in prose that I couldn't do what happens when things move downhill and that's what it's all about, it's just the headwaters to downhill."

I have in the past been a bit dismissive of Leopold's influence on ecology. I see him as something like ecology's gamekeeper, rather than the font of its theoretical wisdom. There is no question, though, that Leopold was a legitimate ecologist; he had served a term as the Ecological Society of America's president. But can it be that all the ecology needed from that science to manage the environment is encapsulated in Leopold's poetic diary? Is it all just "headwaters to downhill," as Likens put it?

Actually, I now suspect that it is. I suspect that the bird watcher in tennis shoes knew all the ecology needed for rational environmental decisions—but for the millions of details that occupy ecologists. Most of those details might better occupy the efforts of botanists, entomologists, ornithologists—whatever biological specialty is appropriate—and geochemists, hydrologists, environmental engineers—whatever physical science fits, than ecologists steeped in community theory.

What would be the advantages? Perhaps we would be able to decouple what are almost religious beliefs attached to the ecology of 1970. We can let Deep Ecology continue its development as a philosophy for life, without it also claiming justification from its being a science. This may be difficult. Darwinism and religion are still clashing in the most enlightened of societies—our own. It can be done, though. Only fundamentalist Christians seem to take issue with the science that comes from applying Darwinian selection to evolutionary change.

Although science can be said to be the invention of philosophers—Descartes, Bacon, Mill—philosophers have had a long history of misinterpreting and misunderstanding the science that they then use in justification of what are strictly moral and ethical issues. There is no doubt, either, that scientists have had to adopt a philosophical point of view in their science, and this goes beyond the methodological niceties that underpinned the philosophical clash we saw in the null model controversy. Consider the scientific advances represented by Heliocentrism, Uniformitarianism, and Darwinism. Clearly scientists were made by each of them to choose a point of view directly at odds with the prevailing philosophy of a dominant religion. There may or may not be something of the same sort of phenomenon going on in ecology, except that the religious view—the quasi-religion of environmentalism—is embedded in the science, rather than fought by it. The anti-GMO fervor of many environmental activists might be a demonstration of it.

The debate over nuclear power is also instructive here. It was (and still is) one of differences in philosophy more than science, although not as clearly so as those scientific revolutions that challenged a religion. There is no scientific theory to chose from that argues that nuclear power is either good or bad. Neither are the facts at issue really in dispute. Points of view—we can do good with nuclear technology or there is no good that can come out of an evil technology—are at the heart of any dispute over the facts of nuclear energy technology. This is not supposed to happen in science. Society needs to be able to get information from science in a dispassionate way. Science, for example, can answer any question the public or policy makers can ask about the characteristics of life, sperm and egg, zygotes, embryos, and fetuses, but it can never answer the question of the morality of abortion. Nonetheless, we would expect that a scientist would present scientific findings impartially, regardless of his personal views on abortion.

Can we expect the same level of dispassion from our environmental scientists, ecologists in particular? Perhaps not to the extent of scientists working in areas less directly affecting public welfare.

I was appalled at about the turn of this century (the 21st) to come across words written by Edgar Fahs Smith, provost of the University of Pennsylvania at the turn of the last century (the 20th).

"Suppose," he had written as his contribution to an imbroglio over academic freedom and a professor with subversive Marxist views, "for illustration, that I, as a chemist, should discover that some slaughtering company was putting formalin in its sausage; now surely that would be none of my business."

These are not words we would wish to be coming out of the mouths of today's professors of chemistry—or any university administrators, whatever their field of expertise may be. Smith's words came just a few years before publication of Upton Sinclair's *The Jungle*. Certainly, we expect more awareness from our scientists of the consequences of their action or inaction. This standard must be severely adhered to by environmental scientists. We won that battle in terms of the environmental and public health consequences of a chemist's actions on that first Earth Day.

But should an ecologist necessarily be an activist?

Almost without exception among practicing ecologists, the answer to that question is yes. In his book, *The Idea of Biodiversity*, David Takacs surveyed biologists who promoted the values of biodiversity. Takacs' goal was to try to better understand why biologists cared so much about biodiversity. His conclusion? They have become advocates for the idea.

"By their ideas, words, and actions, the biologists who promote biodiversity's values are putting an end to the frankly metaphysical notion of an objective, value-neutral search for knowledge," according to Takacs. "Their values shape their worldview and continuously affect their work."

Sounds a bit post-modern—or does it?

Here is how some ecologists have answered the question on activism in their published work.

Yes, according to EO Wilson. We must maintain biodiversity for "a healthful environment, the warmth of kinship, right-sounding moral strictures, sure-bet economic gain, and a stirring of nostalgia and sentiment...." Professor Wilson then went on to say, "Together they are enough to make a compelling case to most people most of the time for the preservation of organic diversity. But this is not nearly enough: every pause, every species allowed to go extinct, is a slide down the ratchet, an irreversible loss for all. It is time to invent moral reasoning of a new and more powerful kind, to look at the roots of motivation and understand why, in what circumstances and on which occasions, we cherish and protect life. The elements from which a deep conservation ethic might be constructed include impulses and biased forms of learning loosely classified as biophilia."

Wilson seemed to have come to a philosophy of Deep Ecology without accepting the ecologically indefensible superorganism concept. He argued that we have a biological need as a species for natural experiences. It was a different path—from the streets of Pensacola, Florida, instead of the fjords of Norway—but it took Wilson to the exactly identical mental place of Arne Naess.

No, apparently according Gene Likens. "There's tremendous public confusion," he has been quoted about the fine line between environmentalism and ecology, "because we often work on the very same things." He tried "very hard not to let my emotions and my personal views color my science."

Yes, according Dan Botkin, but with a caveat. He foresaw active management of the environment, even clear-cutting in forests, if that was what the science showed. Professor Botkin saw a future in which ecologists could be activists without being—in my words—"environmentalists." In his words: "If nature in the twenty-first century will be a nature that we make, then the guide to action is our knowledge of living systems and our willingness to observe them for what they are, our commitment to conserve natural areas, to recognize the limits of our actions, and to understand the roles of metaphor and myths in our perceptions of our surroundings."

Note, in particular, the appeal "to understand the roles of metaphor and myths." Another take on those roles, this one by ecologist Jon Ghiselin in the midst of the profession's anguish over

Calculus and Commoner

NEPA, was: "It is frightening to see all of ecology—if not all of biology—reduced to a part of a chapter, and then presented as a combination of calculus and Commoner."

Yes, according to 20 scientists, a number of whom have had a part in my story, writing in *Science*, out of concern for the human predicament. "Now *all* of our research is done in systems altered by *Homo sapiens*," they bemoaned. It was their view that ecologists should "be ready and willing to devote part of their professional lives to stemming the tide of environmental degradation and the associated losses of biodiversity and its ecological services." Leaving the reader wondering if the scientists really read anything but each other's works or listened to anything other than their own symposia, they charged ecology with a need to teach "the public about those losses." It is out of "a responsibility to humanity" that ecologists should take on those tasks, they concluded. (Not out of protection of their research sites, was left unsaid.)

Their letter to *Science* echoed a more sophisticated piece by a similar number of scientists warning against the ecological consequences of nuclear war that was published in that journal a generation earlier. Some ecologists were authors of both. The need for the large number of authors in each case can only be explained by the psychology of groups.

Yes, was the latest word according to Stuart Pimm, who set up a fictitious ecologist, Dr Brown, to berate on the matter. Dr Pimm found it inexcusable that Dr Brown should consider that he had "no mandate to address the scientific needs of people who protect and manage species and the environment." Pimm's straw professor echoed the past words of the University of Pennsylvania's real professor of chemistry. As an alternative, Pimm suggested that scientists donate a sort of tithe to the environment, at least "a tenth of our time and effort to applied versus academic pursuits." Then there was a call to arms.

"Does all this smack of 'advocacy,' the dirtiest eight-letter word in a scientist's vocabulary?" Pimm asked. "We spend our lives educating our students and our colleagues. What part of our job description prohibits our extending that mission to the public and its political leaders?"

The goal of all of the activism Pimm wanted ecologists to undertake? To "protect more of the Earth's remaining natural ecosystems."

This last statement puts into instant perspective all of the dispassionate statistics, some of which have also been the subject of this book, with which Pimm had built his case in the rest of his book. Pimm's goal was, in fact, the purpose of the Ecological Society of America when Victor Shelford first assembled 22 like-minded people in Philadelphia in 1914. The ESA was started essentially as an organization to protect natural areas so that they could be studied by ecologists. Although the ESA never did become a professional society—the almost 50-year struggle over licensing requirements and the ESA's lax membership requirements attest to that—the obvious conflict of interest in having ecologists protect "the ecology," in a sense, led to the formation, separate from the ESA, of The Nature Conservancy. It arose directly out of efforts made by society members to promote protection of natural areas. In Shelford's words: "Thus the Ecological Society of America has grown out of an attempt to accomplish what may be termed a display of the localities which have served to inspire workers in ecological lines." That goal is important to note. By subscribing to it, you, too, can belong to the society.

Ecologists were able to separate their conservation instincts, at least for real estate, from their science that way, through a separate organization. They were not, however, able to separate their belief in the value of natural diversity from their science. That is too embedded in the science for easy excision. The events I have detailed here give evidence to that. Moreover, unlike the tangible asset of real estate, diversity, as we have seen, is both a very real quantity and an ecological construct. Preserving it is less straightforward than purchasing it. Its preservation requires an intellectual solution.

And that intellectual solution has not necessarily been a scientific one. MacArthur's Three Influential Papers are now something akin to a liturgy for some ecologists. The scientific approach that it had shut out by garnering so much attention and energy from ecologists, if nothing else, was that of Odum's ecosystem. Shorn now of its superorganism implications and the more sophisticated cybernetics (read that as mathematical theory) that had been imputed to it, it is, however, the part of ecology most appropriate to the analysis of the broad environmental issues facing us. For specific cases, the design of a nature preserve or power plant or protection of the spotted owl, for example, there is much science only marginally related to ecosystem processes, but in the end, to understand the workings of our landscapes, analysis of inputs and outputs and the effects of changes to them are unavoidable.

Before the MacArthur Revolution, many ecologists were greatly impressed and influenced by Odum's second edition of *Fundamentals of Ecology*. The copy I have, unfortunately, is the third edition. But it is redeemed by having a name inscribed in it, Bob Rovinsky. It is the name of the mathematician who gave it to me. Not as a gift, but as a way of recycling, in a sense. Bob had shared office space with me as he was finishing up his PhD work in math. David Pimentel had allocated it to him in support of some project or other having to do with pest management. He would not be needing the book, Rovinksy told me. He intended to stick strictly with math.

"Ecology is hard to learn," were his words as I remember them. "You need to know not only biology and math, but you need to know chemistry and physics and a whole bunch of other things."

Yes indeed. Those energy and chemical transformations and rates that drove him back to the "easy" world of mathematics are precisely at the core of what is coming to be known as environmental science. They are also subjects that hold very little interest to many ecologists.

Perhaps it is all for the best that it is so. Perhaps it is all for the best that that thing that had been disparaged as "little more than a point of view" just prior to Rachel Carson's *Silent Spring* awakening the nation to its importance, was easy to learn. I think that, regardless of the disciplinary background of an environmental scientist, we can rest assured that he or she has been suitably exposed to that "point of view" by now.

Perhaps it is also for the best that those environmental issues that get the attention of ecologists are being dealt with by ecologists coming from traditions of experimentation. The ecology stemming from the MacArthur tradition is more of an art than a science. As such, it can stir scientists of any persuasion in any discipline, for better or worse.

"Art," Wilson has said MacArthur had enjoyed quoting, "is the lie that helps us to see the truth." The words are those of Pablo Picasso.

I do not have anyone of equal stature to quote in rebuttal. But take a trip down to your local water treatment plant and see if the personnel there don't agree that it is engineering, in one sense or another, that maintains the water, food, shelter, and transportation systems that we depend on. It might also maintain the needed level of biodiversity.

Pondering those two situations might well help one to find an answer to the question posed by ecologists four decades apart. "What is Ecology?" asked Lee R Dice in 1954 on the occasion of his presidential address to the Ecological Society of America. One would have thought his audience knew the answer to the question. The address was reprinted as a public service, no doubt, in *Scientific Monthly*, then an outlet of the American Association for the Advancement of Science. Judging from it, no one seemed to know what ecology was—or at least there was no general agreement on its meaning even among its participants. To outsiders, it was something of a mystery. Not too many years later, Paul Sears had to explain "The Place of Ecology in the Sciences" in his presidential address to the American Society of Naturalists. The year was 1960. Impending was the explosion of interest in ecology that was set off by Rachel Carson's *Silent Spring*. In 1964, Sears was calling ecology a "subversive" science. By then, everyone seemed to think they knew what ecology was. That was the problem. More recently, a leading Australian ecologist lamented that the term's meaning had been taken away from the science. He asked, once more, in an essay, "What Does 'Ecology" Mean?"

Up to now I have been resisting the temptation to use Lawrence Slobodkin's writing as a sort of Greek chorus. I have not been entirely successful and will now succumb to temptation once again. Slobodkin had some surprising questions to ask in response to calls for ecologists to take on roles of social activism as part of their science. He was specifically responding to one of those multi-authored papers with an all-star cast of ecologists calling for more on some ecological issue. (More meaning attention, funding, action, etc.) Slobodkin likened these efforts to something like a new subdiscipline in ecology that he christened "societal ecology." His questions were: "(1) What legitimate claims can be made for the societal necessity of ecology? (2) What are the objects of ecological research? (3) Is there an independent theory of ecology?" The latter two were unusual questions to be raising about any science that professes to have its roots in the 19th century. Slobodkin provided not an answer to his questions, it seems, but another question.

"Might ecology be the art of environmental engineering with special reference to the organisms and chemical and energy transfers in the environment, and how these are influenced by anthropogenic events?" he asked, then decided that if that were so, it would be good for the science of ecology.

I think that would be good, too, for the environment and for society. That art Slobodkin described, however, was not the ecology that brushed aside the ecosystem concept or refused to come to terms with the scientific method. Neither were would-be practitioners of Slobodkin's useful art necessarily to be found among those 20 names from ecology's *Who's Who* willing to devote their professional lives to the cause of biodiversity.

We should be careful what we wish for.

Afterward

Time itself is apparently a great tool in the process of figuring out history. So, fifteen years after first publishing *Earth Days*, I decided to put out another edition, this time from a more distant perspective and through a new publishing outlet. Historian of ecology, Sharon Kingsland, once pleaded that her own 1985 work, *Modeling Nature: Episodes in the History of Population Ecology*, which included the transition covered here, needed to have been farther removed in time from when those events were unfurling in order to have a proper perspective. Robert McIntosh's encyclopedic, *The Background of Ecology*, which covers even more of the changes in the science of ecology that are the subject of *Earth Days*—and in greater detail in many instances—was also published in 1985. Both works captured the controversies, but not quite how they played out.

So, after more than fifteen years of writing, editing, and translating, all at a distance from what was current in ecology, I threw myself onto producing this revised edition. First, I corrected minor errors and typos (which included several missing sentences), improved some figures and expressions (adding some commas and verbs that my experience with translating suggested were necessary), and updated where needed in individual chapters, which turned out to need remarkably little. Then I turned to working on a new chapter to bring the reader up to date on what the science of ecology has become. Ecology, after all, has now been a mature science for over a generation (two, actually, since the battle over null models) and has transitioned along with the rest of the world through a digital revolution in information technology. I was curious about the changes being brought about, and I thought I could productively use my personal perspective to relate things as they are now to the philosophical and methodological issues covered in *Earth Days*. My hope was of following a "main stream" from the roughly 1950-1980 transition in the science that is the subject of this book.

Fat chance! I could barely find even a few trickles from that "main stream" in the almost impenetrable morass that ecology now seems to me. The Merriam-Webster online dictionary, I find, defines ecology as the "branch of science concerned with the interrelationship of organisms and their environments," a good enough general definition, but about as useful as identical general definitions for both physics and chemistry being the "study of matter and energy," as mentioned previously. Wikipedia does seem to cover all of the current definitions of ecology, noting as it does its practical applications and its overlap with other sciences, one of which is identified as evolutionary biology.

According to Don Strong, a recent, long-time editor of *Ecology*, "What's going on in ecology these days [is] multiple real mathematicians doing real science, real evolutionary ecology with genomics, 'big ecology' using remote sensing and computational tools for appreciating large scales in time and space, cool paleoecology, etc. As for the digital revolution, it means that one can search the literature very quickly." You can find out for yourself, he seemed to imply, by going to the Google Scholar site and googling through pretty much every publication (most of which result from federal funding) for its abstract and, usually, even a free copy. It was a hint that I took him up on.

In general, science, in particular the various fields of biological sciences, has evolved to become remarkably fragmented. The digital revolution has no doubt played a large part in recording, analyzing, and communicating data in ways that ecologists could only dream of around that first Earth Day. There are also more scientists (although not necessarily in the ESA), more sciences, more publication outlets (monthly instead of quarterly issues of *The American Naturalist,* separate subject-matter issues of *Science* and *Nature*, new open-access journals), and more new publications, including something like a tripling of contributions through the ESA. God forbid a need to examine a cv of a mature scientist. Each has more books to his or her name than Darwin and more journal articles than Einstein. It must have to do with more, way more, geniuses being around these days. It causes me to pity the lot of future historians of science. (Will there still be letters on paper from our time for them to peruse? Can scientists' emails in "the cloud" be accessed as easily as Hillary's on her personal computer? And will digital search techniques through digital archives really improve things for those future historians?)

Publication outlets available now and acceptance rates, in particular, of open-access online journals, something new under the sun, do need a bit of looking at here. Molecular biologist John Bohannon reported in *Science* on a sting in which he had submitted a paper under a ludicrously fictitious name (Ocorrafoo Cobange) from a non-existing research institute that "[a]ny reviewer with more than a high-school knowledge of chemistry and the ability to understand a basic data plot should have spotted" as a travesty, with experiments "so hopelessly flawed that the results are meaningless," to test even some of the most influential of commercial publishers. (An echo, perhaps, from the "science wars," in which physicist Alan Sokal had a nonsense paper published in the post-modernist journal, *Social Texts*, in 1996.) Bohannon's fake paper received 157 acceptances out of 255 submissions. A few years later, in 2017, Gina Kolata, a well-respected science writer, wrote stories in the *New York Times* about "predatory journals" that published just about anything for a fee of a few hundred dollars that could be used to pad a cv. Editorships at such journals were also available for cv padding, and she also reported on what she called a "Potemkin village" of a conference at which, "for a hefty fee, you can be listed as a presenter—whether you actually attend the meeting or not." Only in the digital era could that have happened!

Ecology could not be expected to have been excluded, although concern over such things seems limited at present. Beall's List of questionable journals (currently suspended without explanation, but you can still find some older versions of it by googling) had two journals with ecology in their titles. (Six with evolution seemed to have nothing to do with the science. Twenty-nine had environment or environmental in their titles, one of which was the *European Environmental Sciences and Ecology Journal*. A later list has only one with ecology in its title, the *International Academy of Ecology and Environmental Sciences*.)

Afterward

There are also now library issues that make research, especially independent library research, much more comfortable, but nowhere near as much fun. You need no longer travel to university libraries or out-of-the-way repositories of materials (the fun part, sometimes), for you can accomplish more sitting at home at your computer (the comfort part). It is all—or very much of it—on line. Prices, though, keep going up and up for legitimate commercially-published journals. (The illegitimate ones mostly cost nothing, even if they are actually made available. They make their income from fees levied on authors.) *Nature* has printed a special issue on all the associated problems with the proliferation of journals, both real and fake, in their March 28, 2013, issue. You can google it!

However, as the digital storage of titles is making those lovely movable library stacks obsolete, they are also making my trips to nearby university libraries unnecessary. In my opinion, that represents the digital revolution's greatest harm. The stacks, the helpful librarians, the eager undergrads, the view of mountains from the reading room, all were more pleasant than being at a computer in my home office and are missed.

Getting back to that main stream (which many historians who have looked at ecology claim not to exist, although my admitting that would be tantamount to admitting that ecology is but a point of view, rather than a science), I have found in trying to update this volume, that the breadth of my little science to now be so wide and fragmented, the literature so dense, the new jargon so pervasive, that it presents me with a Sysiphean task that could take the rest of my life to accomplish, only to then require another lifetime to update again. So, where I could follow a thread, I incorporated what might be needed of it within the original chapters. The rest, I can not do much more than sketch here, and even then not exhaustively.

Besides evo-devo (evolution and development, working in both directions, I assume), which has to be undoubtedly fragmented by the diversity of organs at issue alone, there is now eco-evo and maybe evo-eco (based on the recognition, from molecular studies, that evolution can be very rapid, on the scale of only a few generations, and interact with the ecology through which an organism evolves) also fragmented, no doubt, by the diversity of organisms and the communities in which they are found. Before abandoning that latter area, once just plain evolutionary ecology, to younger minds and the future, I cannot resist drawing attention to the following comment by Dolph Schluter, whose experimental work on ecological speciation in sticklebacks is now in all the textbooks. He feels that Darwin "would most likely be staggered by the discoveries of genes and molecular evolution" and "chuffed by mounting evidence for the role of natural selection ... in the origin of species." That is perhaps enough said about that, which, in retrospect, is the sort of thing I had once dreamed of doing while in the Andes, "chuffing" Darwin.

    Ecosystems (or communities), the objects of "big ecology," meanwhile, are no longer analyzed mathematically only in search of generalities, as had once been recommended to me by Robert May before he became Sir Robert, but in highly individualized and specific ways, both for management and for basic understanding. Richard Levins's rules on modeling seem to be violated everywhere. Precision and realism seem to be favored over generality. Faster, bigger, better computing devices have much to do with that. The new models deal with patch structure and other nonlinearities (what Sir Robert warned us could lead to chaotic behavior) by adding more and more terms to the same old logistic-type equations to test against the few instances (some now generated with more sophisticated versions of the much-earlier, mites-on-oranges types of set-ups) of real-world data. The whole thing reminds me of being warned not to use too many regression variables back when early computers first allowed sophisticated multi-variate regression analysis. As the argument then went, adding more variables until their number reaches that of the data points will always result in a perfect fit, but with little other meaning to it than what connecting the points with connect-the-dot lines makes (sometimes, a picture, it is true, but not about about the phenomenon being analyzed). It also reminded me of the explanation that I ended a previous chapter with about farmers and their cats. Interestingly, variances in the real and model data still seems to have more significance now to modelers than means.

There are also areas in ecology that are not so much new, but are the subject of new emphasis. New knowledge, gained with new genetic techniques, about soil communities and their interaction with above-ground communities, "the 'brown food web' with the 'green food web,'" as Dan Simberloff has described it to me, is resulting in "much more leading edge research ... at the ecosystem level." It is still entirely the exact kind of research that once drove me from engineering to ecology and then, painfully, away from ecology.

Those new techniques have led to exponential growth in the field of "invasion ecology" (also called "invasion biology" by some and "invasion science" by others, probably in deference to all those techniques borrowed from molecular genetics) and, probably the exponential growth of authors on single papers. Ecology is definitely now a "big science," with big budgets for big studies requiring many researchers. Not as many as the 5,154 authors who collaborated on a recent paper in physics, but substantial groups. They mostly dredge large data bases for patterns (NCEAS and NIMBioS are examples of such), or they result from collaboration by scientists who study disparate aspects of the same subject using different techniques. They also come out of meetings of various working groups that now seem to require a publication to a result. Many such collaborations in ecology have to do with environmental issues, especially global ones.

"Theoretical" from the 1950s to the 1980s, and to some extent still, in ecology and evolutionary biology meant "mathematical." It made for the kind of theory, MacArthur-style theory, that seemingly advanced more rapidly than experimentation could keep up with, particularly in full-scale field experiments. (Laboratory culture experiments *a la Gause* have kept up with mathematical models better.) Simberloff does not now see "much development of MacArthur-type, elegant math," which is hardly obsolete according to him, and MacArthur's ghost is still very much alive, for example, as Island Biogeography Theory (now shortened to IBT in the conservation literature). For at least one of its proponents,

William F Laurance, it is "one of the most elegant and important theories in contemporary ecology, towering above thousands of lesser ideas and concepts." He nevertheless, now that the *sturm and drang* over SLOSS has become irrelevant, finds that "the species-area curve is a blunt tool in many contexts" and "now seems simplistic to the point of being cartoonish" when it comes to management of nature preserves. Still, theory continues apace, multiplying and expanding through analyses of various special conditions and parameters that I suspect even the people in the field are hardly keeping up with. Network theory, which places those elegant logistic-derived equations of the 1970s and '80s into food webs is "the shiny new kid on the block," according to Simberloff, "that everyone wants to play with."

As for those baby boomers who attended that first Earth Day, they are now old fogies. How did they (we) do over their lifetimes? Well, Earth Day itself has become institutionalized as a local cleanup day. Suggested activities in celebration of the day in Vermont (you can google it), aside from road-side cleanup, include learning about composting and recycling, spending time outdoors, fly fishing, hiking, and watching nature, mending, rewiring, and recycling, watching out for what you flush down the toilet, and inflating your tires, among other things, that together amply demonstrate it to have become a day of the environment, as it had in actuality been on the first Earth Day, rather than of ecology, although the word has stuck to those activities like flies once did to the flypaper in the Comstock Hall Fly Lab. Global warming (or climate change, depending on one's politics, apparently) has pushed out ecology (as the study of natural selection in the real world) and possibly replaced it as a part of "the ecology," whatever that term now means. Population growth, once the acknowledged root of all environmental problems, seems also to have faded from concern. (Is there still a chapter of Zero Population Growth extant? Has the movement been driven underground out of political correctness concerns? Can you google that?)

Afterward

Ecosystems and endangered species are about all that is left of real ecology that might come to mind to the "man" on the street, an improvement, perhaps, though, from the days when it simply meant "natural history" to the few who actually knew the term. (The woman on the street, might well have more coming to mind now about it, as will be noted later.) All the new ideas of the Sixties seem to have aged into insignificance for baby boomers in their Mercedes or Humvees—or private jets—whose lives are now spent in heat-generating agedness on fragile waterfront real estate—or in the White House—in which they both aggravate and are threatened by climate change.

We can hope that the younger generations will do things better than we did, but we know it will only be in their own way. After all, they have so much on their plates now.

Genetics in the public arena is now little more than a handmaiden of genealogy and gene therapy, rather than a foundation of evolutionary biology. GMOs (and plastic grocery bags) seem to evoke greater concerns than endangered or disappearing species. (There is a paradox here in the US and in Europe, in which environmental activists follow science blindly on global warming and just as blindly oppose it over genetically engineered organisms.) City dwellers battle global warming by trading inequalities in combustion for composting inequalities (the rich have servants). The microbiomes of our guts and skin may now be more of concern than the ecosystems being harvested in the seas and the forests on fire in the Amazon. Consider the current confusion among those in the news media over Amazon forests being "sinks" or "sponges" for global carbon. (Should I repeat here that the threat of global cooling was expounded by a reputable scientist at the first Earth Day?)

But I promised to follow the threads from then to now. Rachel Carson is one. Carson's reputation has grown in the years since this book was first published. She has been honored in various ways: 100 most important people? Of 20th century? Women? Female Ecologist? *Time* magazine devoted a cover to the most influential scientists of the 20th century that posed Rachel Carson as a bust looking down on

Freud apparently analyzing Einstein. Jean Langenheim listed her as an "Early Pioneer" woman ecologist, albeit without training, in her seminal work on notable women doing research in ecology, while the ESA itself has adopted Carson as an ecologist. (Carson has also been vilified as the murderess of millions of children from malaria. You can google it.) She has now become a legend and is defended as such.

    The only thing new in a recent biography was the conclusion that Carson had learned her ecology from Charles Elton, without ever establishing that she even knew who Elton was as an ecologist. It was based only on the mention of one book by Elton (a good one, *The Ecology of Invasions by Animals and Plants*) that was not his classic text, but an elaboration on talks he gave on the radio, although it can nonetheless be said to have launched the science of invasion ecology.

    Another Carson biography, an anniversary edition, made no mention of what I learned at Aton Forest about Frank Egler and Carson, other than making the strong conclusion that she had never received any help from him. Please read my book, *Magnificent Failure: Frank Egler and the Greening of American Ecology*, sir, read this book, even—or was that passage specifically written in answer to it, anonymously, so to speak?

    Further following that thread, a paper of mine, which is (or had been for several years) used by the ESA as one of three documents recommended on the history of ecology, was turned down three different times by two different journals, with all reviews reading essentially that it was an important document that needed more work. It was never to be enough work, I soon realized. All the reviewers but for one, whom I assume was Linda Lear, who had later volunteered to me that she had reviewed the document, favorably, she thought, had indicated to me by their reviews that they did not want another person on their own, personal, Rachel Carson turf.

    In addition, a famous historian of ecology who is tediously producing an encyclopedic history of ecology from almost prehistoric times to, possibly, the future, cited as the only thing he

Afterward

learned from my research being that Johns Hopkins had three professors who were early members of the ESA. True, but my conclusion was that there was no demonstrable evidence that any of those three had been in any way been involved with Rachel Carson in any way having to do specifically with ecology.

Rachel Carson is perhaps of MOST importance, precisely because she was NOT trained in ecology, as she has admitted (under oath, actually). The legend must be served, though. After all, bridges, schools, hiking trails, nature preserves and areas, even a college, have been named after her.

So, moving along, are ecology majors still lacking in math skills (as in 2002)? I don't know, but I am not hearing complaints.

Stanley Auerbach's emergence from "a culture that did not send students into the field of ecology," may no longer be true. There is now a Black Ecologists Section (of over 30, including Environmental Justice and Inclusive Ecology) in the ESA, but whether that represents an abundant registration or a lack thereof, I did not follow up, but what IS obvious is that women, of which Auerbach made no mention, are now very well represented in ecology, holding positions in the ESA from top to bottom and winning major awards. Their trajectory in ecological research might best be realized from Figure 1 in Bronstein and Bolnick's report on women authors in *The American Naturalist*, which is given further support by Jean Langenheim's aforementioned study.

Among other trends that I can scarcely do more than mention here is the resurrection of Lamackian evolution, at least for certain (very important) molecules at the sub-cellular level. That's new. Also, this is an appropriate place to remind that my description of genetics in the relevant chapters is highly simplified in comparison to the flood of new knowledge coming out, particularly through cancer studies. However, my simplifications have made nothing I have said inaccurate.

"Science Wars?" Deconstructionism? The entire incident having to do with Bjørn Lomborg looks much different from the perspective of a world in which Donald Trump can become President of the United States. Lomborg seemed to have found a similar path to Trump's, but his only raised him to the head of a foundation whose sole purpose seems to be in providing him with a very healthy means of subsistence.

The ESA is still certifying "Professional Ecologists" and still trying to figure out how to do so effectively. Null models are still being philosophized over and tweaked in a manner that is in many ways analogous to my descriptions of computer math models a few pages back. The argument over holism versus reductionism is still around, but has seemingly morphed into one over "top down" versus "bottom up."

The mechanisms underlying species diversity are still under examination. "Multiple reviews have attested to the increase in diversity of species and higher taxa from poles to tropics and cataloged the many hypotheses on underlying mechanisms," according to one rather smallish group (4) of authors, who then cite (with the ellipsis that follows) a current conclusion (2014) by a single representative (Jim Brown) from the era that is the subject of this book that "even as the patterns have become clearer ... the explanations have remained elusive and controversial."

Much of what was of interest in detail then in ecology remains unanswered, or has become old hat or orthodoxy, or has been renamed, or simply no longer remains interesting except to historians. As has been amply demonstrated above, there appear to be better fish to fry for the current generation of researchers, not to mention that people retire or die, taking their ideas with them. Still, this being mainly a personal history, rather than a primer on the subject, some of those ideas remain in my chapters, if for no other reason than a hope that they may draw attention to themselves as lessons in what important portions of the science of ecology have gone through for it to become what it is.

Mathematicians are still tinkering with the logistic equation, community ecologists are still studying communities, evolutionary ecologists are still struggling with levels of selection, while some mathematical types are grappling with how patch structure can be fit into useful theories and molecularly oriented types are seeing all manner of possibilities in new modes of genetic analysis. Meanwhile, applied ecologists are still trying not to misapply what might be applicable.

Hutchinson's "Paradox of the Plankton" has accumulated "more than 2,000 citations" and "continues to accumulate more than 50 citations a year." EO Wilson "is at it again … sending shockwaves throughout the biological sciences," by disavowing kin selection and has called Stephen J Gould a "charlatan." (You can google it!) Meanwhile, Michael Ruse has written a fine book about Gaia, Jared Diamond is still writing controversial best sellers, Marlboro College is once more struggling to survive, applied ecology is no longer the biggest section in the ESA (aquatic ecology is), species are going extinct faster than we can discover them, and firemen are still saving cats. What else could be new?

Dr Kingsland also once reminded me that ascertaining who was right in the end is not the goal of a good history. It is not who won or lost a contest that matters, but how they played, something good to keep in mind.

William Dritschilo
September 22, 2019

Published Sources of Quoted Material

Philadelphia, April 22, 1970, Earth Day

Julian L Simon, "Earth Day: Spiritually Uplifting, Intellectually Debased", (May 1, 1995) Online: http://www.inform.umd.edu/EdRes/Colleges/BMGT/.Faculty/jSimon/Articles/EARTHA5.txt (2/4/02).

Ken Shulman, "The Gospel According to Ian McHarg," *Metropolis* (August 2000) Online: http://www.metropolismag.com/html/content_0800/hrg.htm.

Loren Eiseley, *Darwin's Century* (Doubleday & Company, Garden City, New York, 1958).

The Art of Earth Day

Stewart L Udall, "A Message for Biologists," *BioScience* **14:11** (1964):17-8.

Paul R Ehrlich, *The Population Bomb* (Sierra Club/Ballantine, New York, 1971).

Barry Commoner, *The Closing Circle* (Alfred A. Knopf, New York, 1971).

"Fighting to Save the Earth from Man," *Time* (February 2, 1970).

"Perhaps I Have Already Said Too Much"

Loren Eiseley, *Saturday Review* (September 29, 1962).

Robert L Rudd, *Pacific Discovery* (Nov-Dec, 1962).

Richard S Miller, "Summary Report of the Ecological Study Committee with Recommendations for the Future of Ecology and the Ecological Society of America," *Bulletin of the Ecological Society of America* **46** (1965):61-82.

Linda Lear, *Rachel Carson: Witness for Nature* (Henry Holt and Company, New York, 1997).

The Rachel Carson Student File, Ferdinand Hamburger, Jr., Archives, Johns Hopkins University, Baltimore, Maryland.

Linda Lear, (ed) *Lost Woods: The Discovered Writing of Rachel Carson* (Beacon Press, Boston, 1998).

Published Sources of Quoted Material

Rachel Carson Papers, Beineke Rare Book and Manuscript Library, Yale University, New Haven, Connecticut.

George Johnson, "You Know That Space-Time Thing? Never Mind," *The New York Times Book Review* (June 9, 2002).

Rachel Carson, *The Development of the Pronephros During Embryonic and Early Larval Life of the Catfish (Ictalurus punctatus.)* Master's Dissertation (John Hopkins University, Baltimore, Maryland, 1932).

Great *AEPPS*

WC Allee, AE Emerson, Orlando Park, Thomas Park, and Karl P Schmidt, *Principles of Animal Ecology* (WB Saunders Company, Philadelphia, 1949).

Robert P McIntosh, "Ecology Since 1900," Pages 353-72 in Benjamin J Taylor and Thurman J White, editors, *Issues and Ideas in America* (University of Oklahoma Press, Norman, 1976).

Jane Maienschein, *100 Years Exploring Life, 1888-1988, The Marine Biological Laboratory at Woods Hole* (Jones and Bartlett, Boston, 1989).

Rachel Carson, *The Sea Around Us, Revised Edition* (Oxford University Press, New York, 1961).

Rachel Carson, *Silent Spring* (Houghton Mifflin Company, Boston, 1962).

Rachel Carson Papers.

Frank E Egler, "ESA Needs a Code of Ethics and a Certification Program," *Bulletin of the Ecological Society of America* **53** (1972):2-4.

An Ecologist Reconsiders *Silent Spring* after 17 Years

LaMont C Cole, "Rachel Carson's Indictment of the Wide Use of Pesticides," *Scientific American* (December 1962):173-80.

Egler, "ESA Needs a Code."

Allee, et al, *Principles of Animal Ecology*.

Paul R Ehrlich, "Paul R Ehrlich Reconsiders *Silent Spring*," *Bulletin of the Atomic Scientists* (October 1979):34-6.

Carson, *Silent Spring*.

Autecologists, Synecologists, and Genecologists

Virginia H Dale, Gary W Barrett, Alan T Carpenter, C Ross Hinkle, William J Mitsch, and Louis F Pitelka, "ESA's Professional Certification Program: Let's Make It Work," *Bulletin of the Ecological Society of America* **81** (2000):255-7.

Frank E Egler, "Philip Wylie—Ecologist," *Ecology* **50** (1970):160.

Frank E Egler, "Instant Ecology, in Academia," *Ecology* **55** (1974):691.

William F Cooper, "The fundamentals of vegetational change," *Ecology* **7** (1926):391-413.

Eugene P Odum, *Fundamentals of Ecology, Third Edition* (WB Saunders Company, Philadelphia, 1971).

Robert E Ricklefs, *Ecology* (Chiron Press, Newton, Massachusetts, 1973).

Eiseley, *Darwin's Century*.

Nicholas V Riasanovsky, *A History of Russia* (Oxford University Press, New York, 1963).

Albert Bennett, "What is Physiological Ecology? Definitions and Opinions," *Bulletin of the Ecological Society of America* **63** (1982):341.

Boyd R Strain, Ibid, p 347.

Alston Chase, *In a Dark Wood: The Fight Over Forests and the Rising Tyranny of Ecology* (Houghton Mifflin Company, New York, 1995).

A Tale of Two Brothers and an Uncle

Robert P McIntosh, "The Background and Some Current Problems of Theoretical Ecology," *Synthese* **43** (1980):195-255.

Published Sources of Quoted Material

"Eugene Odum: An Ecologist's Life," Online: http://www.gactr.uga/gcq/gcqspr97/odum.html (2/5/02).

Betty Jean Craige, *Eugene Odum, Ecosystem Ecologist & Environmentalist* (University of Georgia Press, Athens, Georgia, 2001).

Odum, *Fundamentals of Ecology*.

Joel B Hagen, *An Entangled Bank, The Origins of Ecosystem Ecology* (Rutgers University Press, New Brunswick, New Jersey, 1992).

Frederic E Clements, "Nature and Structure of the Climax," *Journal of Ecology* **24** (1936):252-84.

Raymond L Lindeman, "The Trophic-Dynamic Aspect of Ecology," *Ecology* **23** (1942):399-418.

Edward O Wilson, *Naturalist* (Warner Books, New York, 1995).

G Evelyn Hutchinson, *The Kindly Fruits of the Earth* (Yale University Press, New Haven, Connecticut, 1979).

Robert Edward Cook, "Raymond Lindeman and the Trophic-Dynamic Concept in Ecology," *Science* **198** (1977):22-6.

Butterfly Collectors

"Eminent Ecologist for 1962," *Bulletin of the Ecological Society of America* **43** (1962):113.

Stephen Bocking, *Ecologists and Environmental Politics: A History of Contemporary Ecology* (Yale University Press, New Haven, Connecticut, 1997).

McIntosh, "The Background and Some Current Problems."

Charles S Elton, *Animal Ecology* (Sigewick and Jackson, London, 1927).

Robert K Swihart and Peter M Waser, "Gray Matters in Ecology: Dynamics of Pattern, Process, and Scientific Progress," *Bulletin of the Ecological Society of America* **83** (2002):149-55.

Chas B Davenport, "Zoology of the Twentieth Century," *Science* **14** (1901):317.

Thomas S Kuhn, *The Structure of Scientific Revolutions* (University of Chicago Press, Chicago, 1962).

Paul R Needham, "Report of the Committee on Applied Ecology," *Bulletin of the Ecological Society of America* **38** (1957):14-19.

Rexford F Daubenmire, Ray F Smith, and Paul R Needham, "Report of the Committee on Applied Ecology," *Bulletin of the Ecological Society of America* **39** (1958):18-26.

Gene Likens, "Pesticide Pollution of Freshwater Ecosystems," *Bulletin of the Ecological Society of America* **45** (1964):157-9.

Jack Major, "Reports of Committees: Applied Ecology," *Bulletin of the Ecological Society of America* **45** (1964):13-17.

Bulletin of the Ecological Society of America **44** (1963):33.

Udall, "A Message for Biologists."

Robert B Platt and John N Wolfe, "Introduction," *BioScience* **14:7** (1964):9.

Pierre Dansereau, "The Future of Ecology," *BioScience* **14:7** (1964):20-3.

Frank Blair, "The Case for Ecology," *BioScience* **14:7** (1964):17-9.

Jon Ghiselin, "Why is Ecology so near the Lunatic Fringe?" *Bulletin of the Ecological Society of America* **53:3** (1972):13-4.

Stanley I Auerbach, "Ecology, Ecologists and the ESA," *Ecology* **53** (1972):205-7.

James W Curlin, "Courts, Ecology, and Environmental Planning," *Ecology* **53** (1972):373.

Robert A Croker, *Pioneer Ecologist, The Life and Work of Victor Ernest Shelford, 1877-1968* (Smithsonian Institution Press, Washington, DC, 1991).

Michael G Barbour, "Ecologist and Consultant: Do We Have to be Both?" *Ecology* **54** (1973):2.

Published Sources of Quoted Material

Richard Levins, "Ecological Engineering," *Quarterly Review of Biology* **43** (1968) 301-5.

Egler, "Instant Ecology, in Academia."

G Evelyn Hutchinson, "Variations on a theme by Robert MacArthur," Pages 492-521 in Martin L Cody and Jared M Diamond, Editors, *Ecology and Evolution of Communities* (Harvard University Press, Cambridge, Massachusetts, 1975).

The Curious Case of the Superorganism

Odum, *Fundamentals*.

Chase, *In a Dark Wood*.

Eugene Odum, "The Emergence of Ecology as a New Integrative Discipline," *Science* **195** (1977):1289-92.

George Sessions, Editor, *Deep Ecology for the 21st Century* (Shambhala, Boston, Massachusetts, 1995).

Hagen, *Entangled Bank*.

Eugene P Odum, "The Strategy of Ecosystem Development," *Science* **164** (1969):262-70.

Henry Chandler Cowles, "The Work of the Year 1903 in Ecology," *Science* **19** (1904) p. 880.

Bernard C Patten and Eugene P Odum, "The Cybernetic Nature of Ecosystems," *The American Naturalist* **188** (1981):886-95.

Arthur G Tansley, "The Use and Abuse of Vegetational Concepts and Terms," *Ecology* **16** (1935):284-307.

LaMont C Cole, "Sketches of General and Comparative Demography," *Cold Spring Harbor Symposia on Quantitative Biology* **22** (1957):1-15.

MJ Dunbar, "The Evolution of Stability in Marine Environments; Natural Selection at the Level of the Ecosystem," *American Naturalist* **79** (1960):48-67.

Save the Cat!

Charles Darwin, *On the Origin of Species* (John Murray, London, 1859).

Odum, *Fundamentals*.

Allee, et al, *Principles of Animal Ecology,* p. 314.

Raymond Pearl and Sophia A Gould, "World Population Growth." *Human Biology* **8** (1936):399-419. p 313. (As given in above.)

Dennis Chitty, "Population Processes in the Vole and Their Relevance to General Theory," *Canadian Journal of Zoology* **38** (1960):99-113.

Edward O Wilson, *Sociobiology* (Harvard University Press, Cambridge, Massachusetts, 1975).

George C Williams, *Adaptation and Natural Selection: a Critique of Come Current Evolutionary Thought* (Princeton University Press, 1966).

Richard C Lewontin, "The Units of Selection," *Annual Review of Ecology and Systematics* **1** (1970):1-18.

Buried with Full Honors

Allee, et al, *Principles of Animal Ecology*.

David Sloan Wilson, "The Group Selection Controversy: History and Current Status," *Annual Review of Ecology and Systematics* **14** (1983):159-87.

Bernard C Patten, "Environs: Relativistic Elementary Particles for Ecology," *The American Naturalist* **119** (1982):179-219.

Patten and Odum, "The Cybernetic Nature of Ecosystems."

Robert V O'Neill, "Is It Time to Bury the Ecosystem Concept? (With Full Military Honors, Of Course!)" *Ecology* **82** (2001):3275-84.

Carla Cole, "Microbial Microcosm," *In Context* (Winter 1993):18. Online: http://www.context.org/ICLIB/IC34/Margulis.htm.

Daniel McMahon, "The Colonization Hypothesis," *Science* **172** (1971):676.

Published Sources of Quoted Material

W Ford Doolittle, "The Endosymbiont Hypothesis," *Science* **213** (1981):640.

JE Lovelock, *Gaia: A New Look at Life on Earth* (Oxford University Press, Oxford, 1979).

Suzanne Spencer, "Just a Good Guess," *Valley News* (2003):C1.

Richard A Kerr, "No Longer Willful, Gaia Becomes Respectable," *Science* **240** (1988):393-5.

Peter M Vitousek, Paul R Ehrlich, Anne H Ehrlich, and PA Matson, "Human Appropriation of the Products of Photosynthesis," *BioScience* **36** (1986):368-73.

A Mathematician Who Knew His Warblers

G Evelyn Hutchinson, "Homage to Santa Rosalia or Why Are There So Many Kinds of Animals?" *The American Naturalist* **93** (1959):145-59.

Stuart L Pimm, *The World According to Pimm* (McGraw-Hill, New York, 2001).

Edward O Wilson, *The Diversity of Life* (Harvard University Press, Cambridge, Massachusetts, 1992).

Leslie A Real and James H Brown, Editors, *Foundations of Ecology, Classic Papers with Commentaries* (University of Chicago Press, 1991).

Paul LeBlanc, "When a Log is Not Enough," *The Marlboro Record* (Fall 2002).

Leslie Plank, *A Chance to Make Something of Yourself, Marlboro College: A History of Its First Ten Years* (Plan of Concentration, Marlboro College, Marlboro, Vermont, Spring 2002).

Thomas B Ragle, *Marlboro College: A Memoir* (Marlboro College, Marlboro, Vermont, 1999).

Robert H MacArthur, *Mathematical Foundations of Boundary Problems* (Plan of Concentration, Marlboro College, Marlboro, Vermont, 1951).

Stephen D Fretwell, "The Impact of Robert MacArthur on Ecology," *Annual Review of Ecology and Systematics* **6** (1975):1-15.

Daniel I Bolnick, "Letter from the Editor," *The American Naturalist* **191** (2018):iii-v.

Robert K Peet, "The Legacy of MacArthur," *Ecology* **57** (1976):613-4.

James H Brown, "Two Decades of Homage to Santa Rosalia: Toward a General Theory of Diversity," *American Zoologist* **21** (1981):877-88.

Wilson, *Naturalist*.

David Quammen, *The Song of the Dodo: Island Biogeography in an Age of Extinction* (Simon & Schuster, New York, 1996).

The Climax of the Ecosystem

Pimm, *The World According to Pimm*.

Darwin, *On the Origin of Species*.

Do Cats Eat Clover?

Pimm, *The World According to Pimm*.

Lynn White, "The Historical Roots of Our Ecological Crisis," *Science* **155** (1967):1203-7.

Odum, "The Strategy of Ecosystem Development."

Michael E Zimmerman, *Contesting Earth's Future, Radical Ecology and Postmodernity* (University of California Press, Berkeley, 1994).

Carson, *Silent Spring*.

Sessions, *Deep Ecology*.

David Takacs, *The Idea of Biodiversity: Philosophies of Paradise* (Johns Hopkins University Press, Baltimore, Maryland, 1996).

Wilson, *Naturalist*.

Edward O Wilson, *Biophilia* (Harvard University Press, Cambridge, Massachusetts, 1984).

Robert H MacArthur, "Fluctuations of Animal Populations, and a Measure of Stability," *Ecology* **36** (1955):533-6.

Diversity-stability and Other Scams

Ricklefs, *Ecology*.

Robert M May, *Stability and Complexity in Model Ecosystems, Second Edition* (Princeton University Press, Princeton, New Jersey, 1974).

Chaos Comes to Ecology

Robert H MacArthur, *Geographical Ecology* (Princeton University Press, Princeton, New Jersey, 1972).

Robert MacArthur, "Population Effects of Natural Selection," *The American Naturalist* **95** (1961):195-9.

Robert H MacArthur, "On the Relative Abundance Species," *The American Naturalist* **94** (1960):25-36.

Robert H MacArthur and Edward O Wilson, *The Theory of Island Biogeography* (Princeton University Press, Princeton, New Jersey, 1967).

Alfred J Lotka, *Elements of Physical Biology* (Williams & Wilkins, Baltimore, 1925).

May, *Stability and Complexity in Model Ecosystems*.

LB Slobodkin, "On the Present Incompleteness of Mathematical Ecology," *American Scientist* **53** (1965):347-57.

Robert M May, "Biological Populations with Non-Overlapping Generations: Stable Points, Stable Cycles, and Chaos," *Science* **186** (1974):645-7.

Henry S Horn, "Succession," Pages 253-71 in *Theoretical Ecology, Principles and Applications, Second Edition*, Robert M. May, Editor (Sinauer Associates, Inc, Sunderland, Massachusetts, 1981).

Richard Levins, *Evolution in Changing Environments* (Princeton University Press, Princeton, New Jersey, 1968).

Nutches in Nitches, Atoms in Apples

Theodor Seuss Geisel, *On Beyond the Zebra* (Random House, New York, 1955).

Jared Diamond, "Overview: Laboratory Experiments, Field Experiments, and Natural Experiments," Pages 3-22 in Jared Diamond and Ted J Case, Editors, *Community Ecology* (Harper and Row, New York, 1986).

John Cottingham, "Introduction," *Descartes: Selected Philosophical Writings*, Translated by John Cottingham, Robert Stoothoff, and Dugald Murdoch (Cambridge University Press, Cambridge, 1988).

Bertrand Russell, *A History of Western Philosophy* (Simon and Schuster, New York, 1945).

Justin Kaplan, General Editor, John Bartlett, *Familiar Quotations, Sixteenth Edition* (Little, Brown and Company, Boston, 1992).

Alfred North Whitehead, *Adventures of Ideas* (Macmillan, London, 1933).

Karl R Popper, *Objective Knowledge, An Evolutionary Approach, Revised Edition* (Clarendon Press, Oxford, 1979).

William W Murdoch, "'Community Structure, Population Control, and Competition'—A Critique," *The American Naturalist* **100** (1966):219-26.

"Robert Henry Peters," Piero Guilizoni, P. Guilizzoni and F. Oldfield, Guest Editors, *Palaeoenvironmental Analysis of Italian Crater Lake and Adriatic Sediements, Mem. Ist. Ital. Idrobiol., 55: 1-4, 1996*. (Online: http://www.iii.to.cnr.it/pubblicaz/mem55/mem55_01.pdf.)

Gina Bari Kolata, "Theoretical Ecology: Beginnings of a Predictive Science," *Science* **183** (1974):400-1&450-1.

Leigh Van Valen and Frank Pitelka, "Commentary: Intellectual Censorship in Ecology," *Ecology* **55** (1974):925-6.

A Doctor for a Sick Science

Amyan MacFadyen, "Some Thoughts on the Behaviour of Ecologists," *Journal of Animal Ecology* **44** (1975):351-63.

Published Sources of Quoted Material

Roger Lewin, "Santa Rosalia Was a Goat," *Science* **221** (1983):636-9.

Charles AS Hall and DL DeAngelis, "Models in Ecology: Paradigms Found or Paradigms Lost?" *Bulletin of the Ecological Society of America* **66** (1985):339-46.

Glen W Suter, "Ecosystem Theory and NEPA Assessment," *Bulletin of the Ecological Society of America* **62** (1981):186-92.

Jonathan Weiner, *The Beak of the Finch* (Alfred A. Knopf, Inc, New York, 1994).

Quammen, *The Song of the Dodo*.

George W Salt, "Roles: Their Limits and Responsibilities in Ecological and Evolutionary Research," *The American Naturalist* **122** (1983):697-705.

Hal Caswell, "Community Structure: a Neutral Model Analysis," *Ecological Monographs* **46** (1976):327-354.

Garrett Hardin, "The Competitive Exclusion Principle," *Science* **131** (1960):1292-7.

TFH Allen, "The Noble Art of Philosophical Ecology," *Ecology* **62** (1981):870-1.

Marjorie Grene, "A Note on Simberloff's 'Succession of Paradigms in Ecology,'" *Synthese* **43** (1980):41-5.

Daniel Simberloff, "Reply," *Synthese* **43** (1980):79-93.

Richard Levins and Richard Lewontin, "Dialectics and Reductionism in Ecology," *Synthese* **43** (1980):47-78.

Richard Levins and Richard Lewontin, *The Dialectic Biologist* (Harvard University Press, Cambridge, 1985).

Russell, *History of Western Philosophy*.

Wilson, *Naturalist*.

It Wasn't a Barroom Brawl. It Was an Athletic Contest

Lewin, "Santa Rosalia Was a Goat."

Jonathan Roughgarden, "Competition and Theory in Community Ecology," *The American Naturalist* **122** (1983):583-601.

Diamond, "Overview."

James H Brown, Diane W Davidson, James C Munger, and Richard S Inouye, "Experimental Community Ecology: The Desert Granivore System," Pages 41-61 in Diamond and Case, *Community Ecology*.

May, *Stability and Complexity in Model Ecosystems*.

Daniel Simberloff, "A Succession of Paradigms in Ecology: Essentialism to Materialism and Probabilism," *Synthese* **43** (1980):3-39.

Robert H MacArthur, "On the Relative Abundance of Bird Species," *Proceedings of the National Academy of Sciences, USA* **45** (1957):293-5.

FW Preston, "The Commonness, and Rarity, of Species," *Ecology* **29** (1948):254-83.

Martin L Cody and Jared M Diamond, "Preface" Pages *viii-ix* in Cody and Diamond, *Ecology and Evolution of Communities*.

Jared M Diamond, "Assembly of Species Communities," Pages 342-444 in Cody and Diamond, *Ecology and Evolution of Communities*.

Peet, "The Legacy of MacArthur."

Hutchinson, *The Kindly Fruits of the Earth*.

Edward O Wilson and Edwin O Willis, "Applied Biogeography," Pages 522-34 in Cody and Diamond, *Ecology and Evolution of Communities*.

TRE Southwood, "Continuing in the MacArthur Tradition," *Science* **192** (1976):670-2.

William J Platt, "Evolutionary Concepts in the Study of Ecological systems," *Evolution* **30** (1976):857-8.

William L Brown, Jr, and Edward O Wilson, "Character Displacement," *Systematic Zoology* **5:2** (1956):49-64.

Published Sources of Quoted Material

Edward F Connor and Daniel Simberloff, "Species Number and Compositional Similarity of the Galapágos Flora and Avifauna," *Ecological Monographs* **48** (1977):219-48.

Donald Strong, Jr, Lee Ann Szyzska, and Daniel S Simberloff, "Tests of Community-wide Character Displacement Against Null Hypotheses," *Evolution* **33** (1979):897-913.

PR Grant, and I Abbott, "Interspecific Competition, Island Biogeography and Null Hypotheses," *Evolution* **34** (1980):332-41.

John A Hendrickson, Jr, "Community-Wide Character Displacement Reexamined," *Ecology* **35** (1981):794-810.

Donald R Strong and Daniel S Simberloff, "Straining at Gnats and Swallowing Ratios: Character Displacement," *Evolution* **35** (1981):810-2.

Edward F Conner, and Daniel Simberloff, "The Assembly of Species Communities; Chance or Competition?" *Ecology* **60** (1979):1132-40.

Br'er Rabbit and the Tar Baby

Gordon H Orians, "Ecology Evolving," *Ecology* **66** (1985):639-40.

John A Wiens, "On Understanding a Non-Equilibrium World: Myth and Reality in Community Patterns and Processes," Pages 439-57 in Donald R Strong, Jr, Daniel Simberloff, Lawrence G Abele, and Anne B Thistle, Editors, *Ecological Communities: Conceptual Issues and the Evidence* (Princeton University Press, Princeton, New Jersey, 1984).

Michael E Gilpin and Jared M Diamond, "Are Species Co-occurences on Islands Non-random?" Pages 297-315 in Strong, et al, *Ecological Communities*.

Jared M Diamond and Michael E Gilpin, "Examination of the 'Null' Model of Connor and Simberloff for Species Co-occurrences on Islands," *Oecologia* **52** (1982):64-74

Peter M Kareiva and Robert T Paine, "Richard B Root, President," *Bulletin of the Ecological Society of America* **66** (1985):328.

Michael E Gilpin and Jared Diamond, Edward F Connor and Daniel Simberloff, "Rejoinders," Pages 332-43 in Strong, et al, *Ecological Communities*.

Paul H Harvey, Robert K Colwell, Jonathan W Silverton, and Robert M May, "Null Models in Ecology," *Annual Review of Ecology and Systematics* **14** (1983):189-211.

Robert M May, "An Overview: Real and Apparent Patterns in Community Structure," Pages 3-16 in Strong, et al, *Ecological Communities*.

Life is Simpler on an Island

OH Frankel and Michael E Soulé, *Conservation and Evolution* (Cambridge University Press, 1981).

Stephen Jay Gould, "An Allometric Interpretation of Species-Area Curves: the Meaning of the Coefficient," *The American Naturalist* **114** (1979):335-43.

Joel E Cohen, *How Many People Can the Earth Support?* (W. W. Norton & Company, New York, 1995).

MacArthur and Wilson, *The Theory of Island Biogeography*.

Robert M May, "Patterns of Species Abundance and Diversity," Pages 81-120 in Cody and Diamond, *Ecology and Evolution of Communities*.

George Sugihara, "Minimal Community Structure: an Explanation of Species-Abundance Patterns," *The American Naturalist* **116** (1980):770-87.

LB Slobodkin, "Limits to Biodiversity (Species Packing)," Pages 729-38 in Simon Asher Levin, Editor-in-chief, *Encyclopedia of Diversity, Volume 2* (Academic Press, New York, 2001).

EC Pielou, "The Broken-Stick Model: a Common Misunderstanding," *The American Naturalist* **117** (1981):609-10.

George Oster, "Predicting Populations," *American Zoologist* **21** (1981):831-45.

SLOSS: Islands in the Wilderness

MacArthur and Wilson, *Theory of Island Biogeography*.

Joan Roughgarden, "Guide to Diplomatic Relations with Economists," *Bulletin of the Ecological Society of America* **82** (2001):85-8.

Published Sources of Quoted Material

Jared M Diamond, "The Island Dilemma: Lessons of Modern Biogeographic Studies for the Design of Nature Reserves," *Biological Conservation* **7** (1975):129-46.

Daniel S Simberloff and Lawrence G Abele, "Island Biogeography and Conservation: Strategy and Limitations," *Science* **193** (1976):1032.

Jared M Diamond, "Island Biogeography and Conservation: Strategy and Limitations," *Science* **193** (1976):1027-9.

John Terborgh, "Island Biogeography and Conservation: Strategy and Limitations," *Science* **193** (1976):1029.

Robert F Whitcomb, James F Lynch, Paul A Opler and Chandler S Robbins, "Island Biogeography and Conservation: Strategy and Limitations," *Science* **193** (1976):1031.

Daniel Simberloff and Lawrence G Abele, "Refuge Design and Island Biogeographic Theory: the Effects of Fragmentation," *The American Naturalist* **120** (1982):41-50.

Gerald B Goeden, "Biogeographic Theory as a Management Tool," *Environmental Conservation* **6** (1979):27-32.

Michael E Soulé, "Thresholds for Survival," Pages 151-69 in Michael E. Soulé and Bruce A. Wilcox, Editors, *Conservation Biology* (Sinauer Associates, Inc., Sunderland, Massachusetts, 1980).

Bruce A Wilcox and Dennis D Murphey, "Conservation Strategy: the Effects of Fragmentation on Extinction," *The American Naturalist* **125** (1985):879-87.

Do Owls Live on Islands?

John Tunney, *Congressional Record*, 93rd Congress, 1st Session **119** (24 July 1973):25668.

Endangered Species Act of 1973.

Daniel Simberloff, "The Spotted Owl Fracas: Mixing Academic, Applied and Political Ecology," *Ecology* **68** (1987):766-72.

Chase, *In a Dark Wood*.

Still Counting

Bruce A Wilcox, "Insular Ecology and Conservation," Pages 95-117 in Soulé and Wilcox, *Conservation Biology*.

Bjørn Lomborg, *The Skeptical Environmentalist* (Cambridge University Press, Cambridge, 2001).

Up Against Limits

G. Evelyn Hutchinson, *An Introduction to Population Biology* (Yale University Press, New Haven, 1978).

Marylin Chou, David P Harmon, Jr, Herman Kahn and Sylvan H Wittwer, *World Food Prospects and Agricultural Potential* (Prager, New York, 1977).

Cohen, *How Many People Can the Earth Support?*

Ehrlich, *The Population Bomb*.

Ronald Bailey, "Earth Day, Then and Now," http://reason.com/0005/fe.rb.earth.shtml.

Donella H Meadows, Dennis L Meadows, and Jorgen Randers, *Beyond the Limits* (Chelsea Green Publishing Company, White River Junction, Vermont, 1992).

Donella H Meadows, Dennis L Meadows, Jorgen Randers, and William W Behrens, III, *The Limits to Growth: A Report for the Club of Rome's Project on the Predicament of Mankind* (Universe Books, New York, 1972).

GS in "Beyond the Limits," *Earth Island Journal* **7** (Fall 1992):13.

Brian W Arthur, and Geoffrey McNicoll, "Large-scale Simulation Models in Population and Development: What Use to Planners?" *Population and Development Review* **1** (1975):261.

Potatoes Made of Oil

John Stuart Mill, *Principles of Political Economy With Some of Their Applications to Social Philosophy, Two Volumes* (Routledge and Kegan Paul, London, 1848).

Published Sources of Quoted Material

"Fighting to Save the Earth from Man," *Time* (February 2, 1970).

Howard T Odum, *Environment, Power, and Society* (Wiley-Interscience, New York, 1971).

David Pimentel, "Population Regulation and Genetic Feedback," *Science* **159** (1968):1432-7.

Cutler J Cleveland, Robert Costanza, Charles AS Hall, and Robert Kaufman, "Energy and the U. S. Economy: A Biophysical Perspective," *Science* **225** (1984):890-7.

Julian L Simon, "Resources, Population, Environment: An Oversupply of False Bad News," *Science* **208** (1980):1431-7.

Robert M Solow, "The Economics of Resources or the Resources of Economics," *American Economics Review* **64** (1974):1-14.

Michael Grubb, "Relying on Manna from Heaven?" *Science* **294** (2001):1285&6.

Lomborg, *The Skeptical Environmentalist*.

Lone Frank, "Greens See Red Over Revisionist's New Job," *Science* **295** (2002):1817.

David Pimentel, "Exposition on Skepticism," *BioScience* **52** (2002):295-8.

What Are Nature's Services Worth?

George P. Marsh, *Address Delivered Before the Agricultural Society of Rutland County* (Rutland, Vermont, September 30, 1847).

Clean Water Act of 1972.

Walter E Westman and RM Gifford, "Environmental Impact: Controlling the Overall Level," *Science* **181** (1973):819-25.

Walter E Westman, "How Much are Nature's Services Worth?" *Science* **197** (1977):960-4.

Colin W Clark, "The Economics of Overexploitation," *Science* **181** (1973):630-4.

Gretchen C Daily, Susan Alexander, Paul R Ehrlich, Larry Goulder, Jane Lubchenco, Pamela A Matson, Harold A Mooney, Sardra Postel, Steven H Schneider, David Tilman, and George M Woodwell, "Ecosystem Services: Benefits Supplied to Human Societies by Natural Ecosystems," *Issues in Ecology Number 2* (Ecological Society of America 1997).

Robert Costanza, Ralph d'Arge, Rudolf de Groot, Stephen Farber, Monica Grasso, Bruce Hannon, Karin Limburg, Shahid Naeem, Robert V O'Neill, Jose Paruela, Robert Raskin, Paul Sutton, and Masrjan van den Belt, "The Value of the World's Ecosystem Services and Natural Capital," *Nature* **387** (1997):253-60.

Wade Roush, "Putting a Price Tag on Nature's Bounty," *Science* **276** (1997):1029.

Ecological Engineers and Environmental Doctors

LB Slobodkin, "Intellectual Problems of Applied Ecology," *BioScience* **38** (1988):337-43.

TRE Southwood, *Ecological Methods, Second Edition* (John Wiley and Sons, New York, 1978).

Southwood, "Continuing in the MacArthur Tradition."

Roger Revelle and Hans Suess, "Carbon Dioxide Exchange Between the Atmosphere and Ocean, and the Question of an Increase in Atmospheric CO_2 During the Past Decades," *Tellus* **9** (1957):18-27.

Daniel B Botkin, "Forests, Lakes, and the Anthropogenic Production of Carbon Dioxide," *BioScience* **27** (1977):325-31

Bert Bolin, "Changes of Land Biota and Their Importance for the Carbon Cycle," *Science* **196** (1977):613-5.

CF Baes, Jr, HE Goeller, JS Olson, and RM Rotty, "Carbon Dioxide and Climate: The Uncontrolled Experiment," *American Scientist* **65** (1977):310-20.

GM Woodwell, RH Whittaker, WA Reiners, GE Likens, CC Delwiche, and DB Botkin, "The Biota and the World Carbon Budget," *Science* **199** (1978):141-6.

WS Broecker, T Takahashi, HJ Simpson, and T-H. Peng, "Fate of Fossil Fuel and the Global Carbon Budget," *Science* **206** (1979):409-18.

Published Sources of Quoted Material

Charles W Ralston, "Where Has All the Carbon Gone?" *Science* **204** (1979):1345-6.

Richard Levins, "Ecological Engineering."

H Selye, "The Evolution of the Stress Concept," *American Scientist* **61** (1973):692-9

Hans Selye, *Stress Without Distress* (The New American Library of Canada, Scarsborough, Ontario, and Lippincott, New York, 1974).

RH Whittaker, "Evolution and Measurement of Species Diversity," *Taxon* **21** (1972):213-51.

John Cairns, Jr., and Kenneth L Dickson, "A Simple Method for the Biological Assessment of the Effects of Waste Discharges on Aquatic Bottom-dwelling Organisms," *Journal of the Water Pollution Control Federation* **43** (1971):755-72.

NG Hairston, JD Allan, RK Colwell, DJ Futuyma, J Howell, MD Lubin, J Mathias, and JH Vandermeer, "The Relationship Between Species Diversity and Stability: An Experimental Approach with Protozoa and Bacteria," *Ecology* **49** (1968):1091-1101.

MacArthur, "Fluctuations of Animal Populations."

John Cairns, Jr, "Quantification of Biological Integrity," Pages 171-87 in RK Ballentine and LJ Guarraia, Editors, *The Integrity of Water* (U. S. Environmental Protection Agency, Office of Water and Hazardous Materials, U.S. Government Printing Office, Washington DC, Stock Number 055-001-01068-1, 1977).

Joel E Cohen and David Tilman, "Biosphere 2 and Biodiversity: The Lessons So Far," *Science* **274** (1996):1150-1.

A Brook Runs Through It

Likens, et al, *Biogeochemistry of a Forested Ecosystem*.

Frank B Golley, *A History of the Ecosystem Concept in Ecology: More than the Sum of Its Parts* (Yale University Press, New Haven Connecticut, 1993).

Frank H Bormann and GE Likens, *Pattern and Process in a Forested Ecosystem* (Springer-Verlag, New York, 1979).

DW Schindler, "The Impact Statement Boondoggle," *Science* **192** (1976):509.

SI Auerbach, RW Brocken, RB Craig, FO Hoffman, SV Kaye, DE Reichle, and EG Struxness, "Environmental Impact Statements," *Science* **193** (1976):188&248.

MacArthur's Ghost

SJ McNaughton, "Ecology of a Grazing Ecosystem: the Serengeti," *Ecological Monographs* **55** (1985):259-94.

F Stuart Chapin, III, Brian H Walker, Richard J Hobbs, David U Hooper, John H Lawton, Osvaldo E Sala, and David Tilman, "Biotic Control of the Functioning of Ecosystems," *Science* **277** (1997):500-4.

David Tilman, "Biodiversity and Production in European Grasslands," *Science* **286** (1999):1099-1100.

Ann P Kinzvig, Stephen W Pacala, and David Tilman, Editors, *The Functional Consequences of Biodiversity: Empirical Progress and Theoretical Extensions* (Princeton University Press, Princeton, New Jersey, 2001).

FA Bazzaz, "Cooling the Diversity Debate," *Bulletin of the Ecological Society of America* **82** (2001):136-7.

Kevin Shear McCann, "The diversity-stability debate," *Nature* **405** (2000):228-33.

Stephen P Hubbell, *The Unified Neutral Theory of Biodiversity and Biogeography* (Princeton University Press, Princeton, New Jersey, 2001).

M Loreau, S Naeem, P Inchausti, J Bengtsson, JP Grime, A Hector, DU Hooper, MA Huston, D Raffaelli, B Schmid, D Tilman, and DA Wardle, "Biodiversity and Ecosystem Functioning: Current Knowledge and Future Challenges," *Science* **294** (2001):804-8.

F Stuart Chapin III, Erika S Zavaleta, Valerie T Eviner, Rosamond L Naylor, Peter M Vitousek, Heather L Reynolds, David U Hooper, Sandra Lavorel, Osvaldo E Sala, Sarah E Hobbie, Michelle Mack, and Sandra Dia, "Consequences of Changing Biodiversity," *Nature* **405** (2000):234-42.

Published Sources of Quoted Material

Calculus and Commoner

Bernd Heinrich, "We Need Wilderness," *Rutland Daily Herald* (February 7, 2002):A8.

Richard Pease Grant, "Wilderness Enhances Diversity," *Rutland Daily Herald* (February 7, 2002):A8.

Trey Popp, "Prophet of Prosperity," *The Pennsylvania Gazette (*October 27, 2017) repeats the words of Edgar Fahs Smith at http://thepenngazette.com/prophet-of-prosperity/ (accessed on 8/26/2019) that I read in a much earlier edition of the same magazine.

Takacs, *The Idea of Biodiversity*.

Wilson, *Biophilia*.

Jocelyn Kaiser, "Taking a Stand: A Reluctant Warrior," *Science* **287** (2000):1190.

Daniel B Botkin, *Discordant Harmonies, A New Ecology for the Twenty-first Century* (Oxford University Press, New York, 1990).

John Ghiselin, "The Environmental Professional as an 'Interdisciplinarian,'" *Ecology* **57** (1976):1095.

Fakhri Bazzaz, Gerardo Ceballos, Margeret Davis, Rudolfo Dirzo, Paul R Ehrlich, Thomas Eisner, Simon Levin, John H Lawton, Jane Lubchenko, Pamela A Matson, Harold A Mooney, Peter H Raven, Joan E Roughgarden, Jose Sarukhan, G David Tilman, Peter Vitousek, Brian Walker, Diana H Wall, Edward O Wilson, and George M Woodwell, "Ecological Science and the Human Predicament," *Science* **282** (1998):879.

Pimm, *The World According to Pimm*.

Victor E Shelford, "The Ideals and Aims of the Ecological Society of America," *Bulletin of the Ecological Society of America* (1917):1-3.

Wilson, *Naturalist*.

LB Slobodkin, "Proclaiming a New Ecological Subdiscipline," *Bulletin of the Ecological Society of America* **81** (2000):223-6.

Afterward

Dolph Schluter, "Evidence for Ecological Speciation and Its Alternative," *Science* **322** (2009):737-41.

William F Laurance, "Theory Meets reality: How Habitat Fragmentation Research has Transcended Island Biogeographic Theory," *Biological Conservation* **141** (2008):1731-44.

John Bohannon, "Who's Afraid of Peer Review?" *Science* 342 (2013):60-65.

Gina Kolata, "A Scholarly Sting Operation Shines a Light on Predatory Journals," *New York Times (*March 22, 2017).

Gina Kolata, "Many Academics are Eager to Publish in Worthless Journals," *New York Times* (October 30, 2017).

Jean Langenheim, "Early History and Progress of Women Ecologists: Emphasis Upon Research Contributions," *Annual Review of Ecology and Systematics* **27** (1996):1-53.

W Dritschilo, "Rachel Carson and Mid-Twentieth Century Ecology," *Bulletin of the Ecological Society of America* **87** (2006):357-67.

Judith L Bronstein and Daniel I Bolnick, "'Her Joyous Enthusiasm for Her Life-Work …': Early Women Authors in The American Naturalist." *American Naturalist* **192** (2018):664-86.

David Jablonski, Shan Huang, Kaustuv Roy, and James W Valentine, "Shaping the Latitudinal Diversity Gradient: New Perspectives from a Synthesis of Paleobiology and Biogeography." *The American Naturalist* **189** (2017):1-12.

James H Brown, "Why are there so many species in the tropics?" *Journal of Biogeography* **41** (2014):8–22.

Lina Li and Peter Chesson, "The Effects of Dynamical Rates on Species Coexistence in a Variable Environment: The Paradox of the Plankton Revisited," *The American Naturalist* **188** (2016):E46-E58.

Published Sources of Quoted Material

Abraham H Gibson, "Edward O Wilson and the Organicist Tradition," *Journal of the History of Biology* **46** (2013):599–630.

Index

NAME

Abbey, Edward, 209
Abbott, I, 308
Abele, Lawrence G, 272, 277, 306-7, 311, 34-5
Adams, Ansell, 21
Alexander, Susan, 416
Allee, Warder Clyde, 53-7, 100
Andrewartha, HG, 142
Andrews, EA, 38, 50
Auerbach, Stanley I, 108, 113-4, 118-9, 121, 185, 259, 264-5, 285, 312, 434, 462, 493
Bacon, Francis, 254-5, 272, 475
Bailey, Amey, 450-1
Bazzaz, Fakhri, 466
Boessky, Ivan, 414
Bolin, Bert, 428, 431
Birch, L Charles, 142
Blair, Frank,116, 118
Bohannon, John, 486
Bolnick, Daniel I, 494
Bormann, F Herbert, 446-7, 450, 452, 458
Botkin, Daniel B, 243, 249, 426, 429, 432, 478
Bronstein, Judith L, 494
Brower. David, 7, 21-2, 24, 211
Brown, Harrison, 379
Brown, James H, 311, 313, 494
Brown, Lester, 393, 398
Brown, William L Jr, 129, 298-9
Bunker, Ellsworth, 172
Cairns, John, Jr, 439-442
Callicott, J Baird, 211
Carpenter, Frank, 262
Case, Ted J, 270-1, 314
Carson, Rachel, 19-20, 25-6, 31-5, 37-41, 45-63, 66-8, 70-4, 76, 79, 82, 84, 87-8, 91, 108-89, 114-6, 120, 122, 124, 129, 142, 198, 203, 209, 211-2, 352, 367, 369, 379, 388, 394, 396, 419, 428, 442, 461, 481-2, 491-3
Carter, Jimmy, 90
Caswell, Hal, 272-4, 276, 297, 315
Chase, Alston, 89, 126, 193, 195, 352-4, 357-8, 412, 414
Clark, Colin W, 410-2, 414
Clements, Frederick E, 52, 55, 65, 92-3, 107, 118, 126, 132, 138, 190-1, 280
Cody, Martin L, 148. 178, 306, 314
Cohen, Joel E, 289, 328, 369, 371-2, 376-8, 380-1
Cole, Blaine, 345
Cole, LaMont C, 61-2, 64, 66-72, 74, 114, 120, 138, 141, 274-5, 277, 337-8, 367-8, 373
Colwell, Robert, K, 307, 313
Commoner Barry, 9, 26-32, 57, 61, 74, 88, 107, 193, 199, 278, 349, 362, 367-8, 378, 389, 472, 479
Conn, David, 405
Connor, Edward F, 300-1, 311-3, 315-9, 332, 334
Cooper, Arthur, 404
Cooper, William F, 75
Costanza, Robert, 393, 416-8
Cowles, Henry Chandler, 131
Cowles, Rheinhart P, 37-8
Crow, James, 358-9
Daily, Gretchen C, 380, 416
Darwin, Charles, 17, 42-3, 65, 77, 80-1, 87, 93, 101, 105, 117, 130-2, 140-1, 144, 148, 168, 200, 210, 214, 256-7, 268, 289-90, 298, 317, 321-2, 369, 375, 408, 468-9, 485, 488
Dasmann, Raymond, 414
Dawkins, Richard, 154
Dean, James, 18, 182, 345, 389, 421
Descartes, René, 254-5, 257, 466, 475

Index

Diamond, Jared M, 18, 253, 259, 270, 286, 290, 294-8, 301-3, 306, 311-3, 315-9, 333, 340-5, 347, 466, 495
Dice, Lee R, 482
Dubos. René, 7, 17, 74, 367, 372
Eckholm, Erik, 395, 398
Erwin, Terry, 361-6, 406, 417, 438
Ford, EB, 157
Game, Margaret, 345
Gandhi, 126, 194, 209
Gause, Georgii, 52, 227-33, 235, 237-8, 241, 267, 273, 369-70, 381, 422-3, 444, 489
Goulder, Larry, 416
Grier, James W, 394
Einhorn, Ira, 6, 9
Eiseley, Loren, 16-7, 33, 74, 77
Egler, Frank E, 15, 59, 64, 72-5, 116, 120, 492
Ehrlich, Anne H, 197, 380
Ehrlich, Paul R, 6, 18, 23-6, 30, 61, 64, 69-71, 79, 114, 145, 149, 161, 197, 210-1, 353, 367-8, 372-3, 375-6, 378, 380-1, 385. 395, 398, 416, 418
Einstein, Albert, 31, 107, 123, 179, 233, 247, 421, 485, 491
Eldridge, Niles, 157
Elton, Charles S, 26, 52, 58-9, 88, 105, 109, 214, 251, 275, 384, 492
Feynman, Richard, 221
Fisher, Dorothy Canfield, 172
Fisher, RA, 162
Ford, EB, 157
Ford, Ernie, 400
Franklin, Benjamin, 78
Fretwell, Stephen D, 176, 179, 181-2, 184
Friedan, Betty, 209
Frost, Robert, 172
Galileo, 254, 466
Gates, Henry Louis, Jr, 354

Gause, Giorgii, 52, 227-33, 235, 237-8, 241, 267, 273, 369-70, 381, 422-3, 444, 489
Geisel, Theodor Seuss, 252, 387
Ghiselin, Jon 478
Gilpin, Michael E, 270, 311, 313, 315-8, 345
Gleason, Henry A, 190-1
Gleick, James, 241
Goldschmidt, 160
Golley, Frank B, 451-2
Goodman, Daniel, 215-6, 284
Gotelli, Nicholas, 289, 306-8
Gould, Stephen Jay, 157, 210, 495
Gould, Sophia A, 143
Grant, Peter, 275, 298, 300, 303-4, 306, 308-9, 313, 330
Grants, 269, 322
Graves, Gary R, 289
Grene, Marjorie, 278-80
Grinnel, Joseph, 251
Haeckel, Ernst, 65, 76
Hagen, Joel B, 451
Hairston, Nelson G, 257
Haldane, JBS, 168
Hardin, Garrett, 273, 367, 410-1
Hasler, 448
Hawkins, Brad, 319
Heidegger, 210
Hegel, 210, 281-2, 355
Heinrich, Bernd, 473
Hendrickson, John A, 309-11
Herbert, Frank, 7, 17
Hespenheide, Henry, 178, 306
Holm, Richard W, 353
Hubbell, Stephen P, 467-8
Huffaker, Charles, 423
Hurwitz, Charles E, 414-5,
Hutchinson, George Evelyn, 18, 58, 97-105, 108, 124-5, 148, 150, 168, 173-4, 176-7, 182, 187, 200-1, 203, 207, 216, 228, 230, 251-2, 263, 293, 296, 299,

303, 306, 311, 350, 369, 419, 421, 424, 427, 429-30, 438, 452, 467, 495
Hutton, James, 112
Huxley, Aldous Huxley, 211
Janak, 243
Janzen, Daniel, 99, 187, 208-9
Jeffers, Robinson, 211
Jenkin, Fleeming, 132-3
Jennings, Herbert S, 34, 39, 51
Johnson, Lyndon, 379, 389
Johnson, Noye M, 446
Juday, Chancey, 447
Kahn, Herman, 371. 381
Kelvin, Lord, 255
Kenaga, Eugene, 442
Kendeigh, Samuel Charles, 100
Kepler, 31, 231
Kettlewell, HBD, 157
Kingsland, Sharon, 484, 495
Knibbs, GH, 380
Kodric-Brown, 311
Kohn, Alan, 100
Krutch, Joseph Wood, 367
Kuhn, Thomas H, 111-2, 263, 286
Lack, David L, 142, 179, 269, 298-300, 321
Lakatos, 307
Lamarck, Jean Baptiste Pierre de, 131
Lande, Russell, 352-3, 356-60, 422
Langenheim, Jean, 492
Laurance, William F, 490
Lave, Lester 405, 407
Lear, Linda, 49, 492
Lenin, Vladimir I, 179, 182, 277
Leopold, Aldo, 20, 57, 81, 144, 181, 203, 211, 340, 349, 474-5
Leopold, Luna, 7
Leeuwenhoek, Anton van, 378
Levin, Simon A, 237, 264, 386-7
Levins, Richard, 119, 135, 248, 250, 268, 278-84, 297, 306, 352-3, 357-60, 432, 488

Lewontin, Richard C, 149-153, 158, 217, 278-84, 352, 359
Liapunov, AA, 246
Libby, Willard F, 433-4, 444
Liebig, Justus Freiherr von, 377-8, 425
Likens, Gene E, 18, 33, 90, 115-6, 390-1, 404, 427-30, 446-50, 452-61, 463, 471, 474-5, 478
Lindeman, Raymond L, 58, 94-6, 100, 103-5, 107-8, 124, 168, 173, 177
Linnaeus, 77
Lomborg, Bjørn, 24-5, 371, 394-6, 398-402, 494
Lomonosov, Michael, 77-8
Lotka, Alfred J, 230-1, 233
Lovelock, JE, 164-7, 444
Lovins, Amory, 382
Lubchenco, Jane, 416
Lyell, Charles, 112
MacArthur, Betsy, 171, 179-181
MacArthur, John W, 169-72, 177, 180-1, 419
MacArthur, Olive, 172
MacArthur, Robert Helmer, 18, 66, 89, 98, 100, 110, 117, 119, 124-5, 148, 150, 168-87, 203, 205, 210, 213, 215-6, 228, 232-3, 243-4, 247-8, 250-2, 254, 259, 262, 265, 268-9, 271, 273, 275-8, 280, 283, 290. 292, 294, 296-8, 306, 311-4, 316-9, 323-9, 331-7, 342-3, 349-50, 352, 357, 360, 370, 385-7, 419-24, 430, 433, 436, 438, 440, 446, 452, 464, 467-71, 480-2, 489
MacFadyen, Amyan, 263, 288
MacPhee, John, 258
Malthus, Robert Thomas, 375-6
Margaleff, Ramón, 434
Margulis (Sagan), Lynn, 164-166
Marsh, George Perkins, 20, 199, 206, 378, 403-4, 408-9, 416-7
Matson, Pamela A, 197, 416

Index

May, (Sir) Robert M, 187, 218-20, 235-7, 239-40, 244-52, 259, 264, 278, 284, 288-90, 306, 313, 319-20, 330-3, 335, 340, 345, 361, 370, 419, 423, 488
Mayr, Ernst, 133, 153, 157-8, 361
McCoy, Earl, 332, 334
McHarg, Ian, 6, 9, 13, 74
McIntosh, Robert P, 278, 484
McNaughton, SJ, 464-5
Mead, Margaret, 17
Meadows, Dennis L, 376
Meadows, Donella H, 374, 376
Mendel, Gregor, 105, 131, 133
Mill, John Stuart, 378, 475
Miller, Lee, 362
Mitchell, Joni, 23, 397
Mooney, Harold A, 416
Montague, Ashley, 353
Moses, Robert, 6
Muir, John, 209
Murdoch, William, 407
Muskie, Senators Edmund, 6, 404
Nader, Ralph, 7, 9, 60
Naess, Arne, 127, 210-2, 478
Nelson, Gaylord, 7, 9, 33
Newton, Isaac, 31, 78, 123, 162, 230, 240, 247, 254, 289, 466
Nicholson, Alexander J, 142
Norse, Elliot, 204
Ockham (Occam's Razor), William of, 148-9
Odum, Eugene P, 18, 49, 56, 58, 66,75-6, 82, 87-92, 94, 98, 100, 106-8, 110, 114, 116, 118, 121, 124-8, 130, 133-6, 139-40, 142, 145, 149, 160-1, 173, 177, 184-9, 192-3, 196, 199-200, 202, 205, 210-1, 213, 250, 259, 261, 278, 306, 321, 339-40, 367-8, 392, 405-6, 428, 434, 441, 446, 452, 455-9, 462, 481
Odum, Howard T, 18, 56, 58, 90-91, 100, 107-8, 110, 121, 124-6, 139-40, 172-3, 177, 188, 250, 259, 306, 321, 339, 382-3, 389, 391-3, 405-6, 428, 446, 456, 459
Odum, Howard W, 90-91, 127 (in passing)
Odum, William, 456
O'Neill, Robert V, 139. 162-3
Paine, Robert T, 207, 386
Park, Thomas, 57, 103, 386, 422-3
Patrick, Ruth, 311, 330, 335, 436-9
Patton, 311
Pearl, Raymond, 34, 39, 51-3, 57, 67, 142-3, 230, 369, 371, 379
Perrine, Richard (UCLA engineering professor), 8, 412, 433
Peters, Robert Henry, 258-61, 263, 265, 268, 272-3, 278, 301, 304
Picasso, Pablo, 482
Pianca, Erik, 176
Pimentel, David, 18, 367-8, 383-8, 390-3, 398-402, 418, 481
Pimm, Stuart L, 197-8, 313, 361, 400-1, 479-80
Pontecorvo, Giulio, 418
Popper, Karl R, 256-8, 263, 266, 286-7, 304, 307, 466
Porter, Elliot, 22
Postel, Sardra, 416
Powell, Colin, 47
Preston, Frank W, 290-1, 327, 329-30, 333-5, 337, 437-8
Quammen, David, 182, 269-71, 277, 285, 294, 311, 315, 348
Raven, Peter H, 23, 149, 161, 212, 385
Ray, Dixie Lee, 373
Reagan, Ronald, 179, 414, 446
Reed, Lowell, 369, 371
Revelle, Roger, 379, 424
Riasanovsky, Nicholas V, 77
Ricardo, David, 376
Ricklefs, Robert E, 9, 11, 76, 88-9, 178, 382
Rifkin, Jeremy, 14

Root, Richard B, 69, 316
Rosenzweig, Michael, 178-9
Roughgarden, Jonathan (Joan E), 215, 248-9, 284-90, 306, 335, 339
Rovinsky, Robert, 481
Rudd, Robert L, 33, 114-5, 442
Ruse, Michael, 495
Russell, Bertrand, 281-2
Sagan, Carl, 164
Sale, Kirkpatrick, 209
Salt, George W, 134-5, 271-2
Saarinen, Esa, 278
Schindler, David, 452
Schlesinger, William, 185, 432
Schluter, Dolph, 488
Schneider, Steven H, 416
Schoener, Tom 313
Schumacher, EF, 382
Scott, Hugh, 6-7
Sears, Paul, 33, 116, 340, 400, 482
Selander, Robert, 151
Selye, Hans, 434
Shannon, Claude, 438
Shelford, Victor E, 52, 55, 100, 115, 480
Simberloff, Daniel S, 18, 185, 262-6, 268-72, 276-80, 283-6, 289-90, 293-4, 297-8, 300-13, 315-9, 330, 341-5, 347-8, 350, 356-7, 365, 372, 466, 469, 489-90
Simon, Julian, 8, 371, 381, 394-6, 398, 402, 418
Skinker, Mary Scott, 39
Slobodkin, Lawrence B, 257, 335, 419, 422, 483
Smith, Edgar Fahs, 477
Smith, Frederick E, 257
Soulé, Michael E, 346-7, 358
Southwood, TRE, 420
Sparr, MC, 449
Stahl, Andy, 352, 354, 357-60
Strain, Boyd R, 83

Strong, Donald R, Jr, 266, 270-1, 275, 278, 284, 290, 294, 303, 306, 309-12, 339, 485
Suess, Hans, 424
Sugihara, George, 332, 334-5
Suppé, F, 286-7
Suter, Glen, 265
Szyska, Lee Ann, 303
Takacs, David, 212, 477
Tansley, Arthur G, 92, 94, 138, 192
Teale, Edwin Way, 367
Temple, Stanley, 322-3
Terborgh, John, 343-4, 347
Thoreau, Henry David, 20-2, 65, 209-11, 378
Tilman, David, 416, 465-6, 470
Tunney, John, 349
Turgenev, Ivan, 184, 250
Turner, Fred, 244-5
Udall, Stewart, 20-1, 116
Van Valen, Leigh, 42, 280
Verlhurst, Pierre Francois, 369
Vernadsky, VI, 108
Vernadsky, George, 108
Vitousek, Peter M, 197
Von Bertalanffy, Ludvig, 438
Wald, George, 7, 262, 372
Wallace, Alfred Russell, 65, 214, 290
Wallis, 243
Watt, Alex S, 458
Watt, Kenneth, 6, 31, 246
Wegner, 258
Westman, Walter E, 403-7, 415-6, 429
Whelan, Elizabeth, 373
White, Gilbert, 65
White, Lynn, 210
Whitehead, Alfred North, 255-6
Whittaker, Robert H, 100, 108, 120, 373, 388, 404, 427-32, 437-9, 446, 463
Weiner, Jonathan, 268-9
Weitzman, Martin L, 418
Wiener, Norbert, 105, 519, 538

Wiens, John A, 314
Williams, George C, 149-150, 159
Willis, Edwin O, 296
Wilson, David Sloan, 159, 161
Wilson, Edward O, 10-1, 18, 59, 97-8, 124, 149, 168, 179-81, 184, 210, 212, 232, 259, 262, 274, 280, 283-4, 292, 296-9, 323-5, 327, 329-32, 334, 337, 342-3, 357, 361-2, 398, 421, 467, 470, 477-8, 482, 495
Wittgenstein, Ludwig, 287, 466
Woodwell, George M, 108, 367-8, 394, 404, 416, 428, 430-1
Wolfram, Stephen, 43
Wright, Sewall, 160
Wynne-Edwards, 148
Zimmerman, Barb, 348
Zimmerman, Michael E, 209-10, 348

SUBJECT

acid rain, 446, 448, 451, 460, 462
adaptation, 64-6, 72, 80-3, 93, 128-132, 140, 149, 156, 186, 225, 329, 367, 434
Adaptation and Natural Selection, 149, 159
agricultural schools, 61, 352
American Association for the Advancement of Science (AAAS), 46-7, 57, 87, 283, 482
American Peregrin Falcon, 85, 356, 367
Animal Ecology, 58, 251
Applied Ecology Committee (ESA Section), 33, 115-6, 495
A Sand County Almanac, 20, 60, 144, 349, 474
Atomic Energy Commission, 108, 153, 245, 339-40, 428
autecologist, 74, 79, 84

autecology, 83-4, 203, 264, 344, 350, 446
balance of nature, 26, 61, 70, 72, 80-1, 88, 109, 128, 144, 193, 209, 263
Beaufort, North Carolina, 54, 56
Big Ecology, 340, 485, 488
Big Science, 107-8, 118, 185, 340, 427, 489
Big Yellow Taxi, 23, 397
BIODEPTH, 465-6
bio(diversity), 11, 16, 24, 72, 92, 117, 136, 168, 200-5, 207-10, 212, 252, 300, 323, 325, 330, 347, 356, 361, 437, 467-70, (genetic), 329, 355, (habitat), 337, (indices), 173, 216, 362, 364, 437, 439-41, 468, (mechanisms (theories)), 261, 274, 276, 438, 467, 494, (preservation), 11, 16, 473, 477-80, 482-3
bio-ecology, 55-5
biologist, 7, 15, 19, 29, 34, 65, 75, 97, 99, 105, 116, 119, 132, 138, 149, 161, 163-4, 168, 170, 179-80, 212, 227, 235, 239, 241, 245, 251, 262, 267, 275, 278-9, 281, 328, 343-6, 352-3, 423, 427, 454, 477, 486, (conservation), 33, 346, (evolutionary), 129, 153, 157, 161, 261, 294, 297, 315, 353, 422, 455
biology, 12, 17, 51, 66, 75-6, 80, 87, 93-4, 98-9, 105, 133, 162, 178, 236, 248, 253, 262, 264, 274, 281-2, 289, 299, 303, 315, 328, 331-2, 334, 336, 340-1, 359, 423, 479, 481, 489, (conservation), 345, 356, 358. (evolutionary), 64, 120, 140, 232, 271-2, 321, 340, 347, 485, 489, 491, (population), 80, 237, 368, 386
biomagnification, 29, 85, 88, 444
biomass, 188, 198-9, 222, 368, 381-2, 392, 425-6, 431

biome, 73, 92, 117-8, 206, 214, 217, 242, 318, 417, 427
biophilia, 11-2, 80, 212, 478
BioScience, 116, 392, 426, 453
biosphere, 82, 164, 217, 378, 425
black box, 96, 105, 123-4, 188, 374, 447, 454, 458
botanist, 23, 65, 130-1, 475
Brookhaven National Laboratory, 108, 394, 427-8, 430
Bulletin of the Ecological Society of America, 64, 74, 83 99, 465-6
butterflies, 23, 26, 81, 111, 118, 124, 200, 312, 325, 366, 454
California, 23, 81, 84, 113, 191-2, 243, 349, 351, 412-3, 433, 446, (condor), 351, 390, (La Jolla), 424, (San Jose), 8, (Spotted Owl), 356
canonical distribution (log-normal) of Preston, 201-2, 204, 291-2, 312. 327, 329-30, 333, 335, 437
Cape Ann, Massachusetts, 55-6
carbon (dioxide), 165, 188, 199, 222, 378, 406, 424-8, 431-2, 434, 446, 491
carrying capacity (K), 143-5, 227, 230-2, 238, 370, (human), 374, 378, 380-1
Center for the Biology of Natural Systems, 27-8, 362
Central Limit Theorem, 202, 330, 332
chaos, 220, 227, 238-42, 289, 350, 423
character displacement, 298-9, 303-4, 310
chemistry, 12, 71, 74, 76, 82, 87, 94, 111, 161, 186, 203, 218, 253, 262, 396, 423, 454, 477, 479, 481, 485-6
Chicago School, 53, 55, 65, 72, 131
clutch size, 146, 148, 153

coevolution(ary), 24, 71, 149, 161, 323, 385
commercial presses, 44, 318
community, 29, 55-6, 65-6, 72, 82-5, 88-9, 92-5 105, 131-2, 137-8, 150, 156, 161, 177, 195, 197, 200, 207, 216, 236, 252, 272-6, 287, 290, 292-3, 295, 297-301, 303, 305, 308, 312, 314-7, 319-21, 331, 333, 335, 344, 360, 387, 438-41, 458, 464, 488-9, (animal), 55, 87-9, (assembly rules), 295, 297, 301-2, 312, 316-8, 333, 341, 360, (bird), 297, 319, (climax), 92, 107, 131, 192, (diversity), 439, 464, 487, (insect), 312, 314, 319, (plant, 55), 65, 87, 92-5, 138, 424, 437, 458, (theory), 195, 305, 344, 350, 385, 432, 439, 446, 469, 475
competition, 63, 93-5, 112, 133, 141-2, 156-8, 169, 228, 232-3, 254, 268, 271, 273-4, 276, 290, 293, 298-305, 310, 312, 314, 317, 319-21, 368, 370, 380, 422-4, (diffuse), 295, 316-7, (theory), 261, 290, 304-5, 339, 385, 465
Comstock Hall, 69, 386, 400, 490
conservation (species preservation), 16, 20, 87-8, 115, 126, 180, 187, 213, 215, 269, 294, 341-50, 352, 357, 469, 478, 480, 489, (policy), 343, 466, (scientist), 340-2
conservationist, 115, 202, 218, 342, 345-6, 357
controversies in ecology and evolutionary biology, 16, 213, 280, (barroom brawl), 15, 285, 313, 466, (density dependence), 67-8, 140-6, 148, 156, 230, 257, 314, 385, (diversity-stability), 88, 128, 150, 163, 173, 186, 213-7, 225, 228, 235-6, 252, 292, 344, 386, 426-7, 439-40, 464-7, 469, (null model), 272, 274-8,

Index

283, 285-9, 294, 297-8, 301-5, 307-8, 312, 314-6, 318-9, 331-2, 334-5, 339, 342, 439, 447, 465-6, 476, 484, 494, (patterns of distribution), 295, 317, (preservation), 416, 436, 438-9, 465, 468, 470, (refuge design), 294, 346-7, (SLOSS), 337, 348, 490, (species-area), 294, 329-30, 332, 361, (stability/complexity), 186, 228, 235-6, 246, (subspecies), 353-6
Cornell University, 61, 68-9, 97, 187, 221, 237, 243, 260, 316, 340, 367, 373, 384, 386-9, 427, 429-30, 432, 446-7, 454, 456, 460
Cornucopian, 374, 379, 394, 418
Costa Rica, 99, 169, 208
cybernetic(s), 67, 86, 91, 95, 104-6, 108, 136, 141, 165, 189, 218, 225, 230, 385, 462, 474, 481
Darwin's Century, 26, 77
Darwin's finches, 179, 269, 298-301, 303, 308-9, 321, 338-9, 468
data dredging, 310-1
Deep Ecology, 116, 126-7, 164, 210-2, 378, 427, 475, 478
DDT, 20, 23, 29, 58, 63-4, 70, 83-5, 88, 122, 349, 368, 394, 428, 442-3, 461
dodo bird, 189 (dead as a Dodo), 323, 335, 349, (tree), 322-3
Earth Day, 4, 12, 29, 33, 89, 119-20, 205, 260, 262, 372, 396-7, 426, 446, 490 (1970 (first)), 6-10, 12, 23, 29-31, 57, 61, 75-6, 80-3, 86, 88, 91, 100, 113, 118, 145, 162, 185-7, 192, 197, 209, 213-4, 235, 244, 250, 263, 340, 347, 367-8, 370-2, 378, 383, 388-90, 392-4, 396, 400, 403, 410, 415, 418, 433-4, 436, 438, 444, 468, 477, 484-5, 490-1, (1980), 12, 14-20, 161, 264, 394, 396, 412
Earth First!, 126, 212
ecocentrism, 126-8

Ecological Animal Geography, 54
ecological concept, 57-8, 116, 231, 316, 381
Ecological Monographs, 54, 64, 318, 465
Ecological Society of America (ESA), 33, 50, 59, 61, 64, 74-5, 83, 99, 114-6, 119, 121, 204, 262, 404, 416, 430, 465-6, 472, 475, 480, 482, 485, 492-5, (Board of Professional Certification), 74-5
ecologist, 12, 14-6, 21-3, 26, 28-9, 31-33, 44, 50-1, 53, 56, 58-61, 64-9, 72-5, 77-83, 85, 90, 93-6, 99, 104-5, 107, 109-12, 114-21, 125, 128, 130-1, 133, 136-9, 141-2, 147, 149, 151, 159-60, 164, 167-9, 173-4, 176-7, 185, 187-90, 192, 196-200, 202-3, 205-8, 210-2, 214, 216-8, 224-5, 229, 232, 235-6, 239-40, 243-4, 246-7, 250-4, 257-61, 263, 265-8, 270-2, 276-9, 287-9, 291, 294, 297-8, 305, 308, 310-1, 314, 316-7, 319, 321-4, 335-6, 338-40, 343, 346-7, 351, 354, 367-8, 370-7, 380-3, 385-6, 388, 390, 392-3, 395-6, 398, 400, 402-5, 415-6, 418, 421-5, 427, 431, 434-5, 438-42, 444-6, 451, 453-5, 462, 466-70, 473-483, 485, 491, (amateur), 72, 202, (animal), 53, 57, 79, 100, 105, (applied), 428, 440, 495, (bird), 79, 148, 314, (butterfly), 79, 368, (British), 26, 88, 92, 114, 420, (certified professionals), 59, 209, 480, 494. (community), 84-5, 137, 441, 494. (Deep), 166, 211, (ecosystem), 85, 89, 133, 136, 146, 161, 189, 202, 259, 322, (evolutionary), 80, 89, 136, 146, 162, 202, 271, 452, 494, (instant), 74, (plant), 79, 92, 94-5, 108, 131-2, 373, 403-4, (population), 51, 79, 84, 147, 211, 250, 368, 385, (professional), 33,

526

59, 117, 209, 494, (self-conscious(-aware)), 50, 79, (system(s)), 133, 384, 393, 394, 416, 433, 452, (theoretical), 79, 189, 288, 336, 467, (wildlife), 116, 322, 344, 389, (woman), 491-2
Ecology (journal), 44, 64, 86, 103, 120, 173, 185, 260, 306, 318, 327, 362, 440, 485
Ecology (textbook), 88
Ecology and Evolution of Communities, 177
ecology, 7-20, 23, 26-31, 33, 38, 44, 48-62, 64-9, 72-91, 93-8. 100, 103-4, 106, 108-21, 123-8, 130-1, 134-6, 139, 142, 144, 146-7, 150-1, 153, 156, 160-3, 168-70, 172-8, 182-7, 188-90, 193, 195, 197-201, 203, 205, 207, 209-14, 216, 227, 232, 241-54, 256-67, 271-2, 275, 278, 283, 285-6, 288-9, 292-3, 295-8, 301, 304, 306, 308, 311-4, 316-7, 321-4, 329, 335-6, 338-42, 344-5, 347, 349-52, 354, 356-7, 359, 361, 367-8, 372, 375, 381-3, 385, 387-9, 392-4, 397, 399, 403, 406, 415-6, 419-27,430-55, 435-9, 442, 444, 446-7, 452-4, 457-8, 460-3, 468-70, 472, 474-6, 478-9, 481-94, (as a point of view), 4, 14, 50, 72, 79, 83, 109, 384, 435, 481-2, 487, (applied), 115, 127, 200, 203, 359, 432, (community), 84, 446, 458, (definition), 29, 75-6, 87, 106, 484, (ecosystem), 26, 56, 82, (mainstream), 30, 56, 65-6, 74, 392, (plant), 55-6, 189, 319, 429, (population), 23, 61, 80, 84, 203, 227, 235, 385, (radical), 128, 209-10, 469
ecosystem, 10 15-6, 18, 29-30, 56, 58, 65-6, 72, 82, 85, 86, 88-92, 94-6, 104-8, 110-4, 116-8, 120-28, 130-1, 133-40, 149, 151, 156, 159-63, 168, 177, 185, 195, 201, 203, 205-7, 210, 213-4, 217-8, 220-2, 224-7, 235-6, 240-3, 248-51, 254, 259, 261, 264-5, 271-2, 278, 285, 289, 302, 304, 321-3, 339-40, 344, 350, 367-8, 392-4, 398, 404, 406, 416-8, 424-8, 432, 434, 440-2, 444-6, 448, 452-3, 457-9, 461-2, 465, 467-70, 473, 480-1, 483, 488-91, (analysis), 86, 111, 265, 383, 392, 426, 442, 223, (attributes), 188-9, 200, 261, (emergent properties), 134, 136-7, 186, 271, (evolution), 145-6, 149, 188, (forest), 242, 406, 447-8, (function), 222, 406, 415, 417, 464-5, 470, (productivity), 162, 196, 199-200, 213, 465-6, (resilience), 167, 217, 222, 427, 467, 470, (services), 405, 416-8, 445, (stability), 138, 186, 211-3, 216-8, 222, (theory), 163, 195, 474, (trophic-dynamic concept of), 108, 173
ecotage, 126, 128, 211, 415
Endangered Species Act, 203, 349-60, 354-6, 443
energy, 30, 56, 63, 76, 121, 129, 135, 140, 165, 177, 204, 222, 377, 380-3, 388-93, 395, 405, 476, (in ecosystems (environment)), 16, 58, 66, 85, 88, 92, 105, 118, 125-6, 135, 162, 177, 188, 205, 225, 259, 417, 432, 459, 467, 481, 483, 485
Eniwetok, 121, 124, 321,
engineer(ing), 8-10, 12, 14, 86, 113, 119-20, 132, 173, 175, 178, 185-6, 234, 246, 260, 291, 306, 347, 382-3, 388, 419-20, 424, 432-4
environmental, 14, (management), 82, 119, 126, 128, 251, 254, (science), 10, 17, 78, 86, 113, 121, 204, 244, 249, 481
environmentalism, 12-3, 15, 66, 210, 476, 478

Index

environmentalist, 8, 14, 25, 28, 126, 211, 215, 235-6, 372, 375, 396-7, 399, 418, 435, 463, 478

eutrophication, 29, 447

evolution, 16, 62-4, 67, 80-1, 83, 87, 89, 93, 98, 101, 129-33, 135, 137-8, 141, 145-6, 148-9, 152, 155-6, 160-2, 188, 225, 236, 262, 268, 271, 298, 309, 322, 367, 385, 420, 486-8, (Darwinian), 17, 132, 144, 154-5, 229-30, 308, (Lamarckian), 93, 129, 131-32, 230, 299, 493, (theory of), 62, 80-1, 132, 151, 158-9, 257, 261, 347

extinction, 81-2, 85, 157-8, 204-7, 218, 228, 273, 296, 321-3, 325, 327-8, 339-40, 343-4, 347, 349-52, 358, 361, 367, 394, 399-400, 410, 423, 431, 465, 470

falsifiable (hypothesis, falsifiability), 256-8, 261, 272, 278, 287

feedback, 68, 104, 374, 376, 385, (genetic), 385-7, 390

Fish and Wildlife Service, 34-5, 47

Florida State (group), 272, 276, 285-6, 288-9, 298, 304-9, 313-5, 317-8, 341, 345

Florida State University, 266, 268

food web, 26, 29, 82, 200, 214, 468, 489-90

Forest Service, 354, 356-8, 451, 461, 471

Four Laws of Ecology (from Barry Commoner), 26, 29-31, 107, 193, 349, 368

Fundamentals of Ecology (Odum textbook), (49), (76), 87-8, 121, (452), 455, 457, 481

Gaia Hypothesis, 164-7, 444, 495

Galapágos (Islands), 199, 300, 302, 313, 321-2, (finches), 141, 179, 298. 300, (tortoise), 98, 366

genecologists, 74, 79

genetic(s), 34, 64, 81, 133, 146, 151-5, 160-2, 175, 186, 208, 267, 329, 355, 359, 385, 388, 489, 491, 493-4, (diversity), 203, 329, 355, (engineering), 14, 382, (feedback), 385-7, 390, (load), 140, 152-4, 346, 358, (population), 175, 186, 274, 385-6, (variability), 346, 358

geneticist, 132-3, 149-51, 153, 157-8, 160, 164, 186, 358, (population), 149, 151, 153, 160, 164, 277, 335, 358-9, 385

group selection, 140, 145-6, 148, 153, 157-62, 164-5, 225

guild, 189, 315-7, 333

habitat, 33, 72, 79, 81, 85, 116, 158, 169-70, 173, 187, 189, 204, 217-8, 232, 251, 300-1, 303, 322-3, 327, 333, 337, 340-1, 343, 346-7, 350-2, 356, 359, 422-3, 439, (diversity), 203, 332, 337

Harvard, 150, 180, 262, 283, 298, 306, 359, 372, 418, 466

heresy, 75, 129, 140, 145, 148, 165-6, 347

heterozygote, 62-3, 152

heuristic, 125, 237, 264, 284, 333, 371, 384, 415

holism, 51, 58, 91, 109, 135, 280, 282, 494

"Homage to Santa Rosalia," 168, 201, 216, (bones) 299

homozygote, 62-3, 152, 358

HSS, 258, 385

Hubbard Brook, 199, 401, 446-7, 449-53, 456-62, 465, 471, 474

hypothetico-deductive (method), 174, 184, 186, 263, 268, 272, 420

In a Dark Wood, 89, 352, 412, 415

industrial melanism, 63, 156-8

International Biome Program (IBP), 117-8, 427, 434, 451-3, 462
Internet, 43, 105, 179, 295, 372, 417
intrinsic rate of increase (r), 143, 230-1, 237, 328
Johns Hopkins University, 34, 37-41, 49-53, 91, 108, 492
kin selection, 153, 495
JABOWA, 242-3, 249-50, 426, 432, 453, 458
Limits to Growth, 373, 380, 383, 393, 395
little old lady (…) in tennis shoes, 12, 74, 84, 109, 259, 340
logistic equation (S-shaped curve), 51, 142-4, 218, 227-8, 230-2, 234-5, 237-41, 264, 369-71, 374, 379, 381, 383, 396, 403, 416, 419, 432, 488, 490, 494, (r and K), 231-2, 235, 369-70, 385, 426
Los Angeles, 8, 90, 113, 248, 433
Lotka-Volterra equations, 231-2, 235, 246, 273, 289, 350, 470
MacArthurite(ian), 184, 203, 264, 270, 306, 312, 314
macroscope, 91, 446
Maine, 12, 52, 56, 72, 84, 109, 181, 388
Malthusian (neo-), 24, 141, 374-5, 403
Manhattan Project, 107, 113, 118, 379, 434
Marlboro College, 171, 175, 329, 421, 495
Marlboro Circle, 180-1, 280, 283
mice, 13, 137, 151, 158, 200, 206, 237, 339
microbiome, 85, 491
Modern (Evolutionary) Synthesis, 131-3, 146, 153, 160, 348
mutualism(tic), 133, 138-40, 235-6, 322-3, 336

National Academy of Sciences, 173, 176, 283, 358, 390-1, 430
National Environmental Protection Act (NEPA), 117-9, 203, 264-5, 350, 357, 462, 479
National Science Foundation, 107, 113, 308, 451
natural history, 15, 50-1, 55, 84, 94-5, 161, 187, 259, 329, 350, 364, 491
natural resource(s), 20, 91, 115, 209, 368, 375, 395, 404, 407, 411-2, 414, 418
natural selection, 17, 63, 80-2, 87, 89, 130-3, 138, 140-2, 144-6, 148-50, 153-4, 156-7, 159, 163, 188, 208, 214, 236, 256-7, 261, 298, 304, 321, 375, 488, 490
naturalist, 17, 20, 91, 39-40, 84, 124, 157, 174, 212, 214, 290, 340, 437
Naturalist (book), 97, 179
nature, 6, 9, 11-2, 16, 19, 21, 23-4, 30, 34, 37, 50-1, 55-6, 60, 65, 71, 76, 78, 80, 86, 88-9, 94, 104, 106-7, 110, 117, 123-7, 137, 146, 156, 158, 160, 164, 173, 182, 193, 202, 206-12, 215, 227, 240-1, 243, 267-8, 288, 298, 304-5, 319, 326, 331, 365, 368, 370-1, 378, 392, 403, 406, 409, 423, 437, 439, 457, 464, 467, 469, 478, 490, (balance of), 26, 61, 70, 72, 80-1, 109, 115, 128, 144, 193, 209, 212, 263, (p)reserve(s), 340, 344, 481, 490, 493, services (ecosystem), 403-6, 415-8, 432, 445, 470, 479
Nature (journal), 44, 280, 318, 416-7, 470, 485, 487
Nature Conservancy, 115, 472, 480
niche, 72, 82, 85, 150, 173, 189, 200-1, 203, 245, 251-2, 290, 293, 295, 327, 387, 424, (Hutchinsonian), 350, 360, (theory), 200-1, 204, 251
neutral model, 272-4, 319

Index

Oak Ridge National Laboratory, 108, 113-4, 118, 121, 139, 162, 245, 250, 264-5, 285, 428, 434, 451, 462
Oecologia, 44, 318, 320, 353, 356
On the Origin of Species, 81, 200, 385
open-access journals, 45, 318, 485-6
Pacific Lumber Company (Palco), 412-4
paradigm, 111-2, 120, 263
parameter, 95-6, 123, 143, 195-7, 216, 222-3, 225, 227, 231-2, 236-8, 240, 242, 247-8, 251, 329, 334, 369, 425, 427, 490
peer review, 42, 45, 47, 318, 357, 391, (normal process) 176
pesticide(s), 35, 48, 58-9, 70-1, 73, 114-5, 205, 212, 357, 394, 442, 446, (chemical), 20, 60, 439, (DDT), 20, 23, 29, 58, 63-4, 70, 83-5, 88, 122, 349, 368, 394, 428, 442-3, 461, (industry), 48, 394, (manufacturers), 60, 441, (resistance), 62, 67, 70-1, 84, (synthetic), 34, 70
Philadelphia, 6-9, 98, 309, 480
population, 24, 26, 30-1, 51, 62, 64, 66-8, 70, 82, 88-9, 118, 128, 131-2, 134, 136, 138, 141-9, 151-3, 156, 158, 160, 191, 193, 211, 218, 224, 227-8, 230-4, 237-8, 240-1, 247-8, 265, 273-4, 323, 325, 328-9, 331, 336, 342-3, 350-8, 367-81, 383, 385-6, 394-5, 399, 408, 467, (animal), 173, 214, 385, 420, (cycles), 29, 61, 274-5, 277, (density), 141, 143-5, 273, (dynamics), 103, 177, 203, 231, 308, 370, 385-6, 392, 419, 490, (growth), 31, 67, 141-4, 156, 218, 227-31, 237, 328-9, 368-76, 379, 381, 392, 419, (human), 24, 77, 144-5, 152, 370-1, 375-6, 378, (regulation), 140-2, 144, 156, 257, 314, (stable), 143, 218, 370, 467, (stability), 218, 467, (structure), 344, 346, (world), 367, 370, 373, 379-80
post-modernism(ist), 253, 477, 486
Primary Productivity of the Biosphere, 198, 427
Principles of Animal Ecology (AEPPS), 49, 53, 57-8, 103, 126, 142, 230-1
Princeton University, 178, 237, 244, 247, 249, 306, 332, 344
Proceedings of the National Academy of Sciences, 173, 308
productivity, 37, 117, 162, 195, 198, 425-7, 449, 465-7, (measurement), 224-5, 427, (global), 167, 425, 439
rainforest, 10, 92, 124-5, 210, 236
reductionism(t), 135, 282, 494
redwoods, 190-2, 222, 413
revolution(ary), 277, 385, 396-7, 431, 466-7, 476, (counter-), 271, 283, 314, 320, (digital (computer)(, 122, 255, 484-5, 487, (environmental), 20, 203, (in ecology), 15-6, 89, 97, 99, 240, 259, 320, 431, (MacArthur), 385, 419, 481, (scientific), 10, 111-2, (social), 280-1
Salton Sea, 243-4, 248, 264, 385
Science, 27, 44, 47, 68, 127-8, 131, 213, 259-60, 277, 309, 318, 322-3, 341-2, 353, 376, 389, 391-2, 394-6, 404-6, 410, 422, 461-2, 465-6, 479, 485-6
Scientific American, 61, 394
scientific method, 254-5, 263, 266, 272, 286-7, 314, 483
sickle-cell, 62-3, 152
Silent Spring, 19-20, 25-6, 33-5, 39, 41, 48-9, 51-2, 56-9, 61, 63-4, 66-7, 70, 72-3, 75, 79, 81, 83, 88, 115-7, 120, 122, 129, 141, 211, 262, 367, 383, 394, 427-8, 461-2, 481-2
Sierra Club, 21, 24, 378, 414
size ratios, 252, 299-300, 303-4

sociobiology, 11, 155, 298
soil, 31, 79, 92, 94-5, 267, 330, 361, 364, 368, 377, 384, 392, 400-1, 403, 405, 459, 462, 489
species, 24, 54, 56, 65-6, 77-8, 80-1, 84-5, 88, 94, 96, 123-4, 129-30, 132, 137-40, 143, 145-6, 149-51, 153-4, 156, 158, 163, 168-9, 173, 177, 186-7, 189-93, 201-2, 206-9, 211, 213-4, 217-8, 221-33, 235-6, 246, 251-2, 254, 259-60, 267, 273-6, 290-1, 293-6, 298-300, 305, 316, 327, 331, 337-8, 353, 359, 363, 366, 370, 385-7, 394, 406, 422, 426, 432, 440, 442, 444, 464-6, 468-71, 478-9, 488, (abundance distribution (pattern)), 81, 293, 312, 327, 330-3, 335, (competition between), 141, 157, 297, 300, 333, 422, (diversity (index)), 200-1, 203, 300, 436-9, 464, 494, (endangered), 203, 205, 208, 332, 367, 443, 491, (extinction), 157, 206, 296, 323, 325, 337, 343-9, 367, 394, 399, 431, 436, 478, 495, (ground beetle), 364-6, (indicator), 442-3, (island), 295, 301-4, 312, 318, 320-1, 323-31, 337, 342, 344, (K-selected), 231-2, 329, 381, 426, (keystone), 207, 322(number of), 93-4, 173, 201, 361-2, 366, 464, 466-8, 470, (packing), 251, 296, (r-selected), 231-2, 329, 381, 426, (tropical), 190, 192, 214, 225, 316
species-area equation (relationship), 293-4, 324-5, 327, 329-35, 337, 339, 341, 346, 369, 490, (z), 329, 334, 337
spotted owl, 349-52, 356-60, 422, 444, 481
St Louis, 270, 362, 364
Stability and Complexity in Model Ecosystems, 235
standing biomass, 188, 198-9, 425

Stanford University, 197, 248, 284, 306, 416,
superinsects(bugs), 62-3
superorganism, 55, 97, 107, 112, 120-1, 126-8, 133, 135-6, 138-40, 145, 149, 156, 159-65, 186, 188-92, 202, 254, 280, 323, 429, 434-5, 444, 478, 481
symbiosis, 133, 148, 164
synecologists, 74, 79
Synthese, 278-9, 283-4, 317
Syracuse University, 464, 471
systems science, 86, 89, 95-6, 105, 113, 123-4, 166, 188, 203, 375, 411, 446, 452-4
t-allele (tailless), 158-60
Tallahassee Mafia, 306, 308, 311, 319, 332, 350, 465
tautology(ical), 256-7, 273-4, 293, 304, 312
(The American) Naturalist, 44, 134, 138, 174, 176, 185, 259-60, 271, 277, 284-5, 287, 301, 307, 317-8, 345, 347, 353, 485, 493
The Beak of the Finch, 268, 308
The Changing Scenes in the Natural Sciences, 1776-1976, 98
The Closing Circle, 29, 199, 436
"the ecology," 7, 8, 56-7, 61, 127, 200, 480, 490
The Ecology of Invasions by Animals and Plants, 58, 492
The Edge of the Sea, 19, 39-40, 54-5, 57-8
The Population Bomb, 25, 211, 372, 375, 395
The Sea Around Us, 19, 35, 54
The Skeptical Environmentalist, 14
The Song of the Dodo, 182, 269, 311, 319, 348
The Structure of Scientific Revolutions, 111
The Struggle for Existence, 229

Index

The Theory of Island Biogeography, 232, 274, 352, 470
The World According to Pimm, 197, 361
theory of island biogeography, 182, 232, 262, 269, 277, 290, 292-3, 296-8, 307, 330-2, 337. 339-41, 343-9, 352, 357, 361, 467, 489
*Time (*magazine), 9-10, 28, 61,491
Three Influential Papers of Robert MacArthur, 173, 273, 290, 333, 385, 438, 480
tundra, 92, 201, 214
UCLA, 8, 14, 148, 244-5, 270-1, 294-5, 306, 362, 403-5, 412, 454, (Department of Geography), 295, 403, (Environmental Science and Engineering), 86, 294, 362-3, 412, 433-4
Under the Sea Wind, 19, 40
University of California, 44, 306-7, 315, 340, 430, 442
University of Georgia, 90, 126, 161, 306, 467
University of Illinois, 8, 56, 100, 429
University of Michigan, 44, 306
University of Pennsylvania, 6, 208, 477, 479
University of Tennessee, 139, 161
University of Vermont, 308, 473
University of Washington, 207, 386
University of Wisconsin, 14, 322, 340, 455-7, 475
Vermont, 35, 169, 171-2, 175, 177, 180-1, 206, 210, 214, 241-2, 275, 283, 408, 410-1, 472-3, 490
Walden, 22, 210
Wakulla, 313-4
Washington University (St Louis), 27, 270
watershed, 94, 136, 199, 390, 447, 450-1, 457-9

web of life, 26, 59
wilderness, 7-8, 20-3, 110, 180, 210, 409, 412-3, 435, 472-3
Woods Hole Marine Biological Laboratory, 34, 37, 52-3, 56-7, 108, 419, 428-9
World3, 373, 375, 380
Worldwatch Institute, 393, 395, 398
Yale University, 40, 54, 97, 100, 102-3, 105, 108, 150, 169, 172-3, 178, 306, 373, 447
Yellowstone National Park, 30, 162, 193-5, 204, 349

www.ingramcontent.com/pod-product-compliance
Lightning Source LLC
Chambersburg PA
CBHW070615220526
45466CB00001B/4